Quantum Gravity and Computation

This volume argues that concepts from the theory of computation—including information theory, formal languages, and discrete structures—might provide novel paths towards a solution to the problem of quantum gravity. By combining elements of physics with computer science and mathematics, the volume proposes to transform the foundations of spacetime physics and bring it into the digital age.

In recent years, it has become increasingly apparent that a new theoretical framework is needed to solve the problem of quantum gravity. This kind of framework—sometimes referred to as "pregeometry" or even "prephysics"—goes beyond conventional mathematical conceptions of space, time, and matter, seeking their building blocks in more fundamental elements. The essays in this volume explore this approach from a variety of perspectives, including physics-based, mathematical, computational, and philosophical. The new formal frameworks needed to discuss such approaches have their roots in homotopy type theory, formal language theory, and higher category theory; the computational perspective is informed by connections between pregeometric structures and formal proofs and programs. The new philosophical fulcrum supporting these new avenues is inspired by constructivism and meta-structures. By probing at a level of structure beneath the ordinary structures used in general relativity and quantum mechanics, this volume seeks to find new ways of showing how these higher-order structures can be constructed from the deeper elements.

Quantum Gravity and Computation is an essential resource for scholars and graduate students interested in the philosophy of physics, quantum mechanics, and computational science.

Dean Rickles is Professor of History and Philosophy of Modern Physics at the University of Sydney, where he is also Co-Director of the University's interdisciplinary Centre of Time. His recent books include *Dual-Aspect Monism and the Deep Structure of Meaning* (co-authored with Harald Atmanspacher; Routledge 2022); and *Life is Short* (2022).

Xerxes D. Arsiwalla is a Mathematical Physics Fellow at the Wolfram Institute for Computational Foundations of Science, working on quantum foundations, pregeometry, higher category theory, formal language, and consciousness. He is an Academic Director of Physics Programs at Wolfram Research USA and a Co-Founder of the Association for Mathematical Consciousness Science (AMCS).

Hatem Elshatlawy is an interdisciplinary researcher working at the intersection of theoretical physics, computation, and philosophy. Formerly a Research Consultant at Wolfram Research, he has contributed to developments in observer theory, multi-computation, and metamathematics. His interests include constructivist foundations of physics, physics of computation, and the computation of physics.

Routledge Studies in the Philosophy of Mathematics and Physics

Edited by Elaine Landry, University of California, Davis, USA and Dean Rickles, University of Sydney, Australia

The Logical Foundation of Scientific Theories
Languages, Structures, and Models
Décio Krause and Jonas R. B. Arenhart

Einstein, Tagore and the Nature of Reality
Edited by Partha Ghose

A Minimalist Ontology of the Natural World
Michael Esfeld and Dirk-André Deckert

Naturalizing Logico-Mathematical Knowledge
Approaches from Philosophy, Psychology and Cognitive Science
Edited by Sorin Bangu

The Emergence of Spacetime in String Theory
Tiziana Vistarini

Origins and Varieties of Logicism
On the Logico-Philosophical Foundations of Mathematics
Edited by Francesca Boccuni and Andrea Sereni

The Foundations of Spacetime Physics
Philosophical Perspectives
Edited by Antonio Vassallo

Scientific Understanding and Representation
Modeling in the Physical Sciences
Edited by Kareem Khalifa, Insa Lawler, and Elay Shech

The Non-Fundamentality of Spacetime
General Relativity, Quantum Gravity, and Metaphysics
Kian Salimkhani

Bertrand's Paradox and the Principle of Indifference
Nicholas Shackel

Phenomenology and QBism
New Approaches to Quantum Mechanics
Edited by Philipp Berghofer and Harald A. Wiltsche

Quantum Gravity and Computation
Information, Pregeometry, and Digital Physics
Edited by Dean Rickles, Xerxes D. Arsiwalla, and Hatem Elshatlawy

For more information about this series, please visit: https://www.routledge.com/Routledge-Studies-in-the-Philosophy-of-Mathematics-and-Physics/book-series/PMP

Quantum Gravity and Computation

Information, Pregeometry, and Digital Physics

Edited by

Dean Rickles, Xerxes D. Arsiwalla, and Hatem Elshatlawy

Routledge
Taylor & Francis Group
NEW YORK AND LONDON

First published 2026
by Routledge
605 Third Avenue, New York, NY 10158

and by Routledge
4 Park Square, Milton Park, Abingdon, Oxon, OX14 4RN

Routledge is an imprint of the Taylor & Francis Group, an informa business

ISBN: 978-1-032-90094-0 (hbk)
ISBN: 978-1-032-90093-3 (pbk)
ISBN: 978-1-003-54613-9 (ebk)

DOI: 10.4324/9781003546139

Typeset in TexGyreTermesX
by KnowledgeWorks Global Ltd.

Contents

Preface

> [T]here is no necessity to start an explanation of quantum processes in space-time.
>
> Basil Hiley

The problem of quantum gravity consists in trying to unite what we think we know about matter and energy with what we think we know about spacetime. It turns out, that we don't know as much as we thought. The problem remains after over a century of effort by the finest minds.[1] This book adopts the viewpoint that the challenges of quantum gravity, that is of somehow casting quantum theory and general relativity within the same mould, might benefit from an infusion into the current landscape of methods and concepts of computation, information, and ideas from discrete/digital physics. While not necessarily capable of directly resolving all the issues, it has at least the potential to reinvigorate an area that has, if not quite stagnated, slowed down to a worrying degree.

Since the problem of quantum gravity appears to involve a clash of frameworks at the deepest layers of conceptual structure (namely, space, time, matter, and energy), the solution may well lie *beneath* these layers in something else entirely. The task of solving the problem then becomes one of finding substructure/s capable of generating the structures of both quantum theory and general relativity in some appropriate limits or representations. As Basil Hiley nicely put it in the above quotation, while our goal might in some sense *end* in quantum processes in spacetime, an explanation need not lie in those same concepts, but instead in some more primitive, pre-physical structure or pregeometry. We might even require alternative languages for expressing theories or mathematical foundations differing from the usual set-theoretic one.

Why computation and information? Because they stand slightly outside the usual scope of the foundations of physics, yet are becoming ever more integral. Moreover, they are neutral with respect to the controversial, clashing elements at the root of the problems of quantum gravity. Thus, they have the potential for providing appropriate pre-physical/pre-geometric building materials for both spacetime and matter (and perhaps more, such as minds).

The essays in this book come at this basic idea of expanding the scope of possible solutions, in a wide variety of ways, some more radical than others. Some probe the nature of the problem, more or less as traditionally conceived, in slightly new ways, invigorated by the introduction of concepts from information and computation. Other essays focus on the conceptual ramifications of approaches that go down a different path, taking us into fundamentally *discrete* approaches and constructive methods, rather than smooth manifolds. Others try to develop entirely new models based on more primitive processes, or categorical foundations, which also radically impact our notions of time as well as space. A variety of novel epistemological implications also arise as a result of the constructive approach to theory-building, in which completed, infinite structures are often avoided. Likewise, old issues, such as those connected to self-reference and universality, also take on interesting new appearances when they are brought into the fold as structural features of the foundations of a new physics.

The book begins with a chapter by Paul Davies that presents a wide-ranging overview of nature of the problem of quantum gravity, along with the paths that might be taken to resolve it. One of the paths mentioned[2] is to treat general relativity as simply not in need of a quantum description at all, but as part of a hybrid reality that is both quantum and classical. In the final chapter, Daniel Terno presents an approach to this, amounting to the so-called "stochastic gravity" viewpoint. Other paths involve modifying either quantum mechanics or the theory of gravitation, which several essays go down. The position of the editors, and the majority of authors in this book, is that something more radical is required that undercuts both the quantum side and the gravitational side, which corresponds to Davies' 4th path "Replace both with a completely new conceptual framework." Ultimately, Davies (this volume, p. 14) agrees:

> It seems unlikely that an incremental approach to quantising gravity, for example, by refinements to string theory, will produce a decisive breakthrough. Progress on this foundational problem will probably come only from a thorough reconceptualization, such as from discrete spacetime theories, an axiomatic approach or from experimental evidence that quantum mechanics breaks down at some scale of complexity.

However, one decides what is the most appropriate route. What we find very apparent in this book is the opening up of new vistas of exploration as well as both old issues that take on new life in this modified framework and entirely new issues that emerge from the mixing of previously separated domains.

D. Rickles, X. D. Arsiwalla, and H. Elshatlawy

Notes

1. For histories of quantum gravity, see D. Rickles, *Covered in Deep Mist: The Development of Quantum Gravity: 1915-1956* (Oxford University Press, 2000); A. S. Blum and D. Rickles, *Quantum Gravity in the First Half of the Twentieth Century: A Sourcebook* (Edition Open Sources, Max Planck Institute, 2018); D. Rickles, *A Brief History of String Theory: From Dual Models to M-Theory* (Springer, 2014).
2. An approach Davies is well-known for, thanks to his seminal textbook on the subject: N. D. Birrell and P. C. W. Davies, *Quantum Fields in Curved Space* (Cambridge University Press, 1984).

Acknowledgments

This work is supported by the following grants: The Foundational Questions Institute and Fetzer-Franklin Fund, a donor advised fund of Silicon Valley Community Foundation [FQXi-RFP-1817]; the John Templeton Foundation [Grant ID 62106]; and the Australian Research Council [Grant DP210100919]. Thanks also to Darren Dougan and the Big Questions Institute for additional financial support.

Contributors

Xerxes D. Arsiwalla is Mathematical Physics Fellow at the Wolfram Institute for Computational Foundations of Science, working on quantum foundations, pregeometry, higher category theory, formal language, and consciousness. He is an Academic Director of Physics Programs at Wolfram Research USA, a Co-Founder of the Association for Mathematical Consciousness Science (AMCS), and a former Associate Professor at Pompeu Fabra University in Barcelona, Spain.

Mark Bailey is Associate Professor at the National Intelligence University (NIU) and Senior Fellow for Complexity, AI, and Risk at the Center for the Future Mind at Florida Atlantic University. He also directs the Biological and Computational Intelligence Center at NIU, where his work focuses on emergent systems, causal structure, and the epistemic boundaries of prediction. His broader philosophical interests include metaphysical process theories and the explanatory power of information-theoretic models in fundamental physics. He is the author of *Unknowable Minds: Philosophical Insights on AI and Autonomous Weapons* (Imprint Academic, 2025).

David Corfield is Senior Lecturer in the Department of Philosophy at the University of Kent. He is the author of *Towards a Philosophy of Real Mathematics* (Oxford University Press, 2003) and *Modal Homotopy Type Theory: The Prospect of a New Logic for Philosophy* (Oxford University Press, 2020).

Paul Davies is Regents' Professor and Director of the Beyond Center for Fundamental Concepts in Science at Arizona State University. Among his many books are *The Physics of Time Asymmetry* (University of California Press, 1974); *Quantum Fields in Curved Space* (with N.D. Birrell, Cambridge University Press, 1982); *The Goldilocks Enigma: why is the universe just right for life?* (Penguin Press, 2006); *The Demon in the Machine (Penguin Press, 2019) and Quantum 2.0* (Pelican 2025). He was the recipient of the 1995 Templeton Prize for his work on foundational questions in science.

Gemma de les Coves is ICREA Research Professor at the Universitat Pompeu Fabra in Barcelona and Principal Investigator of a research group at the Institute for Theoretical Physics at the University of Innsbruck, Austria. She was awarded a START Prize of the Austrian Science Fund in 2020.

Giuseppe Di Pietra is a DPhil student in Atomic and Laser Physics at the University of Oxford. He graduated with a BSc in Physics and Astronomy from the University of Catania, Italy, and a double MSc in Physics of Complex Systems from Politecnico di Torino, Italy, and Université de Paris, France. His research focuses on a new class of information-theoretic witnesses of non-classicality in hybrid systems made by a quantum sector interacting with one that may not obey quantum theory, such as gravity or biological systems. He is a Graduate Teaching Assistant at the University of Oxford and a mentor for high school and undergraduate students.

Hatem Elshatlawy is an interdisciplinary researcher working at the intersection of theoretical physics, computation, and philosophy. Formerly a Research Consultant at Wolfram Research, he has contributed to developments in observer theory, multicomputation, and metamathematics. His interests include constructivist foundations of physics, physics of computation, and computation of physics.

Renate Loll is Professor of Theoretical Physics at Radboud University in Nijmegen (NL) and also holds a Distinguished Visiting Research Chair at Canada's Perimeter Institute. She is known for her work on quantum gravity and has co-founded Causal Dynamical Triangulations, a modern version of lattice quantum gravity. Loll is a member of the Royal Netherlands Academy of Arts and Sciences (KNAW), an elected Fellow of the International Society of General Relativity and Gravitation and a recipient of the 2025 Amaldi Medal.

Chiara Marletto is Research Fellow working in the Physics Department and Wolfson College, University of Oxford. She is an active member of the Quantum Cluster and of the New Frontiers Quantum Hub. Her research is in theoretical physics, with special emphasis on quantum theory of computation, information theory, thermodynamics, condensed-matter physics, and quantum biology. Some of her recent research has harnessed a recently proposed generalization of the quantum theory of information—constructor theory—to address issues at the foundations of the theory of control and causation in physics. She is the author of a popular book inspired by her research, *The Science of Can and Can't* (Penguin, 2022).

Gerard Milburn is Emeritus Professor in the Department of Physics at the University of Queensland. He is a Fellow of the Australian Academy of Science and the American Physical Society. He is the author of five books including *Quantum Optics* (with Dan Walls: Springer, 2008), *Quantum Measurement* *and Control* (with Howard Wiseman: Cambridge University Press, 2009) and *Quantum Optomechanics* (with Warwick Bowen: CRC Press, 2015).

Daniele Oriti is "Distinguished Researcher" at the Department of Theoretical Physics of Universidad Complutense de Madrid, Spain, where he leads a research group working on quantum gravity and foundations of physics. He has been awarded the A. Stanghellini Prize by the Italian Physics Society (2000), the Sofja Kovalevskaja Prize by the A. von Humboldt Foundation (2008), the Heisenberg Grant by the Deutsche Forschung Gemeinschaft (2018), and the ATRAE Grant by the Spanish Ministry of Education and Research (2023). He is the chair of the International Society for Quantum Gravity, a member of the Nominating Committee of the International Society of General Relativity and Gravitation, and a member of the Steering Board of the International Society of Loop Quantum Gravity. He is editor of *Approaches to Quantum Gravity: Toward a New Understanding of Space, Time and Matter* (Cambridge Universaity Press, 2009).

Dean Rickles is Professor of History and Philosophy of Modern Physics at the University of Sydney, where he is also Co-Director of the University's interdisciplinary Centre of Time. Recent books include *Varieties of Nothingness* (co-edited with Leslie Stein; Chiron, 2024); *Dual-Aspect Monism and the Deep Structure of Meaning* (co-authored with Harald Atmanspacher; Routledge, 2022); and *Life is Short* (Princeton, 2022).

Hisham Sati is Associate Professor of mathematics at NYU Abu Dhabi and lead-PI of the Center for Quantum and Topological Systems; Director, Center for Quantum and Topological Systems (CQTS); Associate Dean for Student Success and Curricular Affairs; Professor of Mathematics; Global Network Professor of Mathematics, Courant Institute of Mathematical Sciences. His research interests include intersection of differential geometry, algebraic topology, and mathematical/theoretical physics.

Susan Schneider is an American philosopher and artificial intelligence expert. She is the founding director of the Center for the Future Mind at Florida Atlantic University where she also holds the William F. Dietrich Distinguished Professorship. Her books include *Artificial You: AI and the Future of Your Mind* (Princeton University Press, 2019) and *Science Fiction and Philosophy* (Wiley-Blackwell, 2009).

Urs Schreiber is Senior Research Scientist of the Center for Quantum and Topological Systems at New York University in Abu Dhabi. Schreiber has worked on developing mathematical foundations for non-perturbative higher quantum gauge fields using tools from algebraic and differential topology, both in high-energy physics and quantum gravity, with application to quantum information and topological quantum materials.

Daniel Terno was born and grew up in Latvia (then part of the USSR). He completed his undergraduate and PhD studies at Technion in Israel with Asher Peres as his thesis advisor. He carried out research as a postdoc at the Perimeter Institute, before joining Macquarie University, where he is Professor of physics. His research mainly belongs to four areas: Quantum information, quantum foundations, relativistic quantum information and black hole physics and quantum gravity.

Vlatko Vedral is Professor of Quantum Information Science at the Physics Department and a Governing Body Fellow of Wolfson College, University of Oxford. He was awarded the Abdus Salam Award (1997), the Royal Society Wolfson Research Merit Award (2007), the World Scientific Physics Research Medal (2009), and the Marko V. Jaric Award (2011). He is a Fellow of the Institute of Physics (2017) and a member of the Academia Europaea (2020). He is the author of several academic textbooks and three popular books, *Decoding Reality: The Universe as Quantum Information* (Oxford University Press, 2018), *From Micro to Macro: Adventures of a Wandering Physicist* (WSPC, 2018), and *Portals to a New Reality: Five Pathways to the Future of Physics* (Basic Books, 2025).

Stephen Wolfram is the creator of *Mathematica*, *Wolfram|Alpha* and the Wolfram Language. He is author of *A New Kind of Science* (Wolfram Media, 2019); *Second Law: Resolving the Mystery of the Second Law of Thermodynamics* (Wolfram Media, 2023); *Metamathematics: Foundations & Physicalization* (Wolfram Media, 2022). He is the originator of the Wolfram Physics Project; and the Founder and CEO of Wolfram Research.

Part 1

Scene Setting

1 Open Questions at the Quantum Frontier

Paul Davies

1.1 Why quantize gravity?

I first began studying quantum gravity as a PhD student when I attended some lectures by Felix Pirani at King's College London in 1968. The mood was upbeat. Twelve years later, Stephen Hawking delivered his inaugural lecture at Cambridge University on the occasion of his appointment to the Lucasian Chair of Mathematics. The title of his lecture was "Is the end in sight for theoretical physics?" The subject was $N = 8$ supergravity, which at the time seemed to offer not merely a consistent theory of quantum gravity, but a complete unification of fundamental physics to boot. Shortly after, $N = 8$ supergravity was eclipsed by string theory, which generalized into M theory and was later joined by loop quantum gravity. In spite of attracting some of the best minds in theoretical physics, all approaches to quantizing gravity remain incomplete decades later. It therefore seems reasonable to question the basic assumptions underlying the specific issue of quantizing the gravitational field and the more general problem of whether a complete unified theory – a so-called "Theory of Everything" – is in any case a well-conceived goal.

A rather basic question is whether gravity needs to be quantized in the first place. How can we be sure it isn't a classical field? There is a total lack of experimental evidence for graviton emission, absorption or scattering; so arguments in favour of quantizing gravity rest on theoretical grounds, and typically involve appeals to consistency of theoretical physics, for example, gedanken experiments that might lead to violations of unitarity or permit superluminal information transfer [1]. However, these investigations are very much a backwater of physical theory and I regard it as an open question as to whether a more

DOI: 10.4324/9781003546139-2

penetrating analysis might yield plausible candidates for a consistent semi-classical theory (i.e. quantum matter coupled to classical gravity). Assuming that gravity does indeed require a quantum description, there would seem to be four possibilities for overcoming the current impasse:

- Modify quantum mechanics
- Modify gravitational theory
- Modify both
- Replace both with a completely new conceptual framework

I shall briefly discuss the first two. The final possibility – a complete conceptual makeover – is discussed in [2] and many chapters in this book.

1.2 Modify quantum mechanics

Quantum mechanics in its standard formulation is the most successful physical theory in history. There are, however, many loose ends, ranging from the philosophical aspects to the technical ones. The most persistent concerns are the cluster of issues surrounding the measurement problem, the quantum-classical transition, the Heisenberg cut and the role of the observer. Moreover, the theory has been tested only under a very restricted range of physical circumstances and not at all in gravitationally relevant situations.

It's possible that quantum mechanics is merely an effective theory with a limited domain of applicability. (The concept of an effective theory is familiar in theoretical physics. For example, Fermi's theory of weak interactions is an effective low-energy theory that must be replaced by GSW electroweak theory at higher energies.) There are two ways one might generalize an effective theory. The first is to embed it into a bigger theory that reduces to quantum mechanics in the domain so far investigated. The second is to postulate departures from standard quantum mechanics (e.g. departures from unitary evolution or the Born rule) above some threshold and provide a bridging theory that connects the quantum realm to the classical.

Let me briefly illustrate examples of each.

1.2.1 Embedding quantum mechanics in a bigger theory with extended non-locality

Non-locality in quantum mechanics arises inevitably from the property of entanglement, and is a defining feature of the subject. It is demonstrated most famously in the experimental tests of Bell's inequality. Curiously, however, quantum mechanics is not maximally non-local. Sandu Popescu and Danny

Rohrlich discovered this while asking [3, 4]: "Is quantum mechanics the unique theory that allows for nonlocal phenomena consistent with special relativity?" They embarked on this investigation to test the hypothesis that perhaps quantum mechanics is indeed defined as the only non-local theory consistent with relativity. What they discovered is that this is not in fact the case: nature could be even more non-local than quantum mechanics predicts, yet be fully consistent with special relativity. Might it therefore be the case that the world is actually more non-local than we hitherto assumed but we have not yet discovered any phenomena that, for example, violate Bell's inequality by a greater margin than quantum mechanics? If quantum mechanics is replaced by a (yet-to-be-formulated) more non-local post-quantum mechanics, usually dubbed a PR theory after Popescu and Rohrlich, then quantum gravity will likewise need to be replaced by post-quantum gravity. It is then possible that this post-quantum gravity theory will avoid the mathematical issues associated with existing attempts at quantum gravity. Of course, it is also possible that it will make matters worse, that is, the post-quantum gravity will run into even more severe mathematical problems. In the absence of a specific PR theory, there is little that can be said about the matter.

1.2.2 Appending a bridging theory

It may be that additional physical processes serve as a bridge between standard (Copenhagen) quantum mechanics and the classical world of everyday reality. There are several proposals along these lines. The best known are the collapse theories according to which an additional physical process with a random action triggers wavefunction reduction, effectively classicalizing quantum states. In the GRW theory, for example, spatially extended wave functions spontaneously and randomly implode to definite spatial locations. Two new physical parameters are introduced: the implosion rate and the localization size, with values bounded by existing observations [5]. Another type of collapse theory, due to Roger Penrose, appeals to gravitation as the trigger for wave function collapse [6]. In a delocalized Schrödinger cat type of state, one may define a gravitational potential energy between spatially separated components in the superposition. Penrose postulates a criterion, akin to Heisenberg's energy-time uncertainty relation, that randomly brings about the abrupt "collapse" of the extended wave function to a single localized component. In this scheme, rather than struggling to quantize gravity, one instead invokes the gravitational field as the bridging mechanism that turns a superposition into an "either/or" outcome.

An alternative set of ideas posits a breakdown of unitarity in black holes. Hawking himself introduced such a theory in 1976, according to which black holes turn pure input quantum states into mixed output states [7]. Hawking described this transition in a much-quoted passage, "Not only does God play dice, but he sometimes throws them where they can't be seen". Later, Hawking abandoned this idea in favour of the preservation of unitarity (and the associated conservation of information) during the black hole evaporation process.

Another suggestion is that departures from unitarity emerge when the system is above a certain level of complexity. This is an old idea that goes back to Schrödinger's famous Dublin lectures entitled "What is Life?" in which he mooted the possibility of "a new kind of physical law" prevailing in living matter, though he was vague about what it might be [8]. The challenge about invoking complexity as a controlling factor in quantum dynamics is that it is an intrinsically systemic property. It would indeed require a new kind of physical law as Schrödinger suggested. There are many definitions of complexity, but one that has been proposed is integrated information [9]. Another involves some measure of entanglement, which rises exponentially with the number of qubits and so represents the most extreme extrapolation of linear superposition, and thus the most stringent test of unitary evolution for a composite system. By way of illustration, a quantum system with only a few hundred entangled qubits – a stated target of the quantum computer industry – would have more branches of the wave function than there are particles in the observable universe. If quantum mechanics breaks down somewhere in sufficiently complex systems, quantum computation involving many entangled qubits is a good place to look.

1.3 Modify gravitational theory

Geometrical theories of gravitation such as general relativity and its modifications are based on the idealization that spacetime is a continuum. However, on general grounds one expects drastic disruptions of geometry (and topology) on scales approaching the Planck length. There is a rich history of attempts to build spacetime out of a foundational substructure, from pre-geometry, twistor theory and spacetime foam to string theory and loop quantum gravity. Irrespective of the specific theoretical details, it seems worth investigating generally how the existence of a fundamental length might impact the formulation of a consistent quantum theory of gravity and, moreover, whether there are any observational consequences.

A crude model of spacetime substructure is to simply posit a fundamental cell size, or pixelation of space and time. This discretization will serve as a regulator of divergent quantities in quantum field theory in general and quantum gravity in particular because it implies a maximum frequency (e.g. the Planck frequency). One such scheme is known as doubly special relativity, in which it is readily shown [10] that if spacetime is pixelated, then free space acquires a refractive index and becomes a dispersive medium for electromagnetic wave propagation, thus opening the way to testing for observational effects, for example, from gamma ray bursts. A fixed length scale breaks local Lorentz

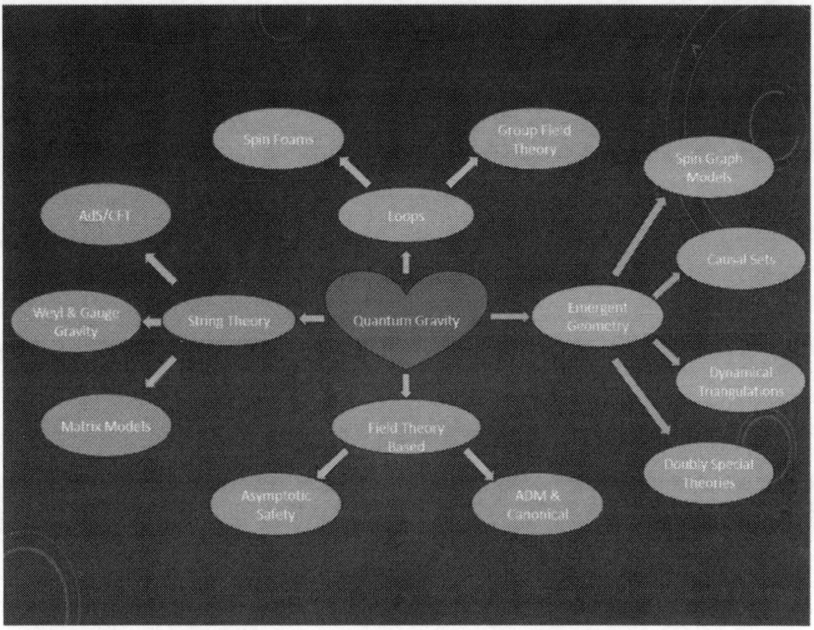

Figure 1.1 A schematic illustration of the broad range of quantum gravity theories.

invariance, but this fundamental symmetry may be restored by introducing curvature in momentum space. The resulting modified propagators may then be used to investigate a range of quantum field effects, such as black hole radiance, moving mirror radiation and Unruh-DeWitt particle detectors [11]. The range of quantum gravity theories with a fundamental length is broad, and illustrated schematically in Fig. 1.1.

1.4 Utility of the semi-classical theory

The absence of a satisfactory theory of quantum gravity has not prevented a large amount of work on a semi-classical theory, in which a classical gravitational field is coupled to quantum matter. Semi-classical electrodynamics is a very useful approximation to a full quantum electrodynamics and recovers a large number of well-known results. A semi-classical approach to quantum gravity has likewise yielded some useful results, such as Hawking's black hole evaporation phenomenon and a description of inflation. The first obstacle in constructing a semi-classical theory of quantum gravity is what to put on the right-hand side of the gravitational field equations, i.e. for the source term. A pioneer of semi-classical gravity, Bryce DeWitt, adopted Schwinger's effective action approach [12]. The effective action is defined as

$$W = i\, ln\, \langle 0_{out} | 0_{in} \rangle \tag{1.1}$$

from which it follows by variation of W with respect to the metric tensor $g_{\mu\nu}$ that

$$2(-g)^{-1/2} \frac{\delta W}{\delta g^{\mu\nu}} = \frac{\langle 0_{out} | T_{\mu\nu} | 0_{in} \rangle}{\langle 0_{out} | 0_{in} \rangle} \tag{1.2}$$

which yields field equations of the form

$$G_{\mu\nu} + \text{ higher order terms in curvature} = \frac{\langle 0_{out} | T_{\mu\nu} | 0_{in} \rangle}{\langle 0_{out} | 0_{in} \rangle} \tag{1.3}$$

where $T_{\mu\nu}$ is the stress-energy-momentum tensor operator corresponding to the quantum field to be coupled to the gravitational field (e.g. the electromagnetic field), while $|0_{in}\rangle$ is the initial quantum vacuum state and $|0_{out}\rangle$ the final quantum vacuum state. The foregoing construction is easily extended to initial states that contain quanta.

De Witt's formulation was adopted by many practitioners during the formative stages of the semi-classical theory [13–15]. However, it was criticized on the grounds that the quantity

$$\frac{\langle 0_{out} | T_{\mu\nu} | 0_{in} \rangle}{\langle 0_{out} | 0_{in} \rangle} \tag{1.4}$$

is not an expectation value and is in general not even a real-valued quantity. An alternative to Eq. (1.3) was suggested and soon widely adopted [16], namely:

$$G_{\mu\nu} + \text{ higher order terms in curvature} = \langle 0_{in} | T_{\mu\nu} | 0_{in} \rangle \tag{1.5}$$

With the benefit of Eq. (1.5), it became possible to calculate the back reaction of the quantum field on the gravitational field, for example, in the Hawking effect, the negative energy flux into the evaporating black hole [17], and in expanding cosmological models, the effect on the expansion rate of the quantum vacuum and any contributions from gravitationally-induced particle creation.

The transition from Eq. (1.3) to Eq. (1.5) leaves open the question of whether the quantity

$$\frac{\langle 0_{out} | T_{\mu\nu} | 0_{in} \rangle}{\langle 0_{out} | 0_{in} \rangle}$$

has a physical interpretation. And as a matter of fact, it does. In 1988, Aharonov, Albert and Vaidman formulated the theory of quantum weak measurements

[18]. In a standard von Neumann projective measurement, a measuring device is coupled to the quantum system of interest and an eigenvalue of an operator A obtained with a probability given by the Born rule. In a weak measurement scheme, the measuring device is only weakly coupled to the quantum system, and the result is usually not an eigenvalue and is compromised by random noise. To compensate, one envisages a large ensemble N of identically prepared systems and computes a statistical average of the measurement results. That average is called the weak value. Because the measurement is weak, the back action of the measurement on the quantum system is reduced. In the limit of large N, the back action is negligible and the weak value is sharply defined. Although all members of the ensemble are chosen to have identical initial states $|a\rangle$, one is free to post-select a sub-ensemble in which the measurements yielded a specific final state $|b\rangle$. In that case the weak value of A is

$$\frac{\langle a|A|b\rangle}{\langle a|b\rangle} \tag{1.6}$$

Weak measurements combined with pre- and post-selection have opened up a new sector of quantum mechanics with widespread experimental confirmation and several practical applications [19]. Inspection of the right-hand side of Eq. (1.2) reveals that it is in fact the weak value of $T_{\mu\nu}$ for systems pre-selected in the quantum state $|\ |0_{in}\rangle$ and post-selected in the state $|0_{out}\rangle$.

Given the poor prospects for semi-classical quantum gravity experiments, the identification of the Schwinger-DeWitt formulation with weak measurements is of little practical value. But it does raise an interesting question in cosmology. A distinctive feature of the universe is its remarkable degree of uniformity (homogeneity and isotropy) on the large scale, a recognized fact that is dignified with the moniker "the cosmological principle". This has led to a focus on the initial conditions of the universe as a state of primordial simplicity, with the universe evolving to greater and greater complexity over time. A specific example of such initial conditions in quantum cosmology is the Hartle-Hawking no-boundary proposal for the wave function of the universe [20]. Primordial simplicity is closely linked to the past hypothesis as the source of the cosmological arrow of time [21]. Thus, the choice of $|0_{in}\rangle$ as an initial state of the electromagnetic or inflaton fields of the universe seems "natural"; $|0_{in}\rangle$ could, for example, be the so-called Bunch-Davies quantum vacuum state [22] at the end of inflation.

What is rarely regarded as "natural" is to impose a final state of simplicity for the universe, such as a vacuum state $|0_{out}\rangle$; although if dark energy remains constant, the universe in the far future will be similar to the de Sitter like inflationary phase, albeit with an expansion rate many orders of magnitude slower. Viewing the current epoch of the universe as a complex phase sandwiched between two simple de Sitter like phases invites the hypothesis the quantum state of the universe is constrained by both initial and final boundary conditions, for example, the vacuum states – $|0_{in}\rangle$ and $|0_{out}\rangle$. In that case, the quantum weak value [Eq. (1.2)] could have operational significance at our epoch as a description of semi-classical quantum gravity, such as black hole radiance [23]. Departures from the predictions of semi-classical theory based on expectation values, as in Eq. (1.5), would depend on how small the quantity $\langle 0_{out}|0_{in}\rangle$ might be. In the case that $\langle 0_{out}|0_{in}\rangle << 1$, dramatic departures dubbed "quantum miracles" by Aharonov [24] might result, with major observational consequences.

1.5 The generalized second law of thermodynamics

Following the discovery by Bekenstein and Hawking that black hole horizon area serves as a measure of entropy, it was possible to generalize the second law of thermodynamics to include the gravitational component. Soon the law was further extended to include cosmological event horizon area, specifically, the area of the de Sitter event horizon. At the time, there was a feeling, articulated explicitly by Penrose [25], that there should exist a measure of "gravitational entropy" for more general spacetimes, which would reduce to horizon area in the limiting cases, but would also quantify departures in spacetime geometry from conformal flatness. One question that arose in relation to this is whether a square centimetre of black hole horizon was "worth" the same as a square centimetre of de Sitter horizon. This can be investigated in two limiting cases. The first is in black hole-de Sitter spacetime geometries, where it can be proved that the exchange of heat between the two horizons never reduces the total area, at least for the spherically symmetric Reissner-Nordström-de Sitter spacetime [26]. The other limit is to consider a Friedmann universe filled with a dilute non-relativistic gas of small black holes and calculate the total horizon area – cosmological + black holes – within a cosmological horizon volume. As the universe expands, black holes drift across the cosmological horizon and the latter area grows as a result of the reduction in density of the black hole gas. Again, the generalized second law is obeyed [27].

In spite of this consistency, the thermodynamic status of cosmological horizons is unclear. A model particle detector in a de Sitter vacuum state responds as if immersed in a bath of thermal radiation, so the de Sitter horizon has a characteristic temperature, which is found to be $H/2\pi$, where H is the Hubble constant. However, the stress-energy-momentum tensor of the de Sitter vacuum is not that of thermal radiation [16]. Rather, it is a simple renormalization of the cosmological constant. This is in contrast to the thermality of a black hole horizon, where there is both a characteristic temperature and a flux of heat energy from the hole. To further determine whether de Sitter heat is "real", one can investigate whether it can be "mined", for example, by a Szilard heat engine.

To accomplish this, it is first necessary to screen out the de Sitter thermal fluctuations with a Casimir-type box to create an oasis at a lower temperature than the de Sitter horizon temperature. A $1+1$ dimensional calculation confirms that it is indeed possible to create a "quiet zone" with zero temperature in de Sitter space; this is a screened region where a particle detector remains unexcited [28]. Therefore, in principle, one could mine the de Sitter thermal fluctuations and transfer the energy into the interior of the box, thus extracting work. This calculation provides evidence that de Sitter heat is "physically real" in spite of the non-thermal nature of its stress-energy-momentum tensor.

The entropic status of cosmological horizons is even less clear when the horizon area becomes time-dependent, as it is for realistic models of the universe. In the case of a time-dependent horizon, a particle detector will respond, but it will not register a thermal spectrum. Thus, there is no characteristic temperature associated with the horizon, although one may approximate with an instantaneous effective de Sitter temperature that evolves with time. In spite of this, it is still the case that the total horizon area increases with time so long as the cosmological fluid obeys the same energy conditions as the area theorem for black hole horizons (the dominant energy condition) [29].

Interestingly, there are some popular cosmological models where the energy condition fails, for example, the so-called big rip model. In that case, the cosmological horizon area shrinks to zero at the big rip singularity [30], which raises the questions of whether the generalized second law should be regarded as paramount, and used to eliminate any cosmological models that violate it. In a famous warning, Eddington wrote [31]:

> If someone points out to you that your pet theory of the universe is in disagreement with Maxwell's equations - then so much the worse for Maxwell's equations. If it is found to be contradicted by observation - well, these experimentalists do bungle things sometimes. But if your theory is found to be against the second law of thermodynamics I can give you no hope; there is nothing for it but to collapse in deepest humiliation.

There is an interesting philosophical issue here. If a field theory (e.g. a proposed Lagrangian) permits a solution that violates the generalized second law, should one merely discard that solution as "unphysical" (as in advanced radiation in electromagnetic theory) or reject the entire theory, on the basis that a sufficiently advanced civilization could in principle construct the said problematic solution and reduce the entropy of the universe – in effect, reversing the arrow of time on a macroscopic scale?

Gravitational theory differs in a crucial respect from other theories in physics, in that unphysical solutions cannot be eliminated via boundary or initial conditions alone. The theory must be augmented by energy conditions, which are introduced ad hoc. In normal analysis, one picks a matter source for the right-hand side of Einstein's gravitational field equations, and then solves the equations to determine the spacetime that the source distribution supports. But one can reverse this procedure and pick an arbitrary spacetime geometry, and then solve Einstein's equations "backwards" to find a matter distribution that will generate it. As is well known, "unpalatable" spacetime geometries, for example, those containing closed timelike curves or naked singularities, will arise unless ruled out by suitable energy conditions (perhaps in combination with boundary conditions). One can draw up a list of "unsuitable" spacetimes on philosophical grounds and then use that as a filter on the permitted types of matter (in effect, eliminating a wide class of field theories). That filter list might include:

- No closed timelike curves (CTCs)
- No naked singularities
- No "naked infinity" (non-globally hyperbolic spacetimes)
- No Boltzmann brains
- No violations of the generalized second law of thermodynamics

1.6 Time machines and cosmic flashing

We notice a curious aspect of the semi-classical theory. Quantum field theory permits states that violate certain energy conditions. For example, the Casimir effect violates the dominant energy condition, opening to the way to traversable wormholes and hence closed timelike curves (CTCs). However, it seems likely (but has not been proved) that the quantum vacuum would blow up on the chronology horizon, stymying any attempt to create CTCs [32].

Another example concerns negative energy fluxes, for example, from accelerating mirrors or from squeezed states. Ford investigated a scenario in which a beam of negative energy radiation was directed at an extreme Reissner-Nordström black hole. By reducing the mass M of the hole but not the charge Q, a sustained flux of negative energy would lead to the condition $M^2 < Q^2$, for which the horizon vanishes, leaving a naked singularity. On careful investigation, Ford found that the total amount of negative energy delivered to the black hole would always be strictly limited by the physical circumstances, and that the condition $M^2 > Q^2$ would be rapidly restored [33]. Ford referred to the mere transitory loss of the horizon as "cosmic flashing". So quantum field theory flirts with disaster – it violates the letter of the above filter, but not the spirit. Why? Why does quantum mechanics save gravitation from producing grossly unphysical spacetimes? Does this point to a deep principle linking quantum mechanics and gravitation that is not apparent in any of the attempts to produce a quantum theory of gravity?

1.7 Quantum cosmology and the birth of the universe

The hypothesis that the universe was born in a quantum process has a deep history going back to Georges Lemaître [34]. The subject of quantum cosmology, in which one assigns a wave function to the entire universe, that is, quantizes the cosmological dynamics itself, is the most ambitious attempt at a quantum theory of gravity. Wheeler and DeWitt [12] laid the formal foundations of quantum cosmology in the 1960s but it was not until the 1980s that there was any serious attempt to construct "the wave function of the universe" (e.g. the work of Hartle and Hawking [20]).

Evidence that the universe did indeed start out in a quantum phase comes from the cosmic microwave background (CMB). The power spectrum of this "big bang afterglow" has imprinted on it the distinctive hallmarks of quantum vacuum noise, for example, from the Bunch-Davies vacuum associated with inflation [35]. If the statistical fluctuations in the CMB are indeed quantum fluctuations writ large and classicalized (by some yet-to-be-determined process), then we have direct evidence for quantum effects on a large scale. At least, it is on a cosmic scale today. But taking the largest scale of inhomogeneity in the observed CMB and evolving back to the epoch of inflation, one may obtain a value for the mass-energy of the much smaller spacetime region giving rise to it. The answer will depend on the energy scale of inflation, but assuming this was the grand unified theory (GUT) scale, one obtains a mass of about 10^{-8} g, which is remarkably close to the current laboratory limit for constructing Schrödinger cat states [36].

1.8 Conclusion

In spite of several decades of effort, a satisfactory theory of quantum gravity remains elusive, although there have been many contenders. General arguments can be made that gravity must be quantum, but direct observational support is still many orders of magnitude away. It is possible that future measurements at Laser Interferometer Gravitational-Wave Observatory (LIGO) [37] or mesoscopic bench-top experiments using entangled states [38] will eventually provide direct experimental support for quantum gravity, but in the interim all we have are plausibility arguments and some overarching principles arising from semi-classical theory, cosmology and horizon area entropy. It seems unlikely that an incremental approach to quantizing gravity, for example, by refinements to string theory, will produce a decisive breakthrough. Progress on this foundational problem will probably come only from a thorough reconceptualization, such as from discrete spacetime theories, an axiomatic approach or from experimental evidence that quantum mechanics breaks down at some scale of complexity.

Acknowledgements

I would like to thank Giovanni Amelino-Camelia, Andrew Jordan, Gerard Milburn, Maulik Parikh, Phil Tee and Jeffrey Tollaksen for helpful discussions on several aspects of this work.

References

[1] Huggett, N., and Callender, C. (2001). Is gravity quantum anyway? And how would we know? *Proceedings of the Philosophy of Science Association*, 2001(3), S382.

[2] Arsiwalla, X. D., Elshatlawy, H., and Rickles, D. (2023). Pregeometry, formal language and constructivist foundations of physics. *arXiv preprint* arXiv:2311.03973.

[3] Popescu, S., and Rohrlich, D. (1994). Nonlocality as an axiom. *Foundations of Physics*, 24(3), 379–385.

[4] Popescu, S. (2014). Nonlocality beyond quantum mechanics. *Nature Physics*, 10, 264–270.

[5] Adler, S. L., and Bassi, A. (2009). Is quantum theory exact? *Science*, 325(5938), 275–276.

[6] Hameroff, S., and Penrose, R. (2014). Consciousness in the universe: A review of the 'Orch OR' theory. *Physics of Life Reviews*, 11(1), 39–78.

[7] Hawking, S. W. (1976). Breakdown of predictability in gravitational collapse. *Physical Review D*, 14(10), 2460–2473.

[8] Schrödinger, E. (1944). *What is Life?* Cambridge University Press.

[9] Albantakis, L., Barbosa, L., Findlay, G., Grasso, M., Haun, A. M., Marshall, W., et al. (2023). Integrated Information Theory (IIT) 4.0: Formulating the properties of phenomenal existence in physical terms. *PLoS Computational Biology*, 19(10), e1011465.

[10] Amelino-Camelia, G. (2010). Doubly-special relativity: Facts, myths and some key open issues. *Symmetry*, 2(1), 230–271.

[11] Davies, P. C. W., and Tee, P. (2024). Quantum vacuum of spacetime with a fundamental length. *Physical Review D*, 110, 025009.

[12] DeWitt, B. S. (1967). Quantum theory of gravity. I. The canonical theory. *Physical Review*, 160(5), 1113–1148.

[13] Dowker, J. S., and Critchley, R. (1976). Effective Lagrangian and energy-momentum tensor in de Sitter space. *Physical Review D*, 13(12), 3224.

[14] Gibbons, G. W., and Hawking, S. W. (1977). Action integrals and partition functions in quantum gravity. *Physical Review D*, 15(10), 2752–2756.

[15] Hartle, J. B., and Hu, B. L. (1979). Quantum effects in the early universe. *Physical Review D*, 20(8), 1757.

[16] Birrell, N. D., and Davies, P. C. W. (1983). *Quantum Fields in Curved Space*. Cambridge University Press.

[17] Davies, P. C. W., Fulling, S. A., and Unruh, W. G. (1976). Energy-momentum tensor near an evaporating black hole. *Physical Review D*, 13(10), 2720–2723.

[18] Aharonov, Y., Albert, D. Z., and Vaidman, L. (1988). How the result of a measurement of a component of the spin of a spin-1/2 particle can turn out to be 100. *Physical Review Letters*, 60(14), 1351–1354.

[19] Jordan, A. N., and Siddiqi, I. (2024). *Quantum Measurement: Theory and Practice*. Cambridge University Press.

[20] Hartle, J. B., and Hawking, S. W. (1983). Wave function of the universe. *Physical Review D*, 28(12), 2960–2975.

[21] Wallace, D. (2023). The local quantum vacuum as the Past Hypothesis. `https://philsci-archive.pitt.edu/22619/1/bunch-davies.pdf`

[22] Bunch, T. S., and Davies, P. C. W. (1978). Quantum field theory in de Sitter space: Renormalization by point splitting. *Proceedings of the Royal Society A: Mathematical, Physical and Engineering Sciences*, 360(1700), 117–134.

[23] Davies, P. C. W. (2014). Quantum weak measurements and cosmology. In D. C. Struppa and J. M. Tollaksen (Eds.), *Quantum Theory: A Two-Time Success Story: Yakir Aharonov Festschrift*. Springer.

[24] Aharonov, Y., and Gruss, E. (2005). Two-time interpretation of quantum mechanics. *arXiv preprint* arXiv:quant-ph/0507269.

[25] Penrose, R. (1979). Singularities and time-asymmetry. In S. W. Hawking and W. Israel (Eds.), *General Relativity: An Einstein Centenary Survey* (pp. 581–638). Cambridge University Press.

[26] Davies, P. C. W., Ford, L. H., and Page, D. N. (1986). Gravitational entropy: Beyond the black hole. *Physical Review D*, 34(6), 1700–1707.

[27] Davies, P. C. W., and Davis, T. M. (2002). How far can the generalized second law be generalized? *Foundations of Physics*, 32(12), 1877–1889.

[28] Davies, P. C. W., Thomas, L., and Zahariade, G. (2021). Heat shield for de Sitter space. *Physical Review D*, 103, 125010.

[29] Davies, P. C. W. (1988). Cosmological horizons and entropy. *Classical and Quantum Gravity*, 5, 1349–1355.

[30] Davies, P. C. W. (1988). Cosmological event horizons, entropy and quantum particles. *Annales de l'Institut Henri Poincaré*, 49(3), 297–306.

[31] Eddington, A. S. (1928). *The Nature of the Physical World*. Macmillan; Cambridge University Press.

[32] Visser, M. (1995). *Lorentzian Wormholes: From Einstein to Hawking*. American Institute of Physics.

[33] Ford, L. H., and Roman, T. A. (1992). "Cosmic flashing" in four dimensions. *Physical Review D*, 46(4), 1328–1339.

[34] Lemaître, G. (1931). The beginning of the world from the point of view of quantum theory. *Nature*, 127(3210), 706.

[35] Akama, S., and Tahara, H. W. H. (2023). Imprints of primordial gravitational waves with non-Bunch-Davies initial states on CMB bispectra. *Physical Review D*, 108(10), 103522.

[36] Bild, M., Fadel, M., Yang, Y., von Lüpke, U., Martin, P., Bruno, A., and Chu, Y. (2023). Schrödinger cat states of a 16-microgram mechanical oscillator. *Science*, 380(6642), 274–278.

[37] Parikh, M., Wilczek, F., and Zahariade, G. (2021). Quantum mechanics of gravitational waves. *Physical Review Letters*, 127(8), 081602.

[38] Bose, S., Mazumdar, A., Morley, G. W., Ulbricht, H., Toros, M., Paternostro, M., Geraci, A., Barker, P., Kim, M. S., and Milburn, G. A spin entanglement witness for quantum gravity. *Physical Review Letters*, 119, 240401.

Part 2

Locality, Non locality, Observers, and Agents

2 Spacetime Events from the Inside Out

Gerard Milburn

2.1 Introduction

Physics alone may not be able to provide answers to the space and time issue. Instead, it is up to neuroscience to address them.

György Buzsáki.

It has been apparent since at least the Chapel Hill conference in 1955 [10] that the world revealed by Schrödinger and Heisenberg is incompatible with the world revealed by Einstein, despite the astounding experimental success of quantum theory and general relativity. The resulting discontent has fuelled a search for a quantum theory of gravity despite the lack of any compelling experimental evidence to do so. There is a general belief that a quantum understanding of space and time will emerge from a quantum theory of gravity.

Or have we missed something? The experimental tests of quantum theory via Bell violations are among the top-shelf achievements of quantum theory and quite consistent with the theory of relativity. Yet there is something mysterious about this consistency. As Gisin puts it, "no story in space–time can describe nonlocal correlations" [11].

Gisin, like many others, describes the quantum correlations revealed in Bell tests as 'nonlocal' correlations. It is hard to describe it any other way, yet it is easy to design expressions which violate the Bell inequalities even for time-like separated observers. In fact, spacetime coordinates of observers play no role at all. It is in this sense that that quantum correlations seem to be coming from outside spacetime itself. What is at stake here is our deep-seated intuition that

DOI: 10.4324/9781003546139-4

an agent, like us, can only act *here and now* according to internal states. This perspective has been emphasized by György Buzaski [12].

In this chapter, we will discuss a version of the Bell tests that uses a single agent and time-like separation of detection events. One might think that this is unlikely to raise issues of nonlocality. However, using the same logic as the standard Bell violation argument, the single agent scenario suggests a surprising interpretation: retrocausation without superluminal signalling to the past.

Prior to quantum theory, we could describe the world as being built from classical variables that describe objective properties of that world. In quantum optics, for example, we perform experiments on the electromagnetic field using sources and detectors. The quantum field is not a classical field like Maxwell's fields. It does not have independent properties like electric or magnetic field magnitudes that have independent causal efficacy. Only measurements' results in a particular experiment context, have causal efficacy. The value of the electric field is established by a particular class of experiments (homodyne and heterodyne detection). If we want to estimate the intensity of the field, we do a very different kind of measurement; we count photons. These are complementary experiments in the sense of Bohr. If we have a single charge in a superposition of two locations in an ion trap, we do not worry that the resulting electromagnetic (EM) field has no classical interpretation. We simply measure what quantum mechanics (QM) predicts.

The classical gravitational field, as revealed in general relativity (GR), is a very different matter. In Einstein's formulation, gravity is represented by the same mathematical object as space time geometry. To measure the gravitational field, we need to find ways to estimate the metric. Einstein used imaginary clocks and rulers. Today, we use a quantum field theory, namely quantum electrodynamics. The objective is to estimate the components of a metric as best as we can given the bounds imposed by quantum uncertainty [6]. A good example is the measurement of perturbations to the Minkowski metric that define gravitational waves, $h_{\mu\nu}$. That is what Laser Interferometer Gravitational-Wave Observatory (LIGO) does, and it certainly uses quantum fields. We tend to think that the gravitational field is an objective fact in the world as it is highly classical. We do not need to worry about graviton statistics. The analogue, in the context of the electromagnetic field, are those states excited by large classical current sources.

Recently, experimentalists have started probing highly non classical sources of gravity. A simple example is a single massive object in a superposition of 'two places' . This only makes sense if there is a background reference frame. The idea of 'two places' presupposes a further background spacetime metric in addition to that produced by the gravitational field of the massive object itself. This is the analogue of the EM field of a charge in a superposition at two places in an ion trap. We could simply claim that there is no objective field until we measure it. But there is a big difference. Gravity has the same mathematical structure as spacetime (the strong equivalence principle). Do we really want to say spacetime is not an objective feature of the world but only revealed, however obscurely, by particular measurement results? This would require us to claim that spacetime events are comprehensively equivalent to measurement results and thus contingent on the kinds of measurement we make. What is at stake in these new experiments is not the reality of gravity or whether gravity is quantum or classical, but the reality of spacetime. It must emerge as an observer-dependent feature by making measurements on something more fundamental. Is this an opening for a better spacetime story of the Bell violations?

2.1.1 Bell tests with three agents

The standard description of photonic tests of Bell inequalities involves two agents and an entangled photon source. I will include an additional agent to make it clear how the required correlation functions are constructed.

A standard Bell test with polarization-entangled photons is shown in Fig. (2.1).

In an ideal experiment, we use single-photon excitations of polarized spatio-temporal modes such that the photons are emitted in opposite directions, with anti-correlated polarization states. The photons are entangled in the polarization. For example, the source could prepare, in every trial, the two-photon state,

$$|\Psi^-\rangle = \frac{1}{\sqrt{2}}(|HV\rangle - |VH\rangle) \tag{2.1}$$

where $|HV\rangle = |1\rangle_{k_A,H} \otimes |1\rangle_{k_B,V}$ and the subscripts label spatio-temporal modes with wave vectors k_A, k_B and corresponding polarization. A single trail corresponds to emitting a two-photon state and counting two photons; one at observer-a and the other at observer-B. If for some reason two photons are not detected, that trial is discarded. As no detector is perfect, this is likely to happen quite often, raising the detection loop-hole.

Let the measurement settings be chosen from a set of discrete rotations $x \in \{\theta_{A,1}, \theta_{A,2}, \ldots \theta_{A,n}\}$ and $y \in \{\theta_{B,1}, \theta_{B,2}, \ldots \theta_{B,n}\}$. Suppose the detector settings are the same, $x = y$, that is to say, the same angle is chosen. We can rotate both angles jointly until we see a perfect anti-correlation at each output. When the photon at A is detected at $a = 1$ channel, the photon at B is detected at $b = -1$ channel, and vice versa. However, from trail to trial, the local measurement outcomes are a random binary numbers ± 1.

In order to see the correlation, an observer must have access to both outcomes in each trial. In the lab this is obvious, as the experimentalist collects all the data from both detectors in each trial. We make this explicit by introducing a 'checker' – labelled C – that receives the data (setting and outcome at each detector) from each observer in a trial in the future light cone of the detection

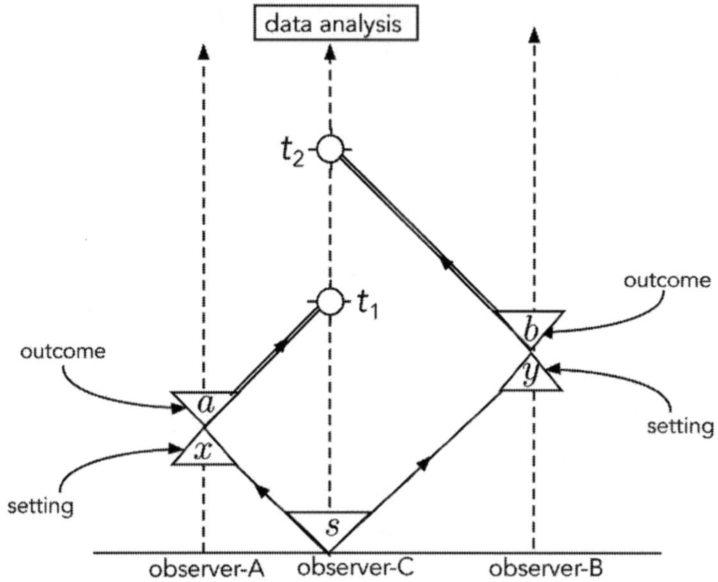

Figure 2.1 A two-party Bell experiment with entangled photons. A source at the origin produces pairs of entangled photons. The photons occupy oppositely directed spatial modes; one goes to detector-A and the other goes to detector-B. Both observers are space-lie separated. After each measurement, the setting and the outcome are sent, over classical channels, to a checker, observer-C, who stores the data for each trial and constructs the appropriate correlation function to check a Bell inequality.

events. For convenience we will suppose the checker is at the same place as the source. Note that the checker receives purely classical information. A space time diagram for the experiment is shown in Fig. 2.1. Observer-C is at rest in the frame of the source and thus can easily synchronize emission and detection events to ensure that data is collected from the right photon pair in each trial.

The analysis of the data is well known. Typically, it involves computing a correlation function known as the CHSH correlation function. Classically this correlation function is bounded by 2. Quantum mechanics predicts that it is bounded by $2\sqrt{2}$ [5], and many experiments have demonstrated that the classical bound is exceeded [11].

2.1.2 Bell tests with one agent

Consider the case depicted in Fig. 2.2. In this case we use the same source as in the usual Bell scenario but now mirrors, *asymmetrically* displaced on either side of the agent, reflect the photons back to a single agent at the source who

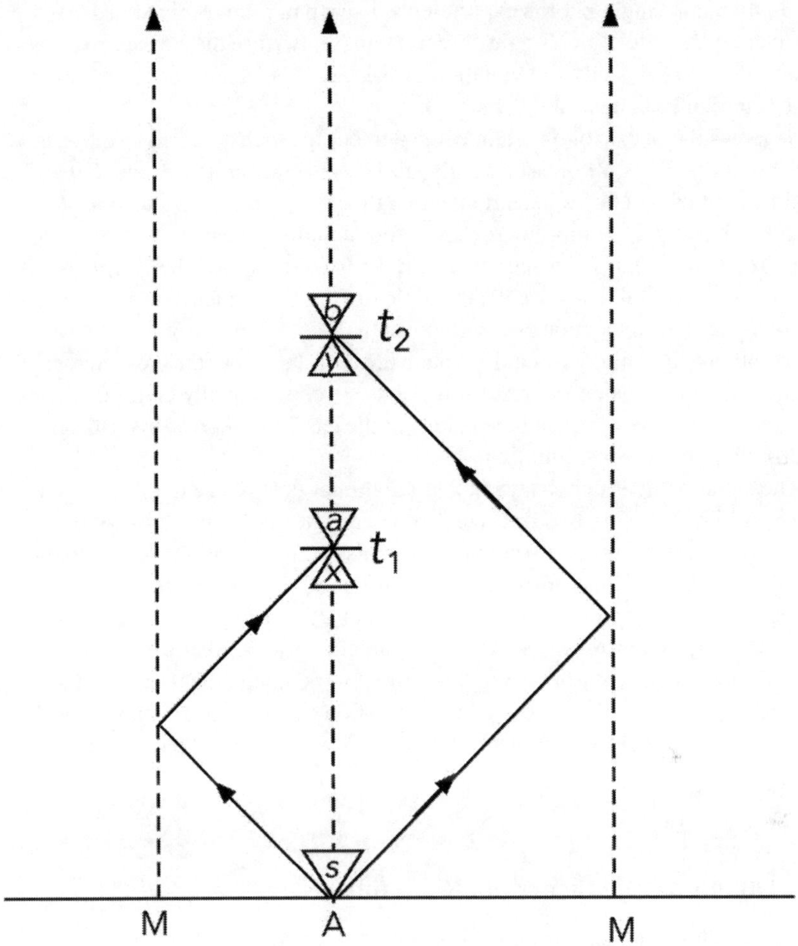

Figure 2.2 A Bell test with a single observer/agent. The agent generates an entangled photon pair, in a known state (indicated by a joint measurement result s) and each photon travels in opposite directions where it is reflected by a mirror at rest in the agent's rest frame. One photon is received back by the agent at the early time t_1 and the other photon is received back at the later time t_2. The dashed lines represent the world lines of the agent and the two mirrors. Variables x, y represent the measurement settings and variables a, b represent the measurement outcomes as in a standard Bell test.

will do the same single-photon experiments on each photon received in a given trial to test a Bell violation. Single observer can easily store the classical results of experiments and verify the violation of the CHSH inequalities at any point in the future of both measurements.

Where is the puzzle of quantum entanglement for such an agent? Nonlocality is not an issue as all measurements are time-like separated. Nevertheless, Bell inequalities will be violated if quantum theory is correct. As quantum correlations are non signalling, there can be no signalling from t_1 to t_2 or vice versa. Yet there remains a deeper puzzle. If we tried to explain the correlations in terms of a local hidden variable, the logic of Bell's argument would imply a symmetric casual connection even without signalling; in other words, retro causation without signalling. We could explain the correlations as the measurement results at time t_1 causing the results at t_2 but we could equally claim that the measurement at time t_2 caused the results at the earlier time, t_1. This is deeply at odds with our classical intuition.

The situation does not change in a gravitational field. Consider the scheme shown in Fig. 2.3. This is a one-agent protocol but one of the mirrors is replaced by a large mass. The single photon pulse travelling to the left is blue-shifted going towards the mirror and red-shifted going away from the mirror. The net effect is simply a delay in the local detection time at the agent. This is the classical Shapiro shift [9]. It has no effect on the degree of violation of the Bell inequality. Even in curved spacetime, the entanglement does not see classical gravity and the retrocausal interpretation remains. It is a purely local phenomenon. Likewise, if the large mass is co-located with the agent at the origin.

2.2 Quantum field theory and spacetime events

In Einstein's formulation, the gravitational field is determined by physical measurements made with clocks and rulers. Unfortunately Einstein was a little vague on just what he meant by local clocks and rulers, and even admitted that this was an inconsistency in the theory [13].

> First, a critical remark on the theory as characterised above. It was noticeable that the theory introduces (besides four-dimensional space) two kinds of physical things, namely 1) rods and clocks, 2) all other things, e.g. the electromagnetic field, the material point, etc. This is in a sense inconsistent; rods and clocks should actually be presented as solutions to the basic equations (objects consisting of moving atomic structures), not as so to speak theoretically self-sufficient beings. However, the procedure is justified by the fact that it was clear from the beginning that the postulates of the

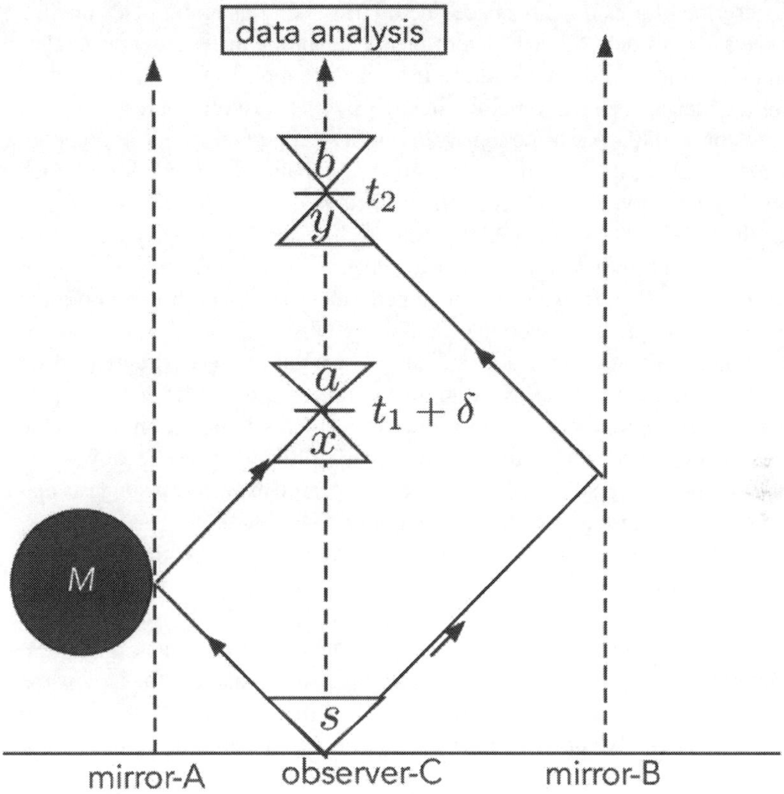

Figure 2.3 A single agent Bell test with gravity.

> theory are not strong enough to deduce from it sufficiently complete equations for physical events sufficiently free of arbitrariness to base a theory of rods and clocks on such a foundation. Unless one wanted to do without a physical interpretation of the coordinates altogether (which in itself would be possible), it was better to allow such inconsistencies—albeit with the obligation to eliminate them at a later stage of the theory. However, one must not legitimise the aforementioned sin to such an extent that one imagines that distances are physical beings of a special kind, essentially different from other physical quantities ("reducing physics to geometry," etc.).

Einstein also made extensive use of classical light pulses to coordinate physical events. We would now replace this with quantum fields. In [6] various schemes were described for estimating spacetime metrics using quantum fields and the ultimate accuracy achievable is determined by quantum uncertainty principles. Kempf has outlined a similar idea [7]. If we deduce metrics from measurement outcomes, then spacetime events are identified with measurement results taking place in some finite spacetime four volume. In a properly formulated quantum field theory, all observers must agree on the probability distribution of such events. Spacetime is objective if a little uncertain.

In the case of flat spacetime, this approach can easily identify an inertial reference frame if we use semi-classsical states of light. However, what kind of spacetime are we to infer using the entangled states in the one-observer protocol described in the previous section? Identifying spacetime with these kinds of measurements already implies a retrocausal structure even in the case of no gravitational field, as we described in the previous section. This would impact the causal set approach to quantum gravity [8]. This is based on taking the causal structure of general relativity as axiomatic, although what is really meant by casual is in fact signalling. The single observer Bell experiment suggests that conflating causal structure with signalling might be unwise.

Problems arise if the quantum fields act back on the gravitational field via the stress-energy tensor. It is relatively easy to see that this must bound the minimum spacetime four volume that can be used to localize a measurement outcome and justify our claim that spacetime events are measurement outcomes. This is because all physical measurements take some time and occupy some three volume. There is a lower bound to this. If a measurement takes place too fast, or is to spatially confined, then the Heisenberg uncertainty principle implies a huge fluctuation of the stress-energy tensor. At some point a black hole is created and the measurement event is causally disconnected from every other observer. This has a physical implication for causal set theory. In that approach, the number of distinct measurement events defines a spacetime volume. If measurement back-action is taken into account, there is a maximum event density in spacetime, and an effective stochastic discreteness to spacetime volumes.

2.3 Gravitational decoherence

The fundamental problem in quantum gravity is this: if a single massive object is prepared in a superposition of two different locations with respect to a fixed coordinate frame, the resulting gravitational field must be non stationary. An example is shown in Fig. 2.4. If instead of the same superposition state for the source mass being used in every trial, we place the mass at one or the other position at random from one trial to the next, the clock will experience the same random red-shifts; however, in this case, it is obvious that the gravitational field is non stationary; we keep changing it from one trial to the next. In more

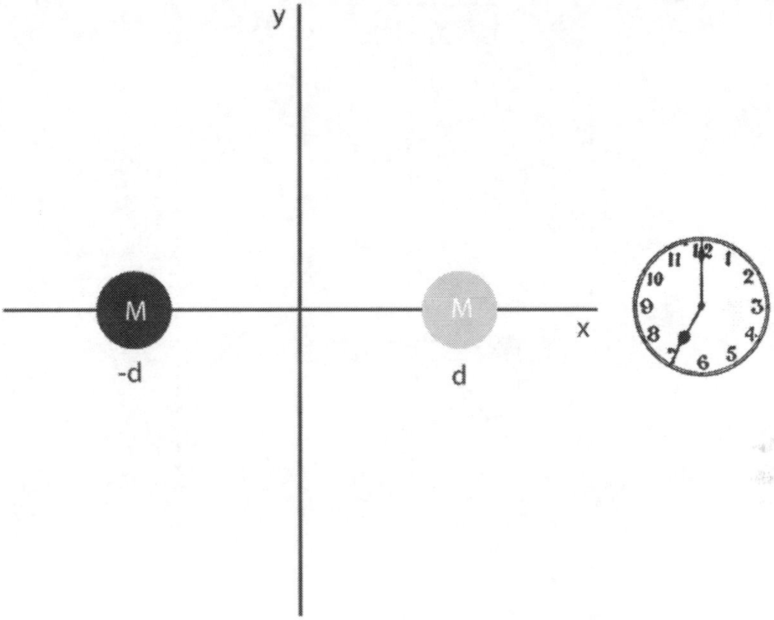

Figure 2.4 A single mass is in a superposition of two locations with respect to a fixed coordinate frame (the earth, say). A very sensitive clock is placed to one side. In a series of repeated trials with exactly the same conditions, the clock will experience two unequal red-shifts, randomly choosing one or the other in each trial. As the clock is the only way we can infer the existence of a gravitational field in this setting, we conclude that the gravitational field at the clock is fluctuating. It is not stationary.

technical terms, the clock experiment cannot distinguish between an initial pure superposition state, which is a zero entropy state, and a maximally mixed state with non zero entropy. We might distinguish the two cases using the names 'pure quantum gravity' versus 'stochastic classical gravity'.

Penrose proposed that we can never actually prepare the pure state required for the first experiment as it will spontaneously collapse into one location or the other as described by the random mixture of the second experiment. He did not give an explanation for how this can happen. Many people have now devised experiments [3] that could in principle distinguish the pure state from the classical mixture. These experiments could distinguish pure quantum gravity from stochastic classical gravity.

The reason is simple: pure quantum gravity can become entangled – quantum correlated – with internal degrees of freedom of particles, whereas

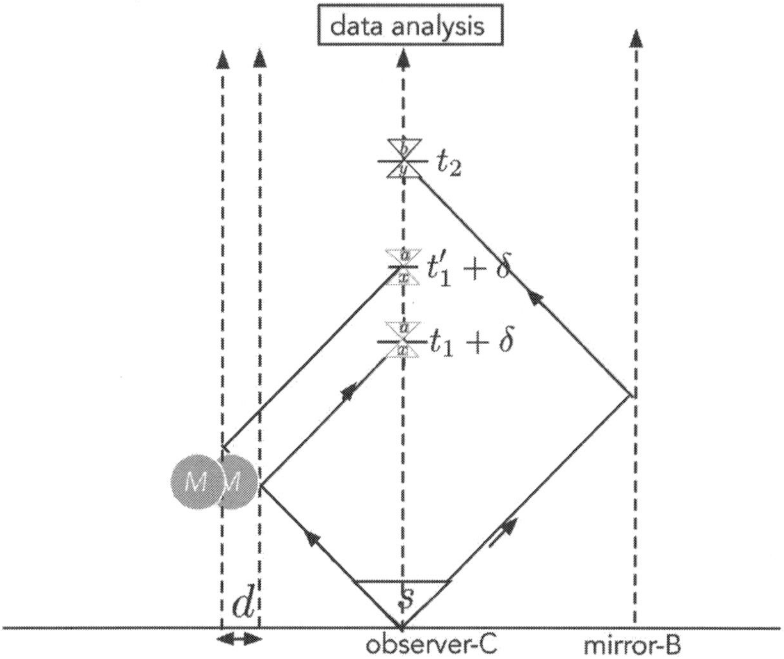

Figure 2.5 A scheme for using time-bin entanglement and a single agent Bell test to search for gravitational decoherence.

stochastic classical gravity cannot; it can only be classically correlated. Conversely, stochastic classical gravity can control quantum systems but cannot entangle them. An entanglement witness can reveal the difference. Any attempt to erase which-path information in an entanglement witness, without acting on gravitational degrees of freedom, will fail.

Let us return to the question of the gravitational field of a non classical source, such as a large mass in a superposition of two displacements with respect to the rest frame of a source of Bell pairs, see Fig. 2.5. The photon travelling to the left will now return to be detected at two possible times. The Shapiro shift is the same as the mass is the same. If the photon samples a fluctuating gravitational field, the Shapiro shift will be stochastic from pulse to pulse. In order to consider the Shapiro effect on a Bell test, we switch to time-bin entangled photons [2] rather than polarization. We can use a version of time-dependent multiplexing to encode a qubit into a sequence of single photon pulses, see Fig. 2.6. There will be no change to a Bell violation, provided the temporal and gravitational degrees of freedom do not become entangled. A time-bin entan-

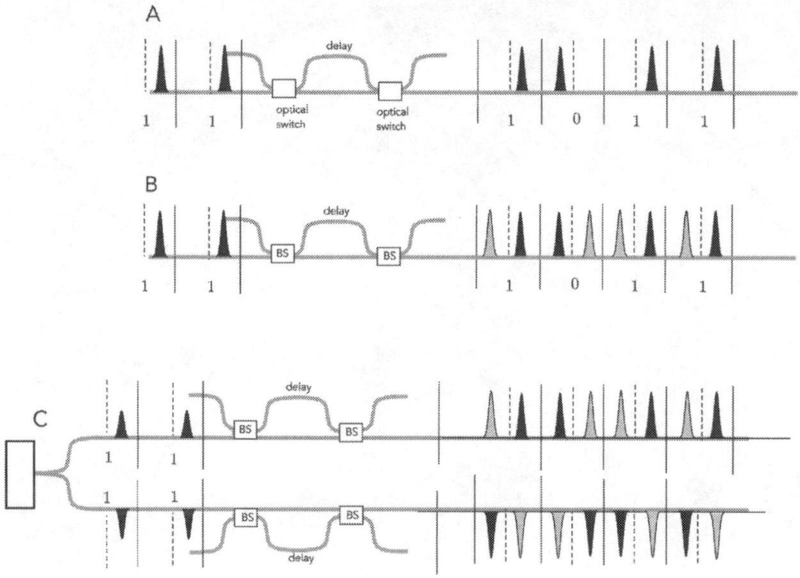

Figure 2.6 [A] A classical optical circuit to encode a bit string into pulse time code. An incoming stream of equally spaced coherent pulses is optically switched onto a direct path or a delay path to encode bits as early or late pulses in each time bin. [B] A time-bin qubit encoder. A sequence of single photon pulses is input to an optical circuit using 50/50 fibre couplers to create an equal quantum superposition of logical bits. This is a time-bin encoded qubit. Note that the total photon number in each time bin is one. [C] A single down conversion source creates pairs of photons simultaneously. Each path passes through a single qubit gate which creates pairs of photons in which one is delayed with respect to the other but we do not know which.

glement Bell test could then be an entanglement witness. If the temporal history of the photons does not become entangled with the gravitational field, the Bell violation can be maximal, and otherwise it is reduced. Note that in this single agent bell test, the operations required by the agent are entirely local.

2.4 Ringworld: A toy model

A common theme in the preceding discussion is that a Bell test can be violated by a single agent with a local clock and local interventions (preparation/measurement). The global structure of spacetime is unknown to the agent;

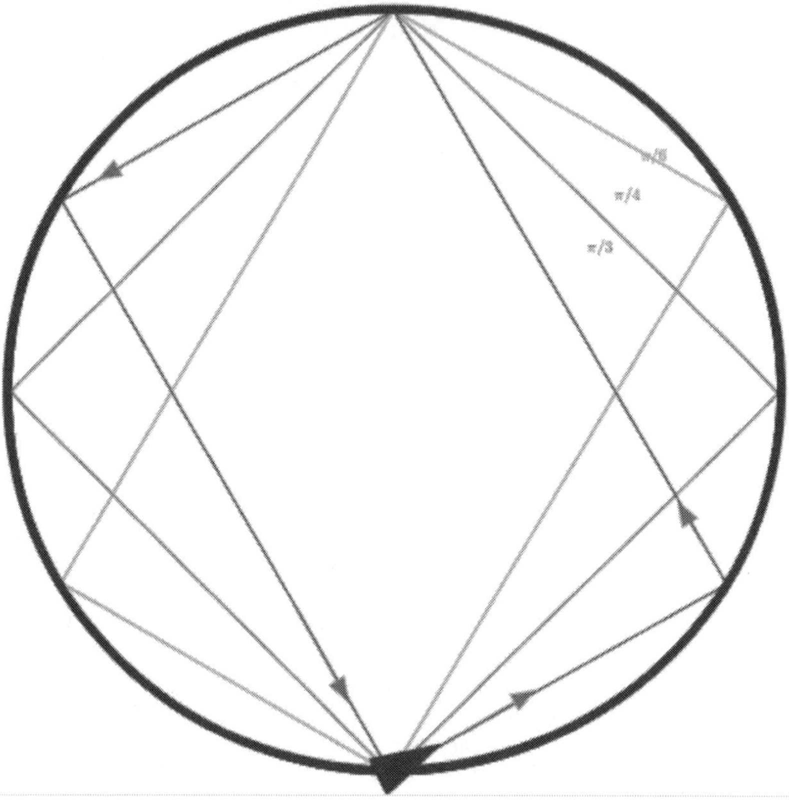

Figure 2.7 An agent on the edge can emit light pulses in different directions. They are reflected from the edge and return to the agent. The agent can control the direction using internal actuators.

it only has a view from the inside. When thinking about such experiments, it is hard to take the inside view of the agent and not import our intuitions of the view from the outside. We will now present an 'intuition pump' which will help you move towards a view 'from the inside out.'

Suppose a learning agent is moving on a unit circle with a reflecting boundary. Inside each agent is a clock and a gyroscope. The agent emits a light pulse at each tick of internal clock. This determines the rate r of pulse emission. The gyroscope estimates the direction of its pointer as a function of the ticks of the internal clock. An example of possible configurations is shown in Fig. 2.7. Thus $\theta = \pi/2$ is along a radius and orthogonal to the tangent to the circle at the agent. We will assume that $\theta = k\frac{\pi}{2K}$, for some integers $k = 0, 1, \ldots K >> 1$. We will

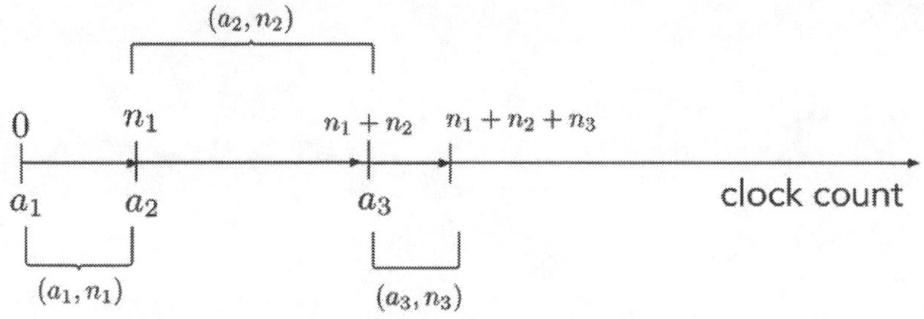

Figure 2.8 The scheme of agent interventions (head angle at emission) and sensations (clock count at pulse reception) versus clock count together with the data training pairs sent to the internal learning machine.

set the speed of light, $c = 1$ and use units of length such that the radius of the circle is unity.

The gyroscope is a simple sensor that measures angular accelerations of the agent's head direction. When combined with a clock, it enables the agent to keep a record of head direction. The agent can only 'know' its internal states indexed by ticks of the internal clock.

We will assume that the agent emits pulse immediately after receiving a pulse: if a pulse is received at a clock count of n, a pulse is emitted at the same clock count of n. The head direction at each step determines the clock count of the *next* emission. Each emission event corresponds to a distinct internal state labelled by a setting of its internal gyroscope a, and a reading of the clock at pulse emission n. Our protocol means that a pulse received at clock count of n is a pulse returning from the previous emission when the head direction was a. The only things the agent has access to are these two things. An internal state is an ordered pair of numbers $S = (a, n)$, where n is the clock count for the next emission and a is the record of head direction at the previous emission. This is summarized in Fig. 2.8

We now equip the agent with an internal learning machine. It works like this. At each tick of the internal clock, the learning machine is sent the head direction a. It then quickly tries to predict when the next light pulse will be received by generating an integer N. For simplicity, we will assume there are only four settings for a. These are labelled with angles given by $0.01\pi, 0.1\pi, 0.15\pi$ and 0.2π, but the agent does not know this. It only knows that there are four different headings, as recorded by its internal gyroscope. At each step one of these four headings is chosen at random.

Assume that the agent is stationary. Given our external god-like view of the agent's world, from the outside in, we can easily give a relation between head

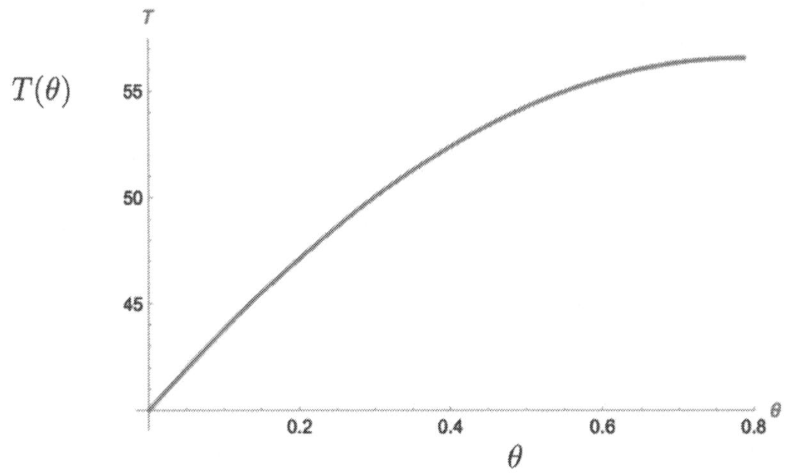

Figure 2.9 A plot of how return time of a pulse varies with head angle. arbitrary units for time.

angle and time taken for a pulse to return to the agent. We assume that spacetime is flat inside the disc. The time taken is then given by $T(\theta) = 40(\cos\theta + \sin\theta)$. We have used an arbitrary scale for time units. The shortest period occurs for a pulse sent along a diagonal and the longest for a pulse sent at 45 degrees to the diagonal. See Fig. 2.9. The agent does not know this function. It simply generates a list of ordered pairs, a label for the head angle and number of ticks of the clock until the pulse returns. The data can be displayed as shown in Fig. 2.10.

The physical machine that implements the learning could be specified in various ways. This can be almost anything, for example, a physical neural network or a physical restricted Boltzmann machine. We will assume that it is a machine that implements a neural network algorithm and that it has a very large number of examples to train on. In Fig. 2.11, we plot an example of the predictions made by this machine once it has been trained using 1000 training pairs. The learned function is stored as physical settings of the physical learning machine inside the agent. In so far as it has learned this functions it has learned a proxy for Euclidean geometry.

In this world, the agent is the only source of light. If there are extrinsic sources of light pulses that do not originate from agents, the training data of the agent is corrupted. Occasionally one of these pulses will be received by the agent. How does the agent distinguish these 'background' pulses from those that the agent itself emits?

Heading	Count
1	41
2	50
4	56
3	54
3	54
3	54
3	54
4	56
3	54
4	56

Figure 2.10 A sample of typical data, actions (head angle) and sensations (count to received pulse), used as inputs to the learning machine. Head directions are settings of an internal gyroscope and chosen at random.

From the agent's point of view, these random emissions are 'background noise'. A key feature of learning is the ability to cope with some uncertainty in the data. In this case, the sensor records are not perfectly correlated with the recordings of head direction. How much noise can be tolerated before learning begins to degrade?

If it receives a random light pulse, the correlation inherent in the ordered pairs (a_j, n_j) is contaminated by independent errors in the n_j. We can represent this by adding/subtracting a small random integer, e_j, to/from each count

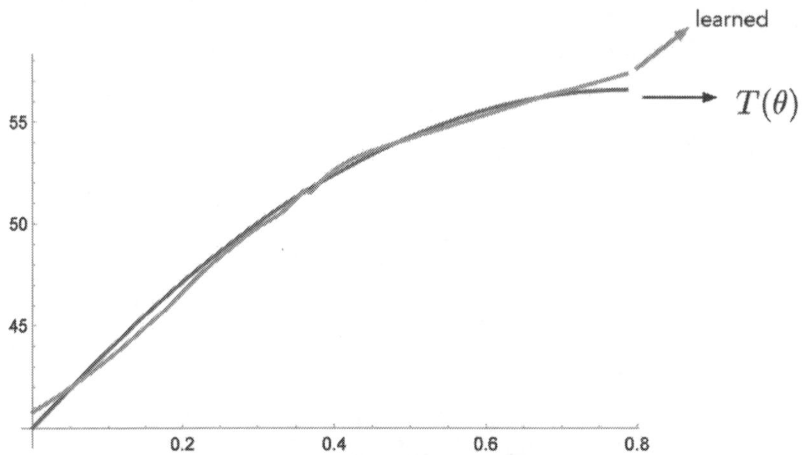

Figure 2.11 An example of the prediction made by a learning machine inside the agent using a large data set of the kind shown in Fig. 2.10. Using only four randomly chosen head directions the $T(\theta)$ curve is the ground truth function that agent is trying to learn.

component.

$$(a_j, n_j) \rightarrow (a_j, n_j + e_j), \quad |e_j| < n_j \tag{2.2}$$

The emission rate from background sources compared to the agent's emission rate is an important fact. If it is small, we might expect the agent can still learn how to control the return time of pulses with changing head direction. Assume e_j is a random integer between -4 and 4. In Fig. 2.12, 1000 trials are used to learn the unknown function. The prediction remains quite good.

We now turn to a non Euclidean example. The analogue of de Sitter space is the Poincaré disc, while the anti-de Sitter space is equivalent to Maxwell's fish eye lens [1]. We treat the latter here. The Maxwell fisheye lens, in the disc of radius R, has a radially dependent refractive index,

$$n(r) = \frac{2}{1 + (r/R)^2} \tag{2.3}$$

We will assume that $R = 1/2$ and thus the refractive index is equal to one on the boundary. Light rays propagating within an infinite two-dimensional plane lens, rays trace out perfect circles. In the case of a disc, rays starting on the boundary are focused at the antipodal boundary point; see Fig. 2.13. Each curve is defined by a head angle as in the flat space case. The situation for a reflecting circular boundary, of radius $1/2$ centred at the origin, is treated in [4].

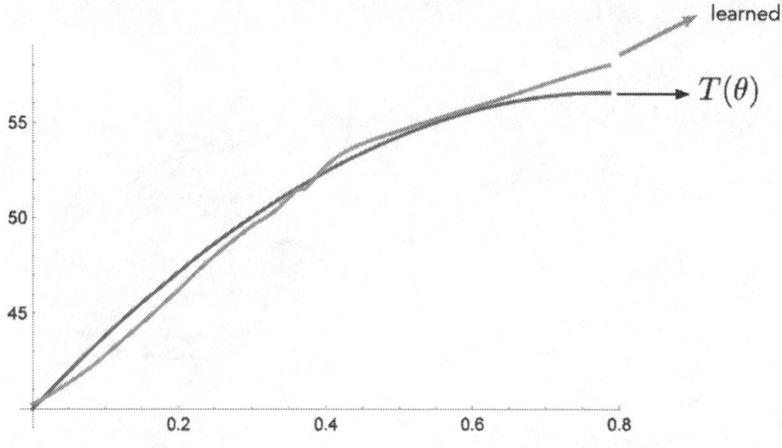

Figure 2.12 A comparison of the unknown function and the learned function when extrinsic light pulses corrupt the training data. The number of training samples is 1000.

In [1] it is shown that the null geodesics on the sphere of radius R correspond to the light rays in the Maxwell fisheye disc, see Fig. 2.13. This implies that the time taken for a light pulse to travel from one side of the disc to the other is independent of the head direction, unlike the flat space case previously discussed.

Each agent is equipped with an internal 'gyroscope' that determines the head direction θ and an internal 'clock' that counts the time taken for a light pulse to be returned. In the case of an empty disc, the time taken depends on the head direction, and the agent can learn the relationship given a simple internal learning machine. In the case of the Maxwell fish eye disc, the time taken is independent of head direction. These are the two 'laws of physics' that the agents learn for each case. In both cases the law can be learned using only local actuators and sensors inside the agent. It knows nothing abut the propagation of light from an external, god-like, view. It projects the learned law 'from the inside out'.

2.5 Conclusion

Einstein constructed general relativity using a profound intuition about how we use local clocks and rulers, yet it was a little vague how such things were built from the fundamental theory. In a quantum world, local clocks and rulers are

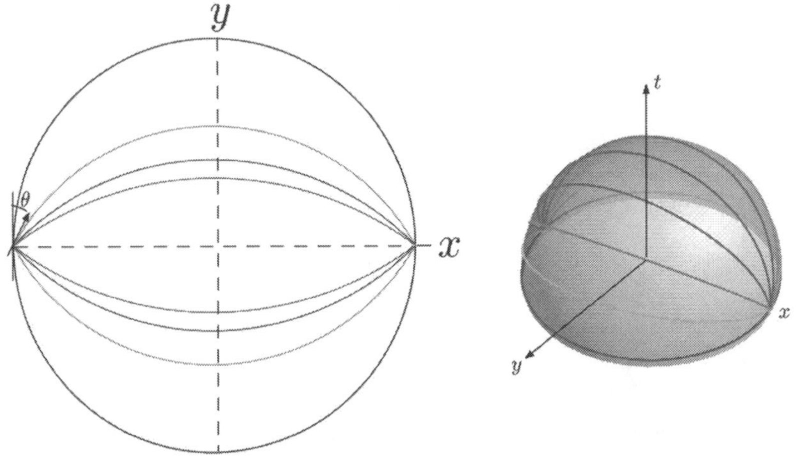

Figure 2.13 The interior of the disc is composed of a spatially varying refractive index that decreases away from the centre. Innermost curve: $\theta = \pi/5$, middle curve: $\theta = \pi/4$, Red=$\theta = \pi/3$.

replaced with local measurements made with quantum sensors. The events from which spacetime is constructed must be classical measurement results made on quantum fields.

If gravity is spacetime, then we must regard gravity as sharing the features of quantum measurements: irreducible stochasticity and contextuality. However, there is a catch here: it is very difficult to conceive of measurements without positing a background spacetime. In the usual test of Bell inequalities with three agents, two of whom are space-like separated, we simply assume that the measurement devices 'have' spacetime coordinates. Yet the observed violation of the Bell inequality is independent of the spacetime coordinates of the detectors involved. We described a single-agent Bell experiment the results of which are the same as the usual three-agent scenario. When the agents are space-like separated, we feel uneasy, but we should also feel uneasy when they are time-like separated as it seems to imply local retrocausality.

We have argued that in thinking about how to make gravity consistent with quantum theory, we should take an inside-out view of the world, an agent centric view in which notions of global time and space are secondary, if not entirely illusory.

Acknowledgements

I wish to thank Peter Evans for useful discussions.

References

[1] Chen, H., Tao, S., Bělín, J., Courtial, J., and Miao, R.-X. (2020). Transformation Cosmology. *Physical Review A*, 102(2), 023528.

[2] Halder, M., Beveratos, A., Gisin, N., Scarani, V., Simon, C., and Zbinden, H. (2007). Entangling Independent Photons by Time Measurement. *Nature Physics*, 3, 692.

[3] Adlam, E. (2022). Tabletop Experiments for Quantum Gravity Are Also Tests of the Interpretation of Quantum Mechanics. *Foundations of Physics*, 52, 115.

[4] Perczel, J., Komár, P., and Lukin, M. D. (2018). Quantum Optics in Maxwell's Fish Eye Lens with Single Atoms and Photons. *Physical Review A*, 98(3), 033803.

[5] Scarani, V. (2019). *Bell Nonlocality*. Oxford Graduate Texts. Oxford: Oxford University Press.

[6] Downes, T. G., van Meter, J. R., Knill, E., Milburn, G. J., and Caves, C. M. (2017). Quantum Estimation of Parameters of Classical Spacetimes. *Physical Review D*, 96(10), 105004.

[7] Kempf, A. (2021). Replacing the Notion of Spacetime Distance by the Notion of Correlation. *Frontiers in Physics*. `https://doi.org/10.3389/fphy.2021.655857`

[8] Surya, S. (2019). The Causal Set Approach to Quantum Gravity. *Living Reviews in Relativity*, 22.

[9] Shapiro, I. I. (1964). Fourth Test of General Relativity. *Physical Review Letters*, 13(26), 789–791.

[10] Morette Dewitt, C., and Rickles, D. (2011). The Role of Gravitation in Physics. In *Report from the 1957 Chapel Hill Conference*. `https://api.semanticscholar.org/CorpusID:125187268`

[11] Gisin, N. (2013). Are There Quantum Effects Coming from Outside Space–Time? Nonlocality, Free Will and "No Many-Worlds." In Antoine Suarez and Peter Adams (Eds.), *Is Science Compatible with Free Will? Exploring Free Will and Consciousness in the Light of Quantum Physics and Neuroscience* (pp. 23–39). New York: Springer.

[12] Buzsáki, G. (2019). *The Brain from Inside Out*. Oxford: Oxford University Press.

[13] Einstein, A. (1970). Autobiographical Notes. In Paul A. Schilpp (Ed.), *Albert Einstein: Philosopher–Scientist* (pp. 1–94). New York: Harper & Row Publishers.

3 On the Role of Locality in the Bose-Marletto-Vedral Effect

Giuseppe Di Pietra, Vlatko Vedral and Chiara Marletto

3.1 Introduction

A particularly promising approach to testing quantum gravity has recently been proposed based on a novel "witness of non-classicality". This witness relies on the entangling power of a given system to conclude that the system has non-classical features. In particular, these tests are based on the so-called general witness theorem (GWT) [2, 25], stating that if a system M (such as gravity) can mediate (by local means) entanglement between two quantum systems, A and B (e.g. two masses), then it must be non-classical [2]. By "local means" here we mean a specific protocol, detailed in [1–3], where A and B must not interact directly with each other, but only via the mediator M, as schematically represented in Fig. 3.1. Interestingly, "non-classicality" is a theory-independent generalisation of what in quantum theory is expressed as "having at least two distinct physical variables that do not commute", which can be expressed within a general information-theoretic framework, the constructor theory of information [24]. Informally, being non-classical means having two or more distinct physical variables that cannot simultaneously be measured to an arbitrarily high degree of accuracy [2]. Due to its generality, the GWT offers a broad theoretical basis for recently proposed experiments that can test quantum effects in gravity at the laboratory scale, based on the generation of gravitational entanglement between two massive probes – the so-called Bose-Marletto-Vedral effect [1, 3]. It also provides a basis for any other experiment that (beyond the case of gravity) intends to show that some system M is non-classical [5], using the effect of its entangling power.

DOI: 10.4324/9781003546139-5

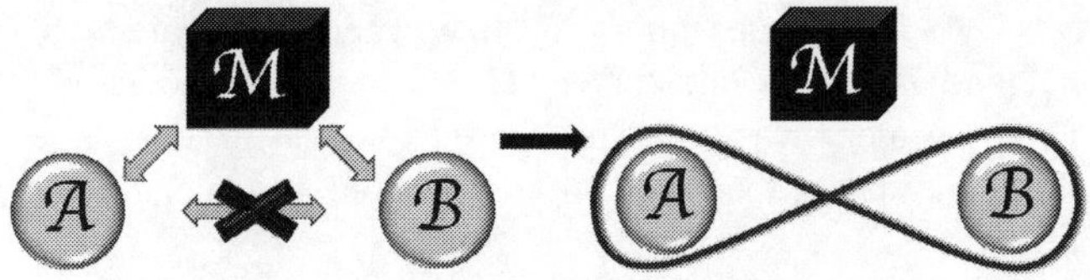

Figure 3.1 Schematic representation of the setup for the general witness theorem. The two space-like separated quantum probes A and B are coupled only via the unknown system M by means of local interactions. Its capability of inducing entanglement between A and B would then be a witness of its non-classicality.

A particularly appealing feature of the GWT is that by using the constructor theory of information, it avoids assuming the usual machinery of quantum information theory, thus extending beyond quantum theory existing results such as the theorems that forbid the creation of entanglement via local operations and classical communication. Moreover, the GWT is proven without assuming the existence of a probability space, in contrast to existing approaches such as generalised probabilistic theories [22]. This generality is particularly important as one wants to use it in a context where the system M may or may not obey quantum theory itself.

This witness relies on the capacity of a system M to generate entanglement between two independent subsystems, initially unentangled with each other. For the witness to be applicable, it is key that the systems A and B are independent – hence it is essential to assume the *principle of locality*. We shall discuss here (1) what the minimal notion of locality on which the witness relies is; (2) how the principle of locality is different and more general than other notions of locality in physics and (3) what known theories satisfy this principle, and hence what the implications of the witness are for those theories. We shall also briefly highlight how the principle of locality is used in the proof of the GWT.

3.2 Summary of the witness of non-classicality

In this section, we recall the key idea of the witness and explain, using an example from quantum theory, how the witness works to conclude that a system is non-classical. A witness of non-classicality is a protocol to probe a system M, whose dynamics is partly unknown, with one or more fully quantum probes Q, to the end of establishing whether M has some quantum features by measuring only the quantum probes.

The scenario we have in mind is one in which M has dynamics that could be fully classical (as in the case of gravity) or affected by dynamical collapse (as in the case of a macroscopic object, e.g., a complex biomolecule).

The witness we shall focus on is the one where there are two quantum probes, A and B, which interact via M – which is a mediator. If one can set up an experiment where the probes interact via M only, and manage to get entangled via M, then the GWT allows one to conclude that M is non-classical.

As we mentioned, the GWT can be proven in a very general framework, without assuming quantum theory's formalism. In this section, however, we shall give a specific example of how the theorem works, using quantum theory. In this example, M being non-classical means that it has at least two variables that do not commute.

Let us assume, for simplicity, that A and B are two qubits. Since our goal is to focus on the role of the principle of locality in this witness, we shall describe the qubits using the descriptors formalism, which stems from the Heisenberg picture of quantum theory [18, 27]. In this formalism, one describes the qubit A with a vector of *descriptors*:

$$\begin{aligned}\hat{q}^{(A)}(t_0) &:= \left(q_x^{(A)}(t_0), q_y^{(A)}(t_0), q_z^{(A)}(t_0)\right) \\ &= \left(\sigma_x \otimes \mathrm{I}_M \otimes \mathrm{I}_B, \sigma_y \otimes \mathrm{I}_M \otimes \mathrm{I}_B, \sigma_z \otimes \mathrm{I}_M \otimes \mathrm{I}_B\right), \end{aligned} \tag{3.1}$$

where σ_k, $k = x, y, z$, are the Pauli operators and I_i, $i = M, B$, is the identity operator on the qubit Q_B and the unknown system M. Similarly:

$$\begin{aligned}\hat{q}^{(B)}(t_0) &:= \left(q_x^{(B)}(t_0), q_y^{(B)}(t_0), q_z^{(B)}(t_0)\right) \\ &= \left(\mathrm{I}_A \otimes \mathrm{I}_M \otimes \sigma_x, \mathrm{I}_A \otimes \mathrm{I}_M \otimes \sigma_y, \mathrm{I}_A \otimes \mathrm{I}_M \otimes \sigma_z\right), \end{aligned} \tag{3.2}$$

is the vector of descriptors of the qubit Q_B. The descriptors allow us to keep track of the evolution of each qubit's *local algebra of observables* while acting on the whole Hilbert space. Notice that $\left[q_k^{(i)}, q_l^{(j)}\right] = 0, \forall i \neq j,\ k, l = x, y, z,$ which implies that quantum observables of two different, non-interacting subsystems must commute.

With this picture in mind, let us now prove the GWT in quantum theory. We recall that being this witness a *sufficient* condition for the non-classicality of the mediator M, we shall assume that at the end of the protocol, the two quantum probes A and B end up entangled.

Considering to the setup shown in Fig. 3.1, let us assume M to be a classical system. Its vector of descriptors will thus have a single component,

$$q_z^{(M)}(t_0) := \mathrm{I}_A \otimes \sigma_z \otimes \mathrm{I}_B, \tag{3.3}$$

following our definition of classicality, to reflect the existence of a single observable for it. We shall identify this observable with the Pauli Z operator. At time t, the quantum probe A interacts with the mediator M. The most general state of the system $A \oplus M$ is:

$$\rho(t) = \frac{1}{4}\Big(\mathrm{I} + \vec{r}_A \cdot \hat{q}^{(A)}(t) + s_z q_z^{(M)}(t) \\ + \vec{t}_A \cdot \hat{q}^{(A)}(t) q_z^{(M)}(t)\Big), \tag{3.4}$$

where $\vec{r}_A$ and $\vec{t}_A$ are real-valued vector, $s_z \in \mathbb{R}$, and $\hat{q}^{(A)}(t)$ and $q_z^{(M)}(t)$ are functions of $\hat{q}^{(A)}(t_0)$ [Eq. (3.1)] and $q_z^{(M)}(t_0)$ [Eq. (3.3)] only, respectively. This state, interpreted as a two-qubit state, is *separable*, meaning that no quantum correlations can be generated between the quantum probe A and a classical mediator M. Thus, it would be impossible to find entanglement between A and B at the end of the protocol as the final state would be separable too, contradicting the assumption of the witness. The reason why $\rho(t)$ in Eq. (3.4) is separable lies in the classicality of M: the mediator needs (at least) another observable that does not commute with σ_z. Thus, observing entanglement between A and B at the end of a protocol performed *by local means only* leads to the conclusion that M must be non-classical, according to our definition of non-classicality.

But why is the locality of the protocol so crucial? The formalism of descriptors in quantum theory allows us to understand immediately the consequences of dropping the locality assumption. Let us modify the GWT setup as shown in Fig. 3.1 and assume that the quantum probes A and B can interact directly in some way. This means that the descriptors of quantum probe A at time t will become a function of the descriptors of quantum probe B at the previous time t_0 in Eqs. (3.1) and (3.2), i.e., $\hat{q}^{(A)}(t) = f(\hat{q}^{(A)}(t_0), \hat{q}^{(B)}(t_0))$. If this is the case, then the state in Eq. (3.4) *can* result in an entangled state for A and B at the end of the protocol even with a classical system M as mediator. Thus, observing the entanglement between the quantum probes will *not* lead to the conclusion that M must be non-classical: there would be no contradiction between observing A and B entangled and the mediator M having only a single variable. In fact, the entanglement in question could have been generated already before A's interaction with M, irrespective of the mediator.

Thus, the assumption of locality is crucial in this example to "force" M in using (at least) two non-commuting variables to entangle A and B: one is used to entangle A and the other to transmit the quantum correlations between the two quantum probes. However, an effective witness of non-classicality must be formulated without relying on the formalism of quantum theory, in order to consider the possibility that the unknown system M may also be described by postquantum theories.

As we shall now explain, the locality assumption can be stated clearly as a general principle, without relying on a specific dynamics, and distinguished carefully from other notions of locality, some of which are dynamics-dependent.

3.3 Locality, no-signalling, microcausality and Lorentz-covariance

Here we distinguish the principle of locality from other principles, such as no-signalling, and properties of specific theories, such as microcausality and Lorentz-covariance. We shall follow an order of generality: locality is the most general property, Lorentz-covariance the least general. We shall refer, for convenience, to a bipartite system, made of two subsystems: A and B.

Principle of locality: The *principle of locality* states that given a partition of a system into subsystems A and B, a dynamical transformation that operates only on A cannot change the *states* of B.

Note that here by "state" we mean the *complete* specification of the state of affairs of a given system, not necessarily what is empirically accessible by measuring observables of that system only. This is also known in the literature as the "ontic state", or "noumenal state", to differentiate it from the "phenomenal state", or "epistemic state", indicating what is observable in the system, [4, 28]. Thus, the principle of locality can also formally be stated as a strict constraint on the states of systems, as follows [24]:

Theorem 1: *The state of a system is a description of it that satisfies two properties: (i) any attribute of a system, at any given time t, is a fixed function of the system state and (ii) any state of a composite system $A \oplus B$ is an ordered pair of states (a, b) of A and B, with the property that if a task is performed on A only, then the state b of the substrate B is not changed thereby.*

In quantum theory, the principle of locality is also called *Einstein's locality*. Often times it is confused with a different notion of locality called *Bell's locality*. Bell's locality refers to the possibility of describing a set of data with a local hidden-variable theory, expressed as a stochastic, real-valued theory, thus based on *c-numbers*. When Bell's inequalities are violated in a given experiment, one can therefore rule out a c-number-based, stochastic description of reality, that is local in the sense of Einstein's locality. Quantum theory is Bell non-local; however, it is a *q-number-based*, deterministic description of reality that satisfies the principle of locality (as stated earlier). Thus, Bell's non-locality of quantum theory should not be misunderstood as a violation of the principle of locality as stated above, but rather as a violation of Bell's locality. From now on, when referring to "locality", we will always mean the property required by the principle of locality, not Bell's locality.

Principle of no-signalling: The *principle of no-signalling* states that a dynamical transformation that operates only on one subsystem A cannot change the *observable properties* of the other subsystem B.

In quantum theory, one can express this principle as follows:

$$[U_A, \rho_B] = 0 \quad (3.5)$$

where U_A is a unitary transformation happening on subsystem A alone, while $\rho_B = \mathrm{Tr}_A(\rho_{AB})$ is the reduced density operator of subsystem B, which can be expressed as a linear combination of the generators of its observables algebra.

For example, if one considers B to be a qubit, then $\rho_B = \frac{1}{2}(\mathbb{I} + \alpha X_B + \beta Y_B + \gamma Z_B)$, where X_B, Y_B and Z_B are the Pauli operators and $\mathbb{I}$ is the identity operator, generators of the qubit's observables algebra. Here, $\alpha, \beta, \gamma \in \mathbb{C}$ with $|\alpha|^2 + |\beta|^2 + |\gamma|^2 = 1$.

The principle of no-signalling and the principle of locality are related. When considering a theory with 1:1 dynamics, including quantum theory, a general theorem shows that the two statements are equivalent [4]. This is because the 1:1 dynamics determines the *completeness* of the theory, which then allows for the description of the complete state of affairs of the system (the ontic state), as required by the principle of locality, starting from its observable properties (the phenomenal state), as mandated by the no-signalling principle. In general, while the principle of locality implies the principle of no-signalling, the converse is not true. This is because the principle of no-signalling only focuses on the locally empirically accessible features of a system. An example of a theory that satisfies no-signalling, but is non-local, is Bohmian Mechanics [26]. This fact becomes important for the witness of non-classicality, which, in its most general form [2], relies on the principle of locality, not on the principle of no-signalling.

Notice also that the above two principles concern notions such as "observables", "variables" and "dynamical transformations", which can be expressed in many different formalisms. Hence, while these principles are both satisfied by non-relativistic quantum theory, quantum field theory, and special and general relativity, they are formulated *independently of* any specific formalism. This is the power of principles as general rules that can constrain particular dynamics, rather than being derived within a particular dynamical law.

Finally, note that the principle of no-signalling can also be phrased, within specific dynamical laws, as requiring a finite speed of propagation for signals between space-like separated systems. Here we have opted for a formulation that is dynamics-independent and does not explicitly refer to speed of propagation.

Axiom of microcausality: In quantum field theory, the property of *microcausality* is often taken as an *axiom*. This property states that the allowed quantum observables (operators) of space-like separated systems must commute. In other words,

$$\left[\hat{Q}_A, \hat{Q}_B\right] = 0, \quad (3.6)$$

for any two quantum observables $\hat{Q}_A$ and $\hat{Q}_B$, respectively, pertaining to two space-like separated systems A and B. Note that when we talk about space-like separated systems, we implicitly refer to a *particular formalism* of special relativity. Moreover, by using a commutator, we are committing to the formalism of quantum theory. In passing, we note that Eq. (3.6) is a property of non-relativistic quantum theory, where the quantum observables of independent subsystems are required to commute with each other. This property implies that both non-relativistic quantum theory and quantum field theory satisfy the principle of locality as stated above: since $\hat{Q}_A$ commutes with $\hat{Q}_B$ and the two theories are complete, operating a dynamical transformation on the observable $\hat{Q}_B$ of subsystem B cannot modify the observable $\hat{Q}_A$ of subsystem A, i.e., its state, as required by the principle of locality. Finally, microcausality, as stated above, is *not* a general principle because it is formulated within quantum theory's and special relativity's formalism.

Axiom of Lorentz-covariance: In special relativity, allowed dynamical variables must satisfy the *axiom of Lorentz-covariance*. This means that they must transform properly under Lorentz transformations so that physical variables must be either scalars, tensors, or spinors. Dynamical laws are then said *Lorentz-covariant* if they are expressed in terms of Lorentz-covariant variables only. A dynamical law with this property satisfies the principle of relativity that all laws must make the same predictions about identical experiments in any two inertial frames. Some physical properties such as scalars are also *Lorentz invariant*, which means that they take the same form in all reference frames. When a law satisfies this property it must also satisfy the principles of locality and no-signalling, as the maximum speed of propagation for dynamical perturbations is the speed of light. The two principles as formulated above refer to "instantaneous" modification of the state or the observable of the other subsystems and do not refer to any speed of propagation or other conditions in space and time. Similar concepts can be defined in general relativity, where one talks about *local Lorentz-covariance* (local in spacetime). The same considerations about the relation with the principle of locality and no-signalling hold in that case.

Once more, Lorentz-covariance is a dynamics-dependent formal requirement, which only holds within special relativity. It is implied by microcausality, but it is less general than the locality or the no-signalling principle as formulated earlier.

Universality: There are also more specific notions of locality, related to the particular structure of dynamical interactions. For instance, there is the property that a given unitary on an n-partite system can be approximated arbitrarily well by a set of unitaries operating on one and two systems. In the field of quantum information, this is called the *universality of two- and one-qubit gates*, which was proven by a sequence of results in the eighties [6–8]. That property can also be proven for continuous dynamics, considering that the latter can be approximated arbitrarily well by the dynamics of discrete systems. These well-known results were also recently termed "subsystem locality" [23]. These properties are, once more, formalism-specific and thus more narrow than the principles discussed earlier.

3.4 The role of locality in the General Witness Theorem

According to the most general proof currently known, locality is one of the two sufficient conditions for the GWT [2]. All viable theories currently trusted to satisfy this principle: in particular, it applies to both special and general relativity.

Let us first state the notion of non-classicality that the witness is meant to assess, in constructor-theoretic terms. We need to recall constructor theory's definition of a superinformation medium, which generalises the concept of a quantum system without relying on quantum theory's formalism.

Central to this definition is that of information variable. An **information variable** X is a set of disjoint *attributes* $\{\mathbf{x}\}$ (i.e., a set of all the states where the system has a given property) for which the following tasks are possible:

$$\bigcup_{x \in X} \{(\mathbf{x}, \mathbf{x_0}) \to (\mathbf{x}, \mathbf{x})\}, \tag{3.7}$$

$$\bigcup_{x \in X} \{\mathbf{x} \to \Pi\,(\mathbf{x})\}. \tag{3.8}$$

The "copying" task [Eq. (3.7)] corresponds to copying the attributes $\mathbf{x} \in X$ of the first replica of the system onto the second, target, system prepared in a blank attribute $\mathbf{x_0} \in X$. The "permutation" task [Eq. (3.8)] describes the possibility of performing a logically reversible computation on variable X, described as a permutation Π on the set of labels of attributes in X. Together, the two tasks [Eqs. (3.7) and (3.8)] indicate that it is possible to perform information processing on variable X. An **information medium** is a system having at least one information variable.

A **superinformation medium** is an information medium with at least two information variables, whose union is not an information variable. This means that the two information variables are complementary to one another, and cannot be copied by the same device. In [24], it is shown that superinformation media have all the qualitative properties of quantum systems. In this framework, it is natural and elegant to express the idea of non-classicality that is relevant to the GWT, as follows.

By a system being **non-classical**, one means an information medium **M**, with maximal information observable Z, that has also another variable V *disjoint* from Z and with the *same cardinality* as Z, with these properties:

1. There exists a superinformation medium $\mathbf{S_1}$ and a distinguishable variable $E = \{\mathbf{e_j}\}$ of the joint substrate $\mathbf{S_1} \oplus \mathbf{M}$, whose attributes $\mathbf{e_j} = \{(s_j, v_j)\}$ are sets of ordered pairs of states, where v_j is a state belonging to some attribute in V and s_j is a state of $\mathbf{S_1}$;
2. The union of V with Z is *not* a distinguishable variable;
3. The task of distinguishing the variable $E = \{\mathbf{e_j}\}$ is possible by measuring incompatible observables of a *composite superinformation medium* including $\mathbf{S_1}$, but impossible by measuring observables of $\mathbf{S_1}$ only.

The above conditions express the fact that a non-classical system has a classical variable (the information variable) that is also an observable; and then it has another dynamical variable that is necessary to mediate the generation of entanglement, which may or may not be directly measurable – its attributes may or may not be all preparable or single-shot distinguishable.

The proof of the GWT goes as follows. First, one considers three systems: A and B (two quantum probes), and M, the mediator, which has a "classical" observable Z, and no other known degrees of freedom. In the case of gravity, the classical basis of the gravitational field would be the number operator or its energy. Let A's descriptors be denoted by $\mathbf{q_A(t_0)}$, B's descriptors by $\mathbf{q_B(t_0)}$, and the mediator's descriptors by $\mathbf{c(t_0)}$ [18]. Let us assume that the observables (or measurable properties) of the three systems at time t, as well as their joint observables, are a *fixed function* of the triplet $(\mathbf{q_A(t_0)}, \mathbf{c(t_0)}, \mathbf{q_B(t_0)})$. In quantum theory, this is given by the trace with the initial state's density operator, which never changes and therefore can be incorporated into the definition of the function.

Assume that initially (at time $t = t_0$) the three systems are *uncorrelated*, for instance (in quantum theory), we would say that they are in a product state where a local observable of each subsystem is sharp with some value. So the mediator will have Z sharp with value say z_0. We assume that it is possible to run the experiment with two distinguishable initial conditions, say for instance, the case where the two masses are prepared in some state s_+, in which case the descriptors shall be labelled as $\mathbf{q^+_{A,B}(t_0)}$, and in another distinguishable state s_-, with descriptors $\mathbf{q^-_{A,B}(t_0)}$.

Suppose that at time t_2 A and B end up in both cases being entangled. This means that at time t_2 they have been prepared in one of two entangled states, call them $\mathbf{e}_+$ and $\mathbf{e}_-$, which are distinguishable too by measuring observables of A and B only. Thus, the following task must be possible:

$$T_E \doteq \left\{ \left(\mathbf{q^\pm_A(t_0)}, \mathbf{c(t_0)}, \mathbf{q^\pm_B(t_0)}\right) \rightarrow \mathbf{e}_\pm \right\}, \tag{3.9}$$

and the goal of the experiments proposed in [1, 3] is to show that T_E is indeed possible upon successfully generating entanglement between A and B. Locality here can be used to conclude that in both configurations, the descriptors of A, M and B are an ordered triplet:

$$\mathbf{e}_{\pm} \equiv \left(\mathbf{q}_{\mathbf{A}}^{\pm}(\mathbf{t}_2), \mathbf{c}^{\pm}(\mathbf{t}_2), \mathbf{q}_{\mathbf{B}}^{\pm}(\mathbf{t}_2)\right). \tag{3.10}$$

Now, we can use the assumption that A and B are not interacting directly, but the interaction is mediated by M. To a first approximation, this means that the interaction happens by letting first A interact with the mediator M at time t_1 and then M interact with the qubit B at time t_2. In terms of tasks, T_E in Eq. (3.9) must thus be made by:

$$T_E^{(1)} \doteq \left\{\left(\mathbf{q}_{\mathbf{A}}^{\pm}(\mathbf{t}_0), \mathbf{c}(\mathbf{t}_0), \mathbf{q}_{\mathbf{B}}^{\pm}(\mathbf{t}_0)\right) \rightarrow \mathbf{r}_{\pm}\right\}, \tag{3.11}$$

performed on $A \oplus M$ only, and:

$$T_E^{(2)} \doteq \{\mathbf{r}_{\pm} \rightarrow \mathbf{e}_{\pm}\}, \tag{3.12}$$

performed on $M \oplus B$ only. Once more using locality, the state of the three systems at time t_1 must be described by one of two triplets, according to whether the initial condition s_+ or the initial condition s_- was used:

$$\mathbf{r}_{\pm} \equiv \left(\mathbf{q}_{\mathbf{A}}^{\pm}(\mathbf{t}_1), \mathbf{c}^{\pm}(\mathbf{t}_1), \mathbf{q}_{\mathbf{B}}^{\pm}(\mathbf{t}_0)\right). \tag{3.13}$$

Locality is here used because we have considered that the descriptor of system B must not have changed since t_0, given that the interaction at t_1 only involves A and M.

The proof proceeds by showing that the descriptors $\mathbf{c}^{\pm}(\mathbf{t}_1)$ are: (1) disjoint (set-wise) from one another, and from the classical states of M; (2) not distinguishable from the classical states of M and (3) not distinguishable from one another, yet the joint state of A and M is (using measurements that involve both A and M). These three properties make M non-classical, in that it has a dynamical variable V, made of the two descriptors $\{\mathbf{c}^{\pm}(\mathbf{t}_1)\}$, which is disjoint from the classical observable and yet is not distinguishable from the latter, just like Heisenberg's uncertainty principle requires. We refer the interested reader to [2] for detailed proof of the above three points.

As we mentioned earlier, it is important to notice that the non-classical variable V of the mediator, unlike Z, may not be an observable, in the sense that $\mathbf{c}^{+}(\mathbf{t}_1)$ may not be distinguishable from $\mathbf{c}^{-}(\mathbf{t}_1)$ in a single shot manner, and that those two states may not be preparable.

It follows from this argument that violating the assumption of locality *invalidates* the witness, giving the appearance that even a classical mediator M can create entanglement between the two quantum probes A and B. This is the case of models using the non-local Newtonian gravity [10], allowing for interactions between A and B [11], or assuming a simultaneous interaction of the mediator M with both the quantum probes, [12]. Moreover, some quantum-classical hybrid models, [13] conceal hidden non-locality in the configuration space dynamics, as noted in [9]. Violating locality means removing the assumption involving the second, non-compatible variable V of M in creating entanglement, as shown in the proof above. This is because the task T_E would be performed in a single step, instead of at least two. Thus, with a non-local model, M would need a single variable to perform T_E, ultimately a classical system. However, as discussed above, both quantum mechanics and general relativity obey the principle of locality in the general form introduced in this work, thus securing its presence among the minimal assumptions of a theorem aimed at proving the non-classical nature of an unknown system.

3.5 Conclusions

The principle of locality as discussed in this work is a plausible candidate for the most general notion of locality in physics, considering its relation to the other relevant notions that we have reviewed in this chapter. It is also the minimal notion of locality on which the witness of non-classicality for the BMV effect [2] must rely.

One interesting open question is in the direction of considering a temporal equivalent of the GWT, with a possible notion of locality in time, not just in space.

Following the pioneering work by Leggett and Garg [15], which demonstrated that quantum systems exhibit a form of temporal correlation that any macro-realistic theories cannot explain, the study of temporal correlations and their relationship to spatial correlations has become increasingly popular in the scientific community [16]. Central to this topic is the idea of *locality in time*, which was explored in [14] and formulated as follows: the results of measurement performed at time t_2 are independent of any measurement performed at some earlier or later time t_1. Despite this formulation, a consensus on the precise meaning of locality in time remains elusive, and it is still debated whether quantum mechanics should be interpreted as adhering to temporal locality [19, 20]. It is therefore an interesting open question how to relate it to the principle of locality "in space", as discussed earlier in this chapter.

A recently proposed witness of non-classicality [17] provides a new framework for probing the quantum nature of an unknown system M by studying the time evolution of a single quantum probe Q – thus providing a "temporal equivalent" of the entanglement-based witness just discussed. Specifically, if a system M induces a quantum coherent evolution in a quantum probe Q while conserving a global quantity of the system $Q \oplus M$, then M must be quantum. This "temporal" witness mirrors the spatial witness of non-classicality proposed in [1, 3], where the ability of M to generate non-classical spatial correlations between two probes A and B is examined. In contrast, the temporal witness focuses on detecting non-classical temporal correlations in the evolution of the quantum probe Q.

Given the crucial role played by the assumption of locality in space in the GWT, it is reasonable to expect that locality in time may play an analogous role in the temporal witness. The connections between these two witnesses suggest that the assumptions underpinning each may be closely related: while locality in space is central to the GWT, the conservation law of a global quantity in the system $Q \oplus M$ could be linked to locality in time in the temporal witness. This raises the intriguing possibility that the conservation law required for the quantum coherent evolution of Q is not only essential for detecting non-classicality but may also offer insights into the nature of temporal locality.

In this direction, following those attempts in quantum theory to treat space and time on equal footing [21], one could interpret the two subsystems A and B in the spatial witness as a *single* system at *two different times* t_A and t_B in the temporal one, and follow the general proof described in the previous section. Can the principle of locality as stated in this paper be applied also in this scenario? Where does the conservation law play a role in the general proof for the temporal case?

We leave the answers to these questions to future work.

Acknowledgements

G.D.P. thanks the Clarendon Fund and the Oxford-Thatcher Graduate Scholarship for supporting this research. This research was made possible through the generous support of the Gordon and Betty Moore Foundation, the Eutopia Foundation and of the ID 62312 grant from the John Templeton Foundation, as part of the `https://www.templeton.org/grant/the-quantuminformation-structure-ofspacetime-qiss-second-phase` "The Quantum Information Structure of Spacetime" Project (QISS). The opinions expressed in this project/publication are those of the author(s) and do not necessarily reflect the views of the John Templeton Foundation.

References

[1] Marletto, Chiara, and Vlatko Vedral. (2017). Gravitationally Induced Entanglement between Two Massive Particles Is Sufficient Evidence of Quantum Effects in Gravity. *Physical Review Letters*, 119(24), 240402.

[2] Marletto, Chiara, and Vlatko Vedral. (2020). Witnessing Nonclassicality beyond Quantum Theory. *Physical Review D*, 102(8), 086012.

[3] Bose, Sougato, Anupam Mazumdar, Gavin W. Morley, Hendrik Ulbricht, Marko Toroš, Mauro Paternostro, Andrew A. Geraci, Peter F. Barker, Myungshik Kim, and Gerard Milburn. (2017). Spin Entanglement Witness for Quantum Gravity. *Physical Review Letters*, 119(24), 240401.

[4] Raymond-Robichaud, Paul. (2021). A Local-Realistic Model for Quantum Theory. *Proceedings of the Royal Society A: Mathematical, Physical and Engineering Sciences*, 477(2250), 20200897.

[5] Raia, Salvatore, Giuseppe Di Pietra, and Chiara Marletto. (2024). The Role of Non-Classicality in Mediated Spatial Quantum Correlations. *arXiv preprint* arXiv:2410.00824. `https://doi.org/10.48550/arXiv.2410.00824`

[6] Deutsch, David Elieser, Adriano Barenco, and Artur Ekert. (1997). Universality in Quantum Computation. *Proceedings of the Royal Society of London. Series A: Mathematical and Physical Sciences*, 449(1937), 669–677.

[7] DiVincenzo, David P. (1995). Two-Bit Gates Are Universal for Quantum Computation. *Physical Review A*, 51(2), 1015–1022.

[8] Barenco, Adriano. (1997). A Universal Two-Bit Gate for Quantum Computation. *Proceedings of the Royal Society of London. Series A: Mathematical and Physical Sciences*, 449(1937), 679–683.

[9] Marconato, Emanuele, and Chiara Marletto. (2021). Vindication of Entanglement-Based Witnesses of Non-Classicality in Hybrid Systems. *Journal of Physics A: Mathematical and Theoretical*, 54(37), 375301.

[10] Marchese, Marta Maria, Martin Plávala, Matthias Kleinmann, and Stefan Nimmrichter. (2024). Newton's Laws of Motion Can Generate Gravity-Mediated Entanglement. *arXiv.org*, January. `https://arxiv.org/abs/2401.07832v1`

[11] Husain, Viqar, Irfan Javed, and Suprit Singh. (2022). Dynamics and Entanglement in Quantum and Quantum-Classical Systems: Lessons for Gravity. *Physical Review Letters*, 129(11), 111302.

[12] Martín-Martínez, Eduardo, and T. Rick Perche. (2023). What Gravity Mediated Entanglement Can Really Tell Us about Quantum Gravity. *Physical Review D*, 108(10), L101702.

[13] Hall, Michael J. W., and Marcel Reginatto. (2018). On Two Recent Proposals for Witnessing Nonclassical Gravity. *Journal of Physics A: Mathematical and Theoretical*, 51(8), 085303.

[14] Brukner, Časlav, Samuel Taylor, Sancho Cheung, and Vlatko Vedral. (2004). Quantum Entanglement in Time. `https://arxiv.org/abs/quant-ph/0402127`

[15] Leggett, A. J., and Anupam Garg. (1985). Quantum Mechanics versus Macroscopic Realism: Is the Flux There When Nobody Looks? *Physical Review Letters*, 54(9), 857–860.

[16] Emary, Clive, Neill Lambert, and Franco Nori. (2013). Leggett–Garg Inequalities. *Reports on Progress in Physics*, 77(1), 016001.

[17] Di Pietra, Giuseppe, and Chiara Marletto. (2023). Temporal Witnesses of Non-Classicality and Conservation Laws. *Journal of Physics A: Mathematical and Theoretical*, 56(26), 265305.

[18] Deutsch, David, and Patrick Hayden. (2000). Information Flow in Entangled Quantum Systems. *Proceedings of the Royal Society of London. Series A: Mathematical, Physical and Engineering Sciences*, 456(1999), 1759–1774.

[19] Adlam, Emily. (2018). Spooky Action at a Temporal Distance. *Entropy*, 20(1), 41.

[20] Cohen, Eliahu. (2023). Quantum Clock Frames: Uncertainty Relations, Non-Hermitian Dynamics and Nonlocality in Time. *Journal of Physics: Conference Series*, 2533(1), 012018.

[21] Zhang, Tian, Oscar Dahlsten, and Vlatko Vedral. (2020). Different Instances of Time as Different Quantum Modes: Quantum States across Space-Time for Continuous Variables. *New Journal of Physics*, 22(2), 023029.

[22] Plávala, Martin. (2023). General Probabilistic Theories: An Introduction. *Physics Reports*, 1033, 1–64.

[23] Christodoulou, Marios, and Carlo Rovelli. (2019). On the Possibility of Laboratory Evidence for Quantum Superposition of Geometries. *Physics Letters B*, 792, 64–68.

[24] Deutsch, David, and Chiara Marletto. (2015). Constructor Theory of Information. *Proceedings of the Royal Society A: Mathematical, Physical and Engineering Sciences*, 471(2174), 20140540.

[25] Marletto, Chiara, and Vlatko Vedral. (2024). Quantum-Information Methods for Quantum Gravity Laboratory-Based Tests. *arXiv*, October. `https://doi.org/10.48550/arXiv.2410.07262`

[26] Goldstein, Sheldon. (2024). Bohmian Mechanics. In Edward N. Zalta and Uri Nodelman (Eds.), *The Stanford Encyclopedia of Philosophy*, Summer 2024 Edition. Metaphysics Research Lab, Stanford University. `https://plato.stanford.edu/archives/sum2024/entries/qm-bohm/`

[27] Bédard, Charles Alexandre. (2021). The ABC of Deutsch–Hayden Descriptors. *Quantum Reports*, 3(2), 272–285.

[28] Spekkens, Robert W. (2007). Evidence for the Epistemic View of Quantum States: A Toy Theory. *Physical Review A*, 75(3), 032110.

4 Toward a Generalized Theory of Observers

Hatem Elshatlawy, Dean Rickles, and Xerxes D. Arsiwalla

4.1 Introduction: Historical perspectives on observers

> *The observer, when he seems to himself to be observing a stone, is really, if physics is to be believed, observing the effects of the stone upon himself. Thus science seems to be at war with itself: when it most means to be objective, it finds itself plunged into subjectivity against its will.*
>
> –Bertrand Russell

The concept of an *observer* has been a core puzzle from ancient philosophy to modern physics, shaping how we interpret measurement, reality, and knowledge. Classical thinkers such as Aristotle placed the observer as a passive spectator to the natural world. But the beginning of modern science came with the demand for the observer to take a more active role, as expressed by Francis Bacon in his New Organon (98th Aphorism Concerning the Interpretation of Nature):

> In the business of life, the best way to discover a man's character, the secrets of how his mind works, is to see how he handles trouble. In just the same way, nature's secrets come to light better when she is artificially shaken up than when she goes her own way.

In modern physics—with the stark differences between reference frames in Einstein's relativity and with Heisenberg's uncertainty principle—the active

DOI: 10.4324/9781003546139-6

and constraining role of the observer came to be seen as not just expedient, but as unavoidable [1, 2].

In quantum mechanics, the observer problem became explicit, ultimately raising questions about consciousness, apparatus, and whether physical law alone dictates the outcome of measurements [4, 5]. As measurement seemingly "collapses" the wavefunction, intervention was recognized as a prerequisite of observation. In the field of cybernetics, a complementary development took place: observation was clearly recognized as a prerequisite for intervention, as observation (of outcomes) provides the input to adjust the actions of an agent in a feedback loop [6, 7]. A similar development can be observed in the fields of biology [8] and artificial intelligence [9], where the observer is increasingly regarded as a feedback-driven agent or subsystem with the capacity to influence and be influenced by its environment.

This chapter aims to consolidate these diverse threads into a generalized observer theory. Specifically, we propose a *formal* definition of minimal observers, discuss their philosophical significance, and explain which concepts (such as measurement, internal vs. external distinctions, and emergent complexity) hinge on the presence of observers. We also connect these ideas to pressing questions in quantum gravity and digital physics [11, 12, 32, 33], where the ultimate structure of reality and computation may depend on observer-oriented frameworks. In addition, we point readers to dual-aspect monism arguments [14] and debates on consciousness that further highlight how observer-internal processes and external meaning may be intertwined.

Furthermore, by a *minimal observer*, we refer to the simplest possible entity that exhibits the core characteristics of observation: the ability to perceive external states, update internal configurations based on input, and generate an action or output, thereby forming a closed feedback loop. This definition, which we formalize in the following sections, ensures that the observer remains functionally distinct from its environment while engaging in meaningful interactions that shape both its perception and responses.

To make our account as rigorous as possible, we include new mathematical developments: theorems on *observer equivalence* and *observational complexity*, explicit *diagrams* to clarify feedback loops, and in-depth comparisons with established theories in physics and philosophy.

4.2 A general theory of observers

4.2.1 The need for a unified model

Despite the variety of fields referencing "observers," there does not exist a single unifying formalism of observers. Quantum physicists define observers in terms of measurement apparatus or "external" classical systems [3, 13], while cognitive scientists see observers as perceiving agents [14], and

computer scientists think of them as abstract data-collecting subroutines [10]. This
domain-based fragmentation of ideas obscures shared principles and hinders cross-disciplinary integration. A generalized observer theory seeks to: identify core *functional* features of observation (sensing, state updating, responding); clarify how these features scale from minimal feedback loops to conscious or socially embedded observers; provide a framework to address both *foundational* questions (quantum measurement, realism vs. anti-realism) and *practical* ones (AI design, ethics, interpretability).

4.2.1.1 Record-keeping and information processing in observer models

A promising avenue for unification lies in recognizing a shared invariant: *observation as an information-recording process*. Hugh Everett's formulation of quantum mechanics conceptualizes observers as *servomechanisms*, automatically functioning machines that register environmental interactions via memory storage. He proposed that observation is best understood as a record-keeping process—observers are not external agents but physical subsystems that *store measurement outcomes* as part of their own state evolution [42]. This insight suggests that observation, at its core, entails a feedback loop in which perception updates an internal record, shaping subsequent responses.

This perspective re-emerges in James Hartle's notion of *Information Gathering and Utilizing Systems* (IGUSs), a generalized framework for modeling observers in physics. IGUSs encompass any system—biological, mechanical, or computational—that collects, stores, and processes information to make decisions or generate outputs [43]. A key feature of IGUS models is their time-sequenced memory, which retains past inputs to inform future states, mirroring Everett's record-keeping servomechanisms. As Bacciagaluppi notes, Everett's concept of servomechanistic observers resurfaces in decoherence-based discussions, where IGUSs formalize the continuous accumulation and processing of data [45]. This suggests that *recorded information*—not subjective awareness—defines an observer's role in physics.

4.2.1.2 Toward a unified observer framework

By highlighting the importance of *record-keeping* across physics and cognitive models, we can extract a fundamental principle of observation. Whether in quantum measurement, artificial intelligence, or biological perception, observers can be described as information-processing units with internal states that encode and update representations of their environment. This shared functionality provides a compelling foundation for a cross-disciplinary observer theory, allowing a unified framework to emerge that reconciles disparate approaches under a common structural paradigm.

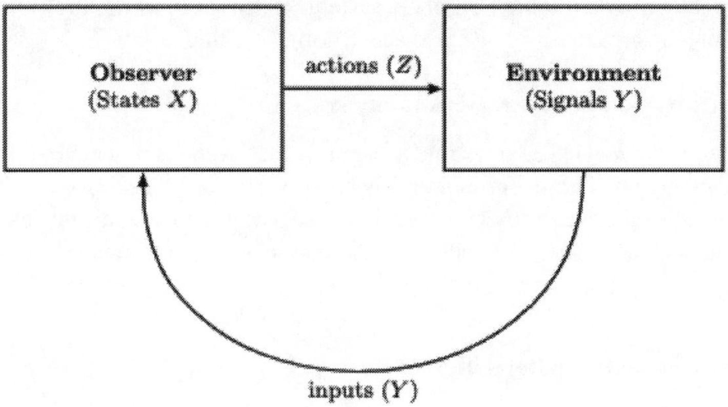

Figure 4.1 Minimal observer-environment feedback loop.

4.3 Minimal observation models in cybernetics

4.3.1 Introduction to cybernetics and observation

Cybernetics, as pioneered by Norbert Wiener [6] and developed by Ross Ashby [7], is the study of communication and control in organisms and machines. Its hallmark is the *feedback loop*, in which a system senses some variable, processes that information, and acts upon its environment, thus influencing future sensory input. Figure 4.1 shows a simple diagram of this loop, highlighting the observer's role.

4.3.2 The sensor-actuator feedback loop

A simplified cybernetic observer consists of three essential components:

1 **Sensor:** Detects environmental or internal states (e.g., temperature, light intensity).

2 **Processing unit:** Interprets sensor data, often by comparing it to a goal or reference.

3 **Actuator:** Executes an action that changes either the system itself or its environment.

4.3.3 Examples of minimal cybernetic observers

THERMOSTAT

A *thermostat* maintains temperature by comparing a set-point to the current temperature sensor reading and toggling a heater. Despite its simplicity, it is widely regarded as a minimal observer system: it senses (temperature), processes (comparing to a set-point), and acts (turning heating on/off).[1]

SIMPLE REACTIVE AGENTS (BRAITENBERG VEHICLES)

Braitenberg vehicles [17] demonstrate how purely reactive sensor-actuator links can yield emergent "intelligent-looking" behaviors. Although lacking complex cognition, they meet basic observer criteria by receiving sensory data, updating motor outputs, and influencing their environment in a feedback loop.

4.4 Second-order cybernetics

4.4.1 Overview

First-order cybernetics keeps the observer *outside* the system: it is a third-person perspective. In *second-order cybernetics* (Fig. 4.2) [15], the observer is integrated *into* the system, allowing for self-reference, thus giving not just a first-person perspective, but one in which the observer enters their own domain.

4.4.2 Key concepts

Observer inclusion: The observer is a part of the feedback process, not a neutral external vantage point.
Self-reference: The observer can observe and modify their own rules, leading to learning or adaptation.
Constructivism: Reality emerges through the observer's activities and interpretations [8, 16].

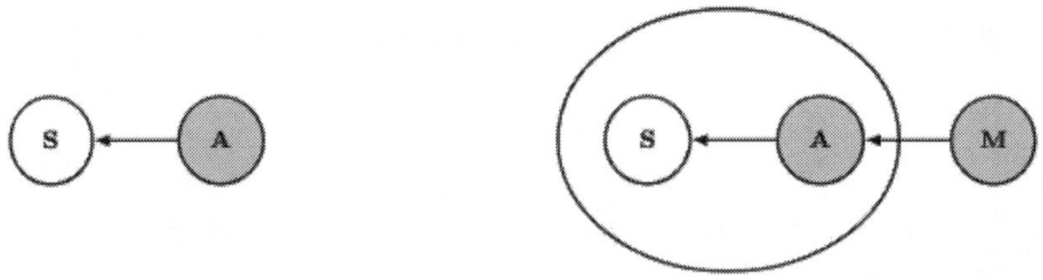

Figure 4.2 First-order vs. second-order cybernetics: (left) first-order observer observes the system, (right) second-order observer modifies the observation process itself.

EXAMPLE: LEARNING NEURAL NETWORKS

A neural network updating its weights W in response to error signals exemplifies second-order cybernetics. The network *observes* (via a loss function) the mismatch between outputs and targets, adjusting W to reduce future mismatch—thus reconfiguring its own internal parameters.

4.5 Developing a meta-model for observers

4.5.1 Core components of the meta-model

Across many examples, we identify the following recurring features:

1 Sensing mechanism (perception)
2 Processing unit (interpretation)
3 Response mechanism (action)
4 Feedback loop (adaptation)
5 Internal model or representation (prediction)
6 Boundary definition (self vs. environment)
7 Self-monitoring (self-observation)

4.5.2 A formal definition of minimal observers

Definition 4.1 (minimal observer): Let O be a system described by the tuple

$$O = (X, Y, Z, f, g, \mathcal{B}),$$

where X is internal state space (finite or countably infinite), Y is input (sensor) space, Z is Output (action) space, f is $X \times Y \to X$: State transition function, g is $X \to Z$: Output function, and $\mathcal{B}$ is a boundary condition demarcating "inside" (the observer's internal states) vs. "outside" (the environment).

Then O is *minimal* if:

$|Y| \geq 1$ (non-trivial sensing),

$|Z| \geq 1$ (non-trivial action),

$|X| > 1$ (non-trivial internal dynamics),

Feedback closure: the observer's actions $g(x)$ alter the environment, which in turn alters subsequent inputs $y \in Y$.

WHAT DOES "MINIMALITY" MEAN HERE?

This definition positions the minimal observer as the simplest system capable of observation in a functional sense: it must sense (Y), update its state (f), and act (g), with a boundary ($\mathcal{B}$) and feedback loop ensuring interaction with an environment. This is minimal in a structural sense—it avoids complexity like memory, learning, or self-awareness, focusing on the bare essentials of observation. However, "minimality" could be interpreted in multiple ways: **structural minimality**, referring to the fewest components needed for observation (e.g., the thermostat example below (Section 4.5.3) fits structural minimality: it exemplifies a basic feedback system with no self-modification or self-production, both of which lie beyond the scope of this minimal model); **functional minimality**, denoting the simplest system that still performs a meaningful role (e.g., distinguishing internal vs. external states); and **ontological minimality**, which identifies the foundational unit from which all observer-like phenomena emerge (possibly tying to autonomy or autopoiesis).

4.5.3 Thermostat as a formal example

As an illustration, let $X = \{\text{ON}, \text{OFF}\}$, $Y = \{\text{Cold}, \text{Hot}\}$, $Z = \{\text{HeaterOn}, \text{HeaterOff}\}$. Define

$$f(\text{OFF}, \text{Cold}) = \text{ON}, \qquad f(\text{OFF}, \text{Hot}) = \text{OFF},$$
$$f(\text{ON}, \text{Cold}) = \text{ON}, \qquad f(\text{ON}, \text{Hot}) = \text{OFF}.$$

and

$$g(\text{ON}) = \text{HeaterOn}, \quad g(\text{OFF}) = \text{HeaterOff}.$$

A boundary $\mathcal{B}$ physically partitions the controller from ambient air. This meets minimal observer criteria: the thermostat senses "Cold/Hot," toggles $\{\text{ON}, \text{OFF}\}$, and acts by turning heat on/off.

4.6 Foundational questions and observer-dependent concepts

4.6.1 Relating the model to foundational questions

MEASUREMENT IN PHYSICS AND OBSERVER ROLES

Quantum theory famously hinges on measurement, prompting debates over whether wavefunction collapse is triggered by consciousness, classical apparatus, or decoherence [4, 5, 13]. Our minimal observer model, although classical, illuminates the *functional* demands of measurement: a system must sense, store, and act upon the input, forging a boundary that designates what is measured (the environment) and what is measuring (the observer). This boundary-centric perspective resonates with relational interpretations of quantum mechanics, where all states and events are observer-relative [28].[2]

COMPUTATION AND COMPLEXITY

Turing machines [10] epitomize minimal universal computation. Our minimal observer is "less ambitious": it does not demand universal problem-solving but ensures a fundamental sensor-actuator feedback. Nevertheless, bridging these models can highlight interesting points, such as whether an observer can, in principle, simulate arbitrary computational processes if given enough states X and a suitable set of transitions f.

CONSCIOUSNESS DEBATES

Whether minimal observers can illuminate the "hard problem" of consciousness [7, 12, 18, 19] remains debatable. Yet the boundary $\mathcal{B}$ and internal modeling h in more complex observers may underlie self-referential processes that are often considered key to subjective experience and conscious phenomenology [14, 20–27]. While minimal observers do not *entail* consciousness, they provide building blocks to analyze how layered observation might scale up to phenomena associated with awareness.

4.6.2 Identifying what would not be defined without observers

Without an observer:

1 There is no clear distinction between **internal vs. external space.**

2 There are no definitive **measurement outcomes.**

3 There are no **reference frames or contextual frameworks.**

4 There is no layering of **hierarchical observation** (e.g., organizations, societies, or multi-level apparatuses).

5 There is no **coarse-graining or hierarchical organization**, since it is observers who impose boundaries that allow phenomena to be described at different scales or levels of abstraction.

Observers provide the partitions that make physics (and meaning) possible. This point resonates strongly with the "ruliological" viewpoint of Stephen Wolfram: in his computational universe picture, reality only becomes tractable once an observer chooses a particular foliation of the underlying causal network, thus

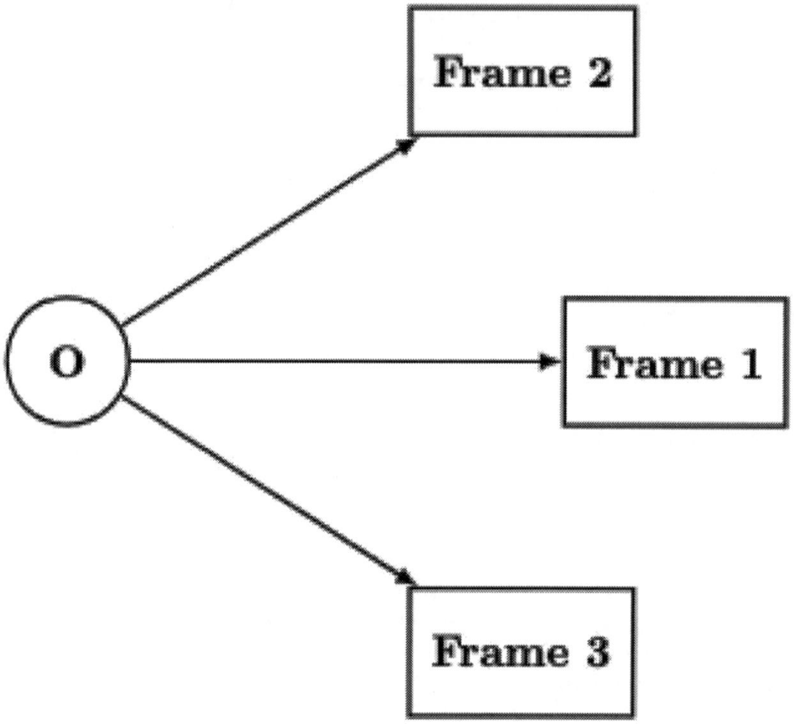

Figure 4.3 Observer defining multiple reference frames in physical or cognitive space.

creating the very coarse-grained structures, frames, outcomes and hierarchical layers, that we then treat as the fabric of experience and scientific explanation [60, 66–68] (Fig. 4.3).

Hierarchical structures, as mentioned earlier, rely fundamentally on observers who define internal states and boundaries at multiple scales. Observers thus serve as "interfaces" that structure interactions between systems and their environments. This perspective resonates strongly with Chris Fields' work, which highlights how boundaries and interfaces created by observers define informational interactions between system components and their surroundings [46]. Without observers, there would be no principled way to partition reality, establish hierarchical frameworks, or define coarse-grained descriptions that enable meaningful layers of interpretation to emerge.

Hence, observation acts not just as a mechanism but as a conceptual foundation for partitioning, measuring, and categorizing phenomena.

4.7 Implications and insights

4.7.1 Case study: Is an electron an observer?

One recurring question is whether fundamental particles (e.g., electrons) qualify as observers. By Definition 4.1, an electron does not meet the minimal criteria. While it interacts with fields, there is no *internal mechanism* that updates "electron states" based on measured input. Instead, quantum mechanics describes its evolution via the Schrödinger equation or quantum field interactions, not a sensor-actuator model with an internal feedback loop. The electron is *observed* but does not itself *observe*, lacking a definable boundary $\mathcal{B}$ that separates "internal states" from "environmental data" in a cybernetic sense.

4.7.2 Quantum gravity and digital physics

In approaches to quantum gravity and pregeometric physics [11, 30, 32–34], space time emerges from underlying discrete structures or informational processes [35–41]. If the universe is essentially computational [12, 31], then observers play a crucial role in defining events or discrete state updates. Specifically, minimal observers could serve as *anchors* that stabilize local measurements, effectively converting "potential states" into classical-like outcomes. They might also define local reference frames or regions of emergent geometry, introducing boundaries into an otherwise unbounded computational cosmos and thus giving localized meaning to information flows.

These considerations align closely with constructivist perspectives linking formal languages and information structures to emergent physical realities [33]. Such minimal observer models not only enrich debates on how classicality or geometry arises from more fundamental substrates but also have implications for philosophical frameworks like dual-aspect monism [14]. Notably, the minimal observer concept might provide fresh insights into long-standing debates concerning the relationship between observers, consciousness, and the fundamental structure of reality.

Indeed, the role of observers as definers of boundaries and informational interfaces offers a bridge to epistemic or informational interpretations of quantum mechanics (such as QBism) and aligns with dual-aspect monism, where mental and physical properties emerge simultaneously from underlying informational substrates [14]. Thus, the notion of minimal observers might serve as a unifying foundation, providing a concrete model to explore how the dual aspects of subjective experience and objective reality might co-arise from a more primitive, informational substrate. As noted by Bacciagaluppi [45], this view can be traced back to Everett's treatment of observers, further suggesting a coherent synthesis of quantum epistemology, constructivism, and digital physics.

4.8 Computational models of minimal observation

4.8.1 *Cellular automata and universality*

RULE 110 AND BEYOND

Rule 110 is a one-dimensional cellular automaton that, despite having extremely simple local rules, is known to be Turing-complete [12]. However, to see it as an *observer*, we must specify how certain cells (or patterns) "sense" local configurations and produce "outputs" that affect the environment. Within such automata, a minimal observer can be implemented as a sub-lattice that monitors local states and changes them according to a rule $f(x, y)$. This demonstrates how emergent complexity might arise from repeated, local observation-based updates (Figs. 4.4–4.6).

4.8.2 *Synthesizing with the meta-model*

When inserted into a larger computational or physical context, the minimal observer definition acts like a module. For instance, in a swarm of robots or distributed computing networks, each node or agent can be viewed as a minimal observer with states, sensor channels, and outputs. By chaining or nesting these observers, we can analyze how *coherent group behaviors*, consensus, or emergent patterns appear in multi-agent systems.

4.9 Ontology of observers in physics

4.9.1 *Internal vs. external spaces*

The distinction between internal and external spaces, and consequently the concept of a boundary, is fundamental in fields ranging from thermodynamics to field theory. Observers explicitly define the boundary B that transforms unstructured "outside" data into structured, measurable signals and internal states [47]. Figure 4.7 provides a schematic of how an observer in a continuum setting might define a region of interest for measurement, resonating strongly with relational or observer-relative approaches in physics [28].

As mentioned, this boundary construction closely parallels Chris Fields' analysis of observers as information-theoretic interfaces that partition physical systems into interacting but distinct subsystems, thereby establishing the

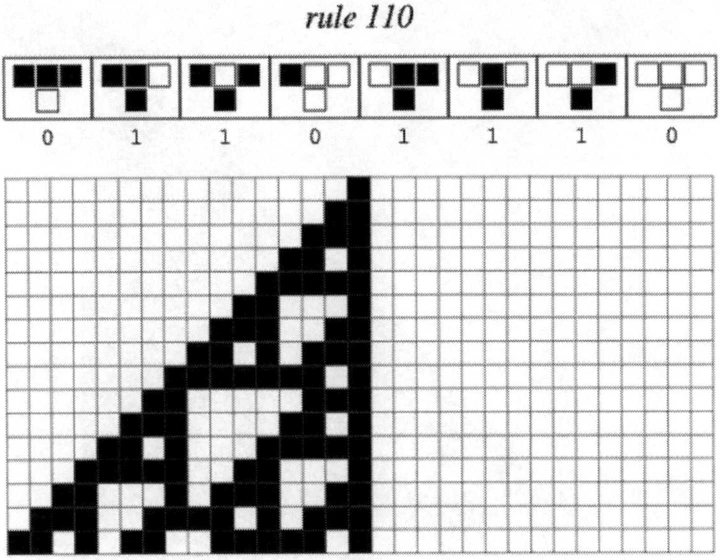

Figure 4.4 Rule 110 is one of the *elementary cellular automaton* rules introduced by Stephen Wolfram [12]. It specifies the next color in a cell, depending on its color and immediate neighbors. Its rule outcomes are encoded in the *binary* representation $110 = 01\ 101\ 110_2$. This rule is illustrated above together with the evolution of a single black cell it produces after 15 steps.

conditions necessary for objective measurement and communication [46]. Fields' interpretation highlights that the observer's choice of interface—analogous to the Heisenberg cut in quantum mechanics—explicitly delineates what parts of the universe are considered "observed systems" versus "observing apparatus," reinforcing that these distinctions are inherently observer-defined and context-dependent [46]. Thus, the act of boundary definition emerges as a central, indispensable feature of any meaningful observer model, linking classical cybernetic observers with foundational quantum-theoretic discussions.

4.9.2 Hierarchies of observers

NESTED OBSERVATION IN SOCIAL SYSTEMS

A corporation may act as an observer by collecting data (markets, consumer feedback), processing internal states (policy decisions), and acting outward (product releases). Individual employees also act as sub-observers, forming a nested or hierarchical structure [29].

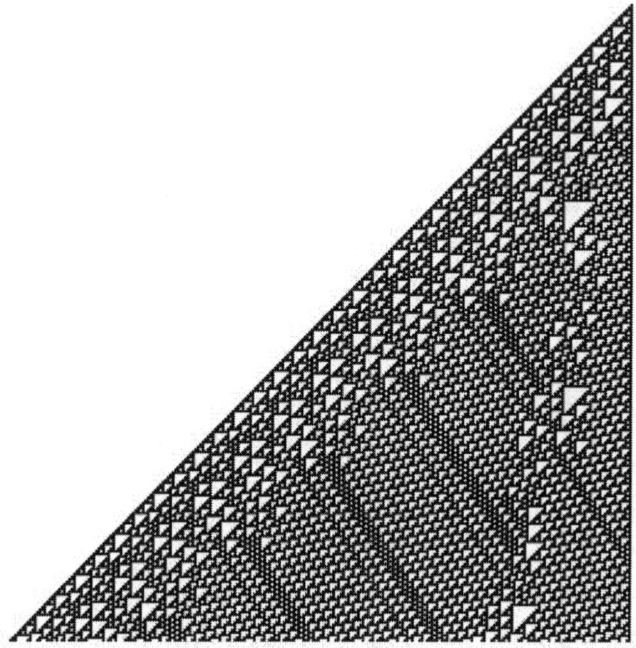

Figure 4.5 250 iterations of CA rule 110.

LAYERED APPARATUS IN PHYSICS

Large experiments (e.g., particle colliders) have multiple layers of detectors, each "observing" sub-events. Their outputs feed into aggregating devices that *observe the observers*, creating a second-order loop that yields final "measurement outcomes." An illustration of a nested observation is shown in Fig. 4.8.

4.10 Expanding the framework: Philosophy, theorems, ethics, and comparisons

4.10.1 Philosophical depth and engagement

4.10.1.1 Kant, Husserl, and Wittgenstein Revisited

KANTIAN TRANSCENDENTAL IDEALISM

Kant posited that the mind actively organizes experience via innate structures (space, time, categories). Observers in our model impose a boundary B and

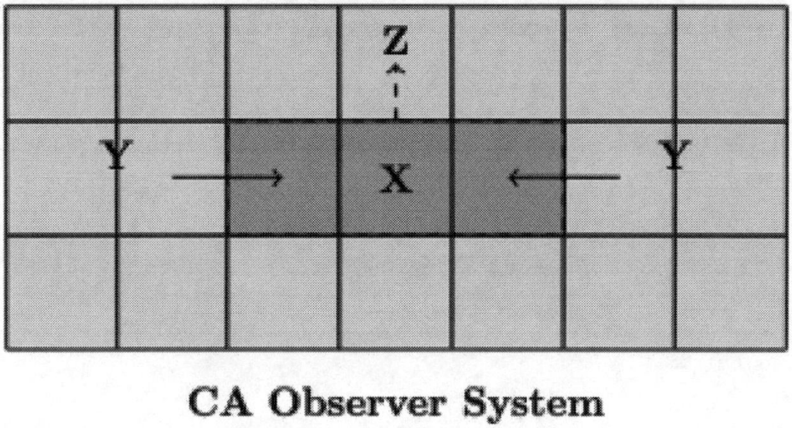

Figure 4.6 A cellular automaton segment where a "block" of cells is designated the observer. The observer's boundary $\mathcal{B}$ encloses internal states X and defines which neighboring cells constitute the input Y. Actions on the local environment serve as Z.

define functions f, g to interpret signals, analogous to Kant's forms of intuition and categories. The environment-in-itself remains inaccessible; what the observer registers is a structured "phenomenal" realm. Formally, one might interpret $f : X \times Y \to X$ as the "transcendental function" shaping raw input Y into the observer's internal "categories" X. This perspective deepens the epistemological stance that "raw data" cannot be known independently of the observer's interpretive structures. Ernst von Glasersfeld's radical constructivism pushes this notion further, proposing that knowledge is not representational of an external reality; rather, it emerges entirely from the observer's self-constructed experiential interface, reinforcing the idea that observers actively construct their own experiential worlds [56]. It also suggests potential alignments with neo-Kantian approaches and contemporary structural realism, where relational structures and observational interactions become more fundamental than intrinsic properties or objects themselves [57, 58].

PHENOMENOLOGY AND LANGUAGE

Husserl's phenomenology and Wittgenstein's later philosophy converge on a single theme that fits naturally into our observer formalism: *meaning arises only through an observer's situated activity*. For Husserl, consciousness is *intentional*—always directed toward some object. A minimal observer's feedback loop realizes a proto-intentionality: sensor inputs Y select a target state, the

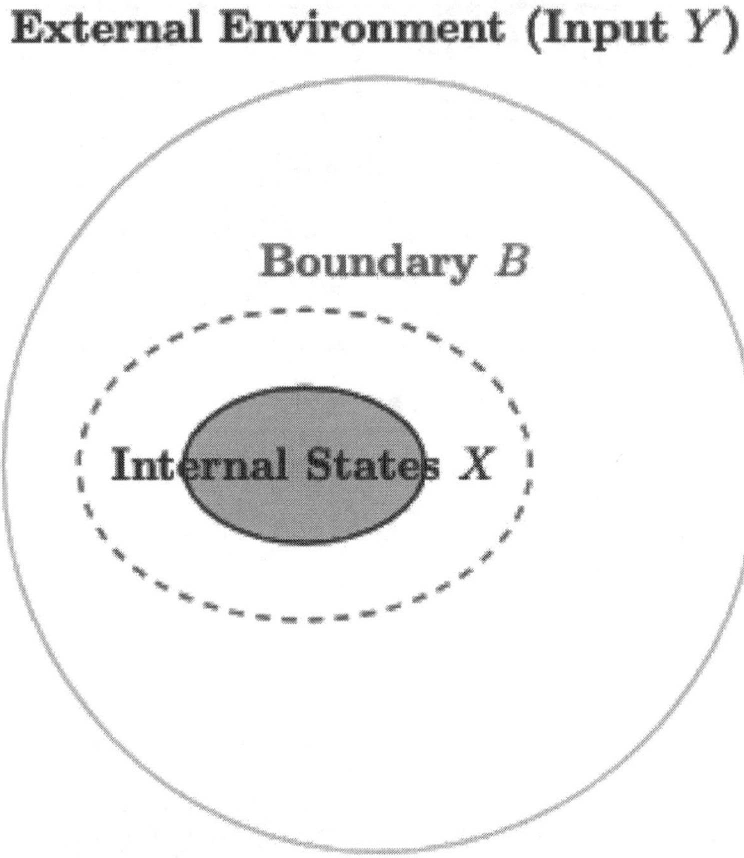

Figure 4.7 An observer in a continuum field scenario defines a boundary $\mathcal{B}$ around certain degrees of freedom (light grey region). The external environment (white) becomes input Y, while internal states (dark grey) belong to X.

internal update f incorporates that target, and the action g projects the observer back toward its object. Husserl's *epoché* then appears as a higher-order modulation of the boundary $\mathcal{B}$, bracketing certain inputs while foregrounding others and thereby re-drawing the line between "internal'' and "external.'' Wittgenstein adds the public dimension: what counts as a meaningful output $g(x)$ depends on the "language game'' played with neighboring observers. Each observer carries a local rule-book-its particular f, g pair—and different rule-books yield different semantic fields. Agreements or conflicts in multi-observer settings thus become questions about whether two

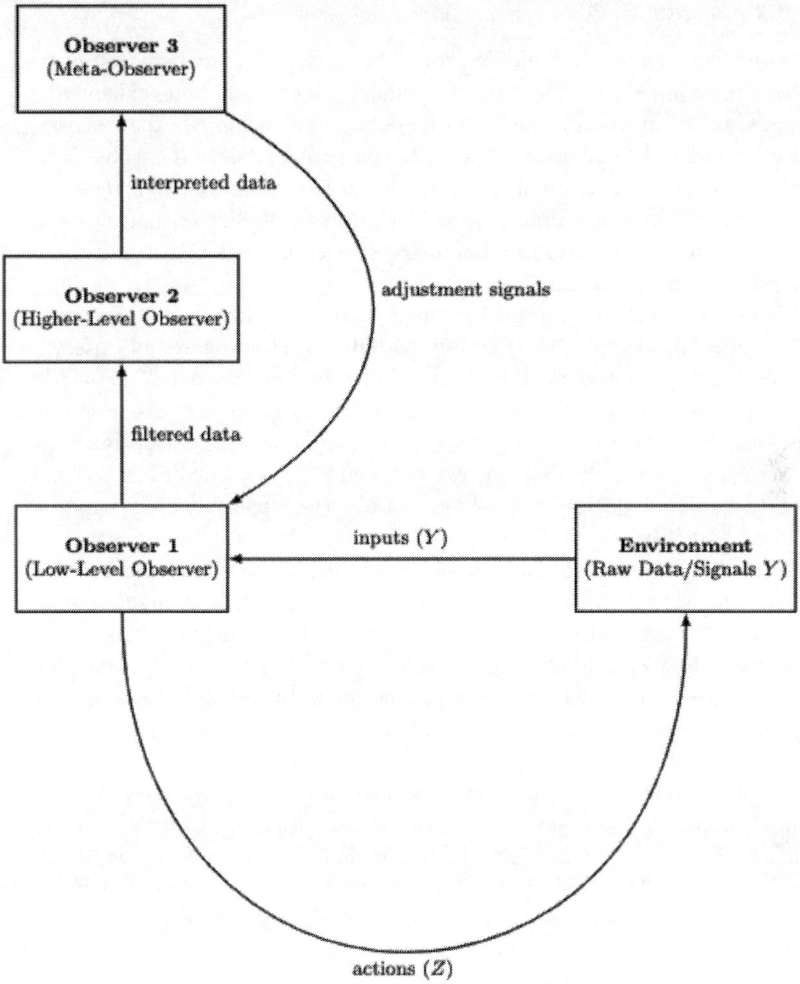

Figure 4.8 Hierarchical observer model: higher-level observers refine and modify the interpretations of lower-level observers, creating a multi-layered observational structure.

feedback loops can be made partially isomorphic. In short, intentional directedness (Husserl) and rule-following meaning (Wittgenstein) are two sides of the same cybernetic coin: both reduce to how an observer carves the world with $\mathcal{B}$ and updates themselves through f, g in dialogue with other agents [44].

4.10.1.2 Realism, anti-realism, and constructivism

Our formalism supports a "middle ground" between strict realism (the environment fully pre-exists observer involvement) and radical anti-realism (the observer creates all facts). Observers co-define phenomena via their action-perception loops; the environment may exist independently, but it is only "measured" or *objectified* once an observer's boundary $\mathcal{B}$ engages with it. Here, we use "measured" in a deliberately qualified sense to emphasize that observation does not merely uncover pre-existing properties but actively participates in structuring the phenomenon observed.

Bhaskar's critical realism [48] posits a stratified reality existing independently, yet acknowledges that it is only partially accessible through observation. Within our framework, the observer's internal state space X explicitly quantifies how much of external reality is filtered by the observer's structure. Thus, our model enriches Bhaskar's viewpoint by providing a rigorous, formal mechanism—namely, the interplay between boundary conditions ($\mathcal{B}$) and internal states (X)—for understanding how observers partially shape what is perceived as reality.

Similarly, the constructivist tradition, exemplified by Maturana and Varela's autopoietic observers, views reality as actively brought forth through ongoing sensor-actuator engagement [8, 16]. Our framework adds depth to this stance by clarifying how exactly observer-boundary interactions evolve dynamically, capturing how reality emerges through a continuous historical process of state updates. It thus provides a computational underpinning to constructivist arguments, showing explicitly how perception-action loops and boundary dynamics lead to the co-creation of observer-environment distinctions over time.

In essence, our observer model bridges realism and constructivism by demonstrating concretely how structured interaction via boundaries and feedback loops can yield an objective yet observer-relative ontology, adding a formal, computational dimension to longstanding philosophical insights.

4.10.1.3 Ethical implications

AGENCY IN AI AND RESPONSIBILITY

When AI systems self-modify their boundaries, the question arises: *who is responsible* if such an observer redefines what data it collects or how it acts? Minimal observers who reconfigure their own $f, g, \mathcal{B}$ approach an autonomy that complicates standard accountability models. As AI systems evolve, they might not only change internal parameters (*weights*) but also alter how inputs and outputs are partitioned or categorized. This redefinition of $\mathcal{B}$—the boundary delineating "system" from "environment"—blurs traditional lines of accountability.

Questions surrounding *moral patiency* and *moral agency* become central [49–51], prompting policymakers and ethicists to reconsider frameworks for attributing responsibility or liability.

ENVIRONMENTAL DEFINITIONS OF VALUE

An ecological observer, as an embodied and situated agent, might sense various biodiversity metrics, interpret them via f, and act to conserve or exploit those measures [53–55]. The boundary $\mathcal{B}$ it adopts—be it an economic perspective or an ecological one—determines how "value" is recognized or ignored. Observers thus shape environmental policy by selecting relevant signals and discarding others. This perspective suggests a dynamic interplay between how we define "resources" or "assets" and the ways we measure them. If an observer is geared toward short-term economic gain, its f function might ignore longer-term ecological consequences [52]. Conversely, a sustainability-focused observer would incorporate broader temporal scales and systemic interdependencies, thereby reshaping policy outcomes. Integrating observer theory into environmental ethics could guide more holistic decision-making processes, revealing blind spots in purely market-driven valuations.

4.10.2 Comparison with existing observer theories

4.10.2.1 QBism: Formalizing belief updates with an observer's boundary

THE QBISM'S VIEW

Quantum Bayesianism (QBism) treats the wavefunction as an observer's subjective belief about a quantum system. Rather than describing objective system properties, quantum states reflect an agent's personal degrees of belief. Importantly, QBism does not prescribe belief updates via the Born rule as probability updating (diachronic), but rather employs the Born rule to detect inconsistencies within an agent's probability assignments at a given moment (synchronic coherence) [68]. QBism similarly does not mandate Bayesian updating as the sole coherent strategy—belief updates depend fundamentally on each agent's particular belief conditions [68, 69].

While QBism clarifies measurement interactions conceptually, it deliberately avoids providing a third-person or physical dynamical mechanism explaining how belief states physically interact with measurement outcomes. This is intentional: QBism seeks to dissolve the measurement problem by reframing quantum mechanics as a decision-theoretic tool for agents, thus avoiding explanations via another physical process [70]. Nevertheless, clarifying the agent–environment boundary remains conceptually beneficial, particularly in explicit modeling scenarios.

STRUCTURED OBSERVER–SYSTEM INTERACTION

Our minimal observer framework explicitly formalizes these concepts by defining an observer as a structured entity $O = (X, Y, Z, f, g, \mathcal{B})$, with internal states X representing subjective beliefs, and boundary $\mathcal{B}$ explicitly distinguishing the observer from its environment. Measurement formally corresponds to sensory inputs $y \in Y$ crossing the boundary, updating the observer's internal state via a general transition function $f : X \times Y \rightarrow X$. Crucially, f need not represent solely Bayesian updating but can reflect any coherent belief-adjustment strategy consistent with the agent's prior conditions. Thus, rather than resolving ambiguities in QBism, this formalism concretizes QBist principles in an operational manner, explicitly modeling the interface between subjective belief and external quantum events.

PHENOMENOLOGICAL SCENARIO (QBism)

Consider a quantum coin-flip scenario. Initially, an observer encodes subjective uncertainty regarding outcomes. Upon measurement, the observed outcome crosses the boundary $\mathcal{B}$ into sensory input space Y, prompting state-transition f to yield a new internal belief state. This concrete depiction operationalizes QBism's subjective belief perspective, explicitly modeling informational flow across agent-system boundaries. Such explicit modeling, while not required by QBism, offers clarity and enables exploration of scenarios involving multiple interacting observers.

PHENOMENOLOGICAL ILLUSTRATION (QBism)

Imagine two observers with distinct priors observing the same quantum coin toss. Each observer, represented by internal state sets X, independently updates beliefs after measurements. Subsequent communication (interactions crossing boundaries $\mathcal{B}$) allows observers to exchange information; however, alignment of subjective beliefs is not guaranteed by mere communication alone, as Bayesian agents require substantial prior agreement (e.g., overlapping priors or common sample spaces) to reach consensus [69]. This explicitly modeled observer framework thus allows rigorous and nuanced exploration of QBism's relational epistemology, emphasizing the complexity and depth inherent in subjective belief alignment.

4.10.2.2 Relational quantum mechanics: Observer-relative states via internal structure

THE RQM VIEW

Relational Quantum Mechanics (RQM) asserts that the state of a system is not absolute; it only exists relative to a given observer or reference frame [28]. Different observers can have different accounts of a sequence of events, and there is no "God's-eye-view" wavefunction for the whole universe. In RQM, any physical interaction can play the role of an observation—even an inanimate object can be an "observer" in the sense that the object has a state relative to another system. However, RQM as originally formulated (e.g., by Carlo Rovelli) provides a conceptual framework rather than a detailed operational model: it says each observer might have their own Hilbert space of information, but it doesn't specify how an observer is structured or how exactly one defines when a fact becomes relative to a particular observer. For instance, RQM contends that if observer A measures system S, then S has a definite outcome state for A, but a second observer B who hasn't interacted may still describe the composite $A + S$ as in a superposition. This raises the question: what precisely counts as an "observer," and what is the criterion for an event being realized relative to that observer? The minimal observer model addresses this by giving a rigorous operational definition of an observer and the moment when something becomes a fact for that observer.

FORMALIZING THE RELATIONAL ROLE OF OBSERVERS

In our model, the boundary $\mathcal{B}$ explicitly delineates the division between an observer's internal degrees of freedom and the external system. This provides a concrete way to implement RQM's core idea that "state is relative to the observer." According to the model, a physical interaction becomes an observation when some input y crosses $\mathcal{B}$ and updates the internal state X of the observer. At that moment, we say the external system has a definite property relative to that observer, encoded by the observer's internal record. Each observer $O = (X, Y, Z, f, g, \mathcal{B})$ thus sees a different state of the world, as each maintains its unique internal record reflecting its interaction history. Our framework makes this relational nature explicit: if a quantity about the system is not encoded in X (because the interaction hasn't occurred), then from that observer's perspective, no definite fact yet exists. Conversely, once X is updated with some outcome, that outcome becomes a fact relative to that observer, even if for another observer the outcome remains unrealized. Thus, the model does away with the need for a universal, absolute wavefunction; the states recorded internally by each observer are sufficient. Practically, this suggests assigning quantum states (or classical data) to observer-system pairs rather than systems alone—the observer's internal state X thus becomes part of the system's description. What the minimal observer model adds is an operational criterion for observer-relativity: an event or state is observer-relative if and only if it is recorded in (or can be inferred from) the observer's internal state X as a result of an interaction crossing boundary $\mathcal{B}$.

THE OBSERVER'S INTERNAL STRUCTURE (X, Y, Z) IN THE RQM CONTEXT

Another novel contribution of our model is that it explicitly distinguishes roles within an observer. Traditional RQM treats observers as monolithic entities relative to which states are defined. Our model introduces internal structure: X (memory), Y (sensor inputs), and Z (actions). For instance, the input space Y precisely specifies which environmental interactions the observer detects, while the internal state space X explicitly limits the observer's informational capacity, reflecting the partial and coarse-grained nature of observed facts. This fine-grained structuring clarifies RQM's implicit assumption that observers record partial information. For example, an observer device with just two internal states ("event happened" vs. "didn't happen") represents the simplest possible relational scenario: relative to this minimal observer, reality reduces to binary distinctions. By contrast, richer state spaces allow more detailed descriptions, capturing progressively finer-grained relational states. The presence of action outputs Z further enriches the RQM view by showing observers as active agents capable of influencing their environments. Hence, our formalism clarifies how observers interact through boundary crossings and internal updates, rigorously tracking relational facts and eliminating ambiguity about their realization and communication.

PHENOMENOLOGICAL SCENARIO (RQM)

Consider the well-known Wigner's friend scenario. Alice (Observer O_A) is inside a lab, measuring an electron's spin, while Wigner (Observer O_W) remains isolated outside, yet plans to measure the combined Alice-electron system later. According to RQM, the electron's spin state is definite relative to Alice but indefinite (entangled) relative to Wigner. The minimal observer model explicitly represents this scenario by assigning each observer a distinct boundary and internal state: Alice's boundary $\mathcal{B}_A$ includes her measuring apparatus, and upon observing a result (spin-up), her internal state X_A is updated to encode this fact. For Wigner, whose boundary $\mathcal{B}_W$ has not yet interacted with Alice's lab, no input has entered his state space X_W, so from his perspective, no definite outcome has occurred. Only after Wigner interacts across his own boundary $\mathcal{B}_W$ (by observing Alice's record) does the spin become a definite fact relative to him. This provides a concrete instantiation of RQM's central thesis that facts emerge relationally upon interactions. Our model, thus, operationalizes

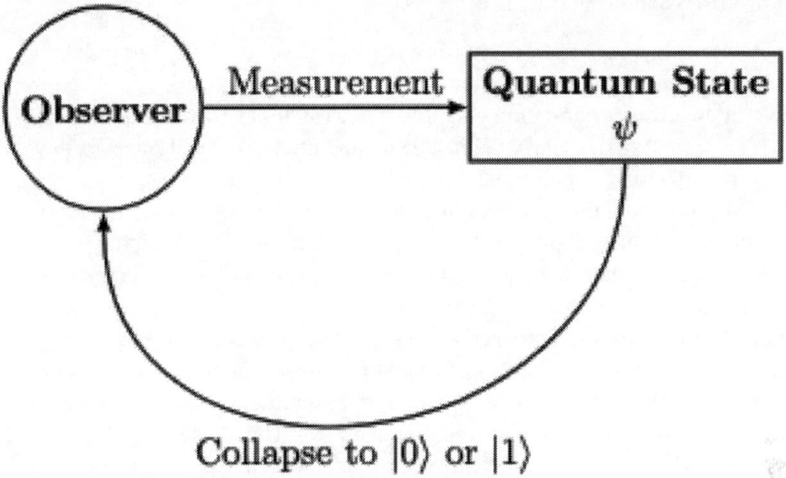

Figure 4.9 Quantum observer measuring a quantum state.

and concretely tracks how and when relational states arise—adding significant operational clarity and formal rigor to the philosophical and conceptual assertions of RQM.

4.10.2.3 Copenhagen interpretation: A concrete classical–quantum cut via X and $\mathcal{B}$

THE COPENHAGEN INTERPRETATION

The Copenhagen interpretation emphasizes the fundamental division—the *Heisenberg cut*—between the quantum system, described by a wavefunction, and the classical measuring apparatus, that records outcomes. Traditionally, the precise location of this cut is flexible, provided the apparatus and observer remain classically describable while the measured system is quantum. Measurement collapses the wavefunction, producing definite classical outcomes (Fig. 4.9), but Copenhagen interpretations leave the exact nature and placement of this boundary unclear. Debates persist regarding whether the cut involves conscious observers or macroscopic apparatuses, resulting in well-known puzzles such as Schrödinger's cat and Wigner's friend scenarios. The minimal observer model clarifies this issue by explicitly embedding the classical–quantum cut within a formal structure of the observer itself.

THE MINIMAL OBSERVER AS THE CUT

In our framework, an observer explicitly embodies the classical apparatus along with its information-storage structure. The internal state space X is defined to be classical by design, consisting of stable states (memory bits or pointer positions). The boundary $\mathcal{B}$ clearly delineates the quantum-classical interface: everything external to $\mathcal{B}$ remains quantum, while internal states in X are classical records. Measurement thus involves two clear stages: (1) a quantum interaction at boundary $\mathcal{B}$ produces input $y \in Y$ (e.g., photon hitting a photographic plate), and (2) the internal state X is updated to a definite outcome state via the transition function $f : X \times Y \to X$. Before measurement, the system may remain in superposition from the observer's perspective; afterward, X holds a single definite classical outcome. Thus, our model provides a mathematically rigorous criterion for the condition when quantum measurement occurs, eliminating the need for vague notions of wavefunction collapse. The Heisenberg cut's position becomes explicitly adjustable through placement of $\mathcal{B}$—from immediate detector boundary to encompassing an entire laboratory—offering a quantitative foundation for Copenhagen's traditionally qualitative statements.

EMERGENCE OF CLASSICAL RECORDS

The minimal observer model further elucidates how quantum events yield classical records. A classical record in Copenhagen terms is an unambiguously readable outcome, such as a pointer reading or detector click. In our framework, classical records correspond precisely to stable internal states X. Before measurement, the observer's internal state is undetermined; measurement interactions update X to a stable classical state. Once recorded, this classical state persists, consistent with classical robustness ensured in experiments by decoherence and amplification. Hence, the quantum-to-classical transition explicitly corresponds to boundary-crossing inputs updating internal observer states. The model thus provides a transparent, physically grounded account of quantum state collapse without invoking consciousness or special collapse rules—only finite internal storage and boundary-based interactions.

PHENOMENOLOGICAL SCENARIO (COPENHAGEN)

Consider a Stern–Gerlach experiment: a beam of silver atoms (quantum spins) passes through an apparatus, splitting into spin-up or spin-down components. According to Copenhagen, the atom remains in a superposition until striking a detector. Our minimal observer model clarifies this scenario explicitly. The detector plus its readout electronics constitutes an observer O, with boundary $\mathcal{B}$ defined at the detector's interaction surface. The observer's internal state space X initially encodes a neutral "ready" state. When an atom with unknown spin interacts at the detector surface (boundary $\mathcal{B}$), an input $y \in Y$ crosses into the detector, triggering f to update X into a stable, definite state—either "spin-up recorded" or "spin-down recorded." This clearly defines when the quantum system transitions into a classical record. Prior to interaction, the spin state is indefinite within observer O; post-interaction, X explicitly encodes a single classical result. Thus, the minimal observer model operationally identifies the exact moment and mechanism of quantum collapse into classical reality, addressing longstanding ambiguities within Copenhagen interpretations and providing new experimental criteria for the quantum-classical boundary.

4.10.3 *Intrinsic space of observers and background-independent frameworks*

4.10.3.1 Concept of an intrinsic observer space

Intrinsic observer space refers to treating the set of all possible observers as a space with its own structure, rather than assuming a single "God's eye" view. In physics, one example is the manifold of all observer states in spacetime. For instance, in general relativity one can consider the 7-dimensional manifold consisting of every possible future-timelike unit tangent vector at every point—essentially "all possible observers" (each point in this observer-space corresponds to an observer with a location and velocity) [62, 63]. The goal is to describe physics from within this space of observers, using only relationships between observers, without appealing to an outside absolute frame. This is an intrinsic approach because the description is given in terms of quantities an observer can define internally (their own clock, ruler, measurement records, etc.), rather than coordinates defined by an external backdrop.

In an intrinsic observer-space formalism, one often introduces equivalence classes of observers and observer-dependent coordinates: two observers might be regarded as "the same" in some intrinsic sense if there is a transformation mapping one's observations to the other's. There is a philosophical subtlety here. One view ("view from nowhere") treats different observer frames as mere perspectives on an underlying invariant reality, identifying them via equivalence classes. Another view ("view from everywhere") holds that each observer's frame defines its own reality without needing to be quotiented out [64]. In practice, most physical theories define an observer-space along with transformations relating observers, so that one can either quotient out those differences (to find invariant physics) or consider each perspective as fundamental. Below we formalize these ideas by defining observer equivalence and coordinate structures, and then discuss how going background-independent—not only in spacetime but in the space of observers itself—leads to new formulations in category theory, differential geometry, and physics.

4.10.3.2 Equivalence classes of observers

Observer versus frame: In the relativistic portions of this section we momentarily use "observer'' in the *kinematical* sense of a time-like world-line equipped with a local orthonormal frame (or tetrad). That frame can be dragged by a Lorentz transformation in special relativity, or by a spacetime diffeomorphism in general relativity; the full, feedback–loop observer of Section 4.1 is therefore represented by an *orbit* (gauge class) of such frames rather than a single coordinate chart.

Observer equivalence means that under certain transformations, two observers are considered physically or functionally the same. In relativity, for example, the relativity principle states that all inertial frames are fundamentally equivalent—no preferred inertial frame exists. This can be phrased as: "all reference systems are equivalent," reflecting a democratic principle that physics should not single out one observer over another [65]. Any two inertial observers moving at constant velocity relative to each other are related by a Lorentz transformation; they form a single equivalence class under these symmetry transformations. More generally, in general relativity any two observers (inertial or not) can be related by a diffeomorphism (a smooth coordinate transformation) acting on the spacetime manifold. If a diffeomorphism maps one observer's worldline and measurements to another's, the two are considered the same physical situation due to general covariance (diffeomorphism symmetry). In gauge-theoretic language, choosing a different observer or reference frame is often akin to a gauge transformation—a change of description that does not alter physical content [62, 64]. Thus, an equivalence class of observers can be defined by the symmetry transformations of the theory: all observers related by a valid change-of-frame (Poincaré transformations in special relativity, or arbitrary diffeomorphisms in general relativity) belong to the same class, representing one "physical situation" viewed from different perspectives.

In more abstract terms, one can formalize observers and their transformations using groupoids or categories. We may define a category of observers where each object is an observer (or an observer's coordinate system) and each morphism is a change-of-observer transformation (such as a coordinate change or Lorentz boost). This category is naturally a groupoid (every transformation is invertible), encapsulating the idea that all observers are on equal footing and can transition into one another's perspective [65]. Within this framework, an isomorphism (invertible morphism) between two observer objects indicates that they are effectively the same observer in a relational sense. Put differently, an isomorphism is a dictionary translating one observer's internal descriptions into another's. Using this idea, one can rigorously define an equivalence relation: $O_1 \sim O_2$ if and only if there exists a bijective structure-preserving mapping (an isomorphism) between observer O_1 and O_2. It can be shown that this $\sim$ indeed satisfies reflexivity, symmetry, and transitivity (i.e. it is an equivalence relation). In practical terms, $O_1 \sim O_2$ means the two observers differ only by a relabeling of their relevant structures (such as coordinate labels or internal states) while preserving all relationships—they are "the same observer described in two different languages."

For example, in a formal observer-as-system model (discussed more in the computational context below), one theorem states that two observer systems are equivalent if there is a bijection between their internal state sets, input channels, and output channels that makes their transition and observation dynamics coincide; this bijective homomorphism defines an observer isomorphism. In other words, two observers are "equivalent" if they differ only by relabeling their internal states, inputs, and outputs in a way that preserves the structure of their feedback and state-transition dynamics. Thus we obtain equivalence classes of observers as sets of all observers isomorphic to each other.

By factoring out (i.e. identifying) observers related by these transformations, physics focuses on the invariant content. However, one may also study the space of inequivalent observers as a manifold or set of distinct vantage points. Clearly defining the equivalence relation is crucial—whether it's "same physical trajectory up to re-labeling," "same dynamics up to relabeling of states," or some other criterion depending on context. Moreover, one must decide whether to adopt the view that only the equivalence class has objective meaning (the view-from-nowhere stance) or that each equivalence class member may be treated as having its own reality unless/until related to another (the view-from-everywhere stance) [64].

4.10.3.3 Observer-dependent coordinates and frames

Each observer in the intrinsic space typically comes equipped with their own coordinate system or frame of reference. That is, an observer defines how to measure space and time (or other quantities) relative to themselves. Two different observers generally have different coordinate descriptions of the same events. Formalizing this, one can assign to each observer a coordinate chart or a basis of measurement. For instance, an observer in spacetime has a natural choice of time coordinate (their proper time along their worldline) and space coordinates (defined by some simultaneity convention, like projecting events orthogonally to the observer's worldline). This leads to different splittings of spacetime into "space" and "time" for different observers.

A simple example: in special relativity, two inertial observers moving relative to each other have time axes that mix into each other's space axes—what one calls "now" is a mixture of "now" and "later" for the other, due to the relativity of simultaneity. In general relativity or accelerating frames, the differences are even more pronounced. A notable case is the Rindler observer (uniformly accelerated) vs. an inertial Minkowski observer: each has a different foliation of spacetime into space+time. The accelerated Rindler observer uses Rindler coordinates, in which their constant-time slices are hyperbolas that an inertial observer would describe as accelerated trajectories. This difference in "observer's space" has concrete physical effects—each observer literally perceives a different version of space and time. For example, the Unruh effect can be interpreted as arising from this difference: an accelerated observer (Rindler frame) sees a thermal bath of particles, while an inertial observer sees vacuum, because the very notion of what constitutes a "particle" (or even a vacuum state) depends on the observer's space–time splitting [66]. In cosmology as well, an "observer's space" can be defined at each moment for an arbitrary worldline via a chosen simultaneity prescription; this space generally differs from any globally preferred slicing (e.g., an inertial observer in an expanding universe defines space in a way that is neither homogeneous nor isotropic, even if the universe has a homogeneous slicing) [66]. All these examples underscore that coordinate structures distinguishing observers are fundamental—one must carefully specify which observer's frame is being used when assigning values to measurements.

Mathematically, one way to encode observer-dependent frames is with the frame bundle or observer bundle on spacetime. The bundle of orthonormal frames at each spacetime point includes all possible orientations of an observer's axes at that point. Picking a specific frame (a basis of time+space directions) is like choosing an observer at that event. In differential-geometric terms, the space of observers can be modeled as this bundle: each point in observer-space might be labeled by (x, u), where x is a spacetime point and u is a unit timelike 4-velocity at x (the observer's local velocity or time-direction). One can introduce coordinates on this observer-space: for example, coordinates $(t, x, y, z, v_x, v_y, v_z)$ could label an observer's spacetime position (t, x, y, z) and their 3-velocity components (v_x, v_y, v_z) (or analogous parameters like rapidity and orientation angles). Such coordinates on observer-space describe how observers differ by location and state of motion. The transformations between observers (e.g., Lorentz boosts, rotations, or general coordinate changes) will act on these coordinates. By studying the observer-space as an entity, we can attribute geometric structures to it (such as a metric or connection defined on O). The key idea is that an observer's coordinate system is not global and absolute; it is attached to the observer. When comparing two observers, we either transform one's coordinates into the other's or work in the larger space containing both and map between their coordinate patches.

One concrete example of defining observer-dependent coordinates is the construction of a canonical reference frame for any given observer in curved spacetime. Lachièze-Rey [66] provides a prescription to foliate spacetime according to an arbitrary observer's notion of simultaneity. This yields a unique slicing (and thus a set of spatial coordinates at each instant) corresponding to that particular observer, even if the observer is accelerating or in a general cosmological model. The result is that different observers have different "slices" of space—different 3D surfaces that they consider to be happening "now"—and thus they decompose spacetime differently. The Langevin observer (in the twin paradox) or a rotating observer, for example, will assign a different geometry to space (non-Euclidean, etc.) than an inertial one [66]. All these constructions are ways of giving each observer an intrinsic coordinate map. Such a coordinate structure is crucial for observer-dependent physics (like defining particle horizons or energy as measured by that observer).

In summary, the intrinsic space of observers comes with a myriad of possible coordinate choices, one per observer. The relations among these coordinate systems (through transformations) are what connect one observer's measurements to another's. This viewpoint emphasizes that many quantities (length, time interval, even the vacuum state of a field) are not absolute but depend on the observer's frame. By formalizing an observer-space, we keep track of these dependencies systematically and can define what it means for a quantity to be covariant or invariant under change of observer. An invariant would be something all observers agree on or that transforms trivially (e.g., the spacetime interval, or an abstract action integral), whereas a covariant quantity has a well-defined transformation law that allows any observer to calculate what another would see.

4.10.3.4 Extending background independence to observer space

Background independence traditionally means a theory does not presume a fixed spacetime structure—the spacetime geometry is dynamical or undetermined, and only relationships (e.g., causal or metric relations) that satisfy the equations have physical meaning. General relativity is the prime example: there is no fixed background metric; the geometry (metric field) is a part of the solution. Extending this idea beyond spacetime to the space of observers means the theory also does not assume a preferred or fixed observer/frame of reference. In other words, not only is spacetime relative, but the arena of all observational viewpoints is itself treated without prior structure. All observers are on equal footing, and the laws of physics must be formulated without secretly choosing a special observer or coordinate system in advance. This is essentially the principle of relativity and general covariance taken to the next level—it suggests that the space of observers is taken as fundamental, and spacetime (with its events and distances) might even be secondary or emergent from relations among observers [62].

One concrete realization of this idea comes from Cartan geometry and frame bundle techniques. In a fully background-independent view, one can start with the space of observers as a primary object and later extract spacetime as a derived concept. For example, consider again the observer manifold O (all events + a unit timelike direction at each event). On this 7D manifold, we can define fields and a Cartan connection such that no *a priori* spacetime metric is needed—the only input is the local symmetry group (the Lorentz group) which acts on the fibers. Gielen and Wise propose exactly this: "taking observer space as fundamental" and formulating gravity in terms of an observer-space geometry [62]. In their approach, spacetime can be recovered as a quotient of the observer space if certain integrability conditions hold (essentially, if a global 4D slice through the 7D space can be consistently defined). If those conditions fail, it suggests a scenario in which an absolute spacetime cannot be stitched together—instead, physics might only be definable in terms of overlapping observer perspectives, making spacetime an observer-dependent, relative concept [62]. This radical possibility is a form of fully relational ontology: the fundamental description is a network of observer states and their relations, with no single universal spacetime backdrop for all events.

Even without going to such extremes, insisting on observer-background independence means our formalism should be covariant under change of observer. In practical terms, any statement or equation should be valid in any frame—or transform appropriately between frames—without relying on a fixed background frame. Category theory provides a natural language for this: one can require that the theory is formulated as functorial or natural with respect to the category of observers. For instance, in a category-theoretic formulation of quantum physics or gravity, one proposal is to assign to each observer their own state space (Hilbert space) and then relate these via functors. Crane suggests that a "state for quantum gravity" could be described as a functor from the category of observers to the category of vector spaces (Hilbert spaces) [67]. What this means is that each observer (object in the category) gets a vector space of states, and an observational transition (morphism) between observers (representing a change of reference frame or perspective) induces a linear map between their state spaces. Physical laws would then be invariant under such changes if they arise as natural transformations between these functors (so that the state assignments to each observer are consistent with evolving or transforming the system). This categorical construction embodies background independence at the level of observers: there is no single "preferred" state space or single observer's coordinates in which the theory must be formulated. Instead, the principle of relativity is built in—any observer's description can be transformed to any other's systematically.

Another way to enforce no preferred observer is to treat the full groupoid of observers as the arena for physics, instead of a single spacetime. As mentioned, one can view all frames as connected by morphisms. The laws of physics should be expressible in a way that is invariant under moving along these morphisms.

In effect, this is like saying the fundamental formulation is done "up on O" (the observer space) rather than down on M (spacetime) with a chosen frame. When done properly, this yields all the same physics but with a manifest guarantee of symmetry. A simple analogy is how one formulates Maxwell's equations or other laws in tensor form—by writing them in a covariant form, one shows they hold in any coordinate system (observer). Here we elevate that idea: we formulate the entire theoretical framework on a structure that does not bias any particular observer or coordinate system. Only relations between observers (like relative velocities, or intersection events) might enter.

In practical terms, achieving full background independence including observers might require additional constraints or symmetry principles. For example, requiring that no global structure (like a global time coordinate or global inertial frame) appears, forces the introduction of fields or connections that compensate for shifting observer perspective. The observer-space approach introduces something like an "internal observer group" gauge symmetry—shifting an observer's 4-velocity direction (i.e. moving along the fiber at fixed spacetime point) is a symmetry operation. Demanding invariance under this symmetry can lead to new conservation laws or conditions (just as gauge invariance leads to conserved currents via Noether's theorem). In short, extending background independence to observers means that the theory's degrees of freedom and constraints should be described in a way that does not require fixing an observer. This is a natural generalization of general covariance: not only are coordinates unphysical but even the choice of an observing frame is unphysical until an interaction (measurement or communication) relates two observers.

The payoff of this extension is conceptual clarity and universality. It becomes clear which aspects of a theory are genuinely invariant and which are convention-dependent. It also helps bridge physics with information theory and computation, where the role of the observer (or agent) is crucial. By not pinning down an observer, we keep the theory general enough to apply, say, to any agent (human, machine, or particle) that could be making observations. The theory then has to supply rules for how different agents' accounts compare—which is precisely the role of the transformations in observer space or the natural transformations in the categorical approach. This fully relational perspective is at the heart of many modern discussions in quantum gravity, quantum foundations, and even philosophy of science, where one tries to eliminate any lingering absolute structure, including the abstract "observing subject," and replace it with a web of relations.

4.10.3.5 Category-theoretic perspectives on observers

Category theory offers a high-level, structural way to describe observers and their interrelations. We already touched on the category of observers idea, where observers are objects and transformations between observers (changes of frame or perspective) are morphisms. Because any observer should in principle be able to transform to any other (given the appropriate coordinate transformation or data translation), this category is typically a groupoid (every morphism is invertible). Formulating physics in this language makes symmetries and equivalences explicit. For example, the collection of all inertial frames in special relativity can be seen as the objects of a category, with a morphism for each Lorentz transformation mapping one frame to another. That category has the structure of the Poincaré group action (essentially a one-object category if we identify all inertial frames as instances of the same abstract frame, or a groupoid if we treat each frame as a separate object). The advantage of the categorical view is that one can then impose conditions like functoriality or naturality to ensure physics is consistent across different observers.

A striking use of category theory in observer-related physics is in quantum gravity and quantum foundations. The work of Crane (1993) and others envisioned a scenario where each observer has their own Hilbert space of quantum states, and physics is a kind of many-object generalization of quantum mechanics. In this approach, consistency between observers is maintained by categorical structures. Specifically, Crane described that "a state for quantum gravity is given by a functor from the category of observers to the category of vector spaces" [67]. In plain terms, this assigns to each observer a vector space (typically a Hilbert space) such that when you have a morphism (an observation change or reference-frame transformation) from observer A to observer B, the functor provides a linear map between A's vector space and B's vector space. The physical state of the universe would then not be a single vector, but rather the entire functor—which consistently assigns state vectors to each observer and ensures that if two observers are related by a transformation, their state descriptions are related by the corresponding linear map. Time evolution or other processes can be described by natural transformations (mapping functors to functors) which play the role of dynamics in this schema. This categorical formalism is closely related to the idea of a topological quantum field theory (TQFT), where a state is associated to boundaries (observers can be thought of as "boundaries" between observed system and observer) and consistency conditions must hold for gluing boundaries (analogous to communicating or transforming between observers).

Another category-theoretic approach is the groupoid model of relativity. As mentioned, instead of focusing only on a symmetry group (which usually has one object/state and many automorphisms), one considers the category of all observers with morphisms as allowed transformations. This can handle cases that a single group cannot (e.g., when only certain observers can directly transform to each other, or when composition of transformations has path-dependence as in general relativity's gravitational holonomies). In a groupoid, each object has its own little group of automorphisms (its symmetry group of leaving that observer invariant), and the whole structure can encode both symmetry and the equivalence relation between different frames. Oziewicz proposed to "formulate the physics of relativity in terms of the groupoid category of observers, keeping strictly the most democratic interpretation of the Relativity Principle that all reference systems are equivalent" [65]. This means rather than starting with a fixed space and one group acting on it, one starts with the many-object category where each object is a reference frame and each morphism is a change of frame, and one builds the theory there. Doing so can reveal hidden assumptions of the usual approach—for instance, the breakdown of a single-group picture when considering non-inertial observers or gravity can be naturally accommodated by a groupoid (since accelerating frames might not be related by a single global Lorentz transformation, but can be related piecewise). The category approach also generalizes to observers of different types (imagine a category that includes classical and quantum observers as different kinds of objects, and morphisms that describe interactions or translations between their descriptions).

Category theory also provides tools for hierarchies of observers. One can consider 2-categories or higher categories where morphisms between observers themselves have morphisms (think of one observer observing a pair of other observers in communication—this could be a 2-morphism in a higher category of observers). These abstractions can formalize complex scenarios like "observer A watches observers B and C conduct an experiment." While these are mostly theoretical at this stage, they hint at a unified language for multi-observer interactions.

In less abstract terms, categories help enforce that physics is independent of the *choice* of observer by design. If the theory is formulated as a functor on the *category of observers*, then by definition it assigns equivalent data to *isomorphic* observers—so every member of an equivalence class receives the same physics. This is a powerful way to encode observer-independence: physical predictions arise as functorial assignments that do not discriminate between isomorphic viewpoints. Any quantity that is *strictly* observer-invariant will factor through this functor to the *quotient category of equivalence classes*. Importantly, that quotient is not a return to a single, objective "view from nowhere"; it is itself defined *relationally*, via the mappings generated by observers (a stance fully aligned with second-order cybernetics [15]). By contrast, quantities that remain observer-relative live entirely in the original category and transform non-trivially along its morphisms.

To summarize, category-theoretic perspectives treat observers as fundamental objects and changes of perspective as fundamental morphisms. They naturally encode equivalence (via isomorphisms) and can enforce that the theory treats all observers without favoritism (via functorial assignments). This is a very general framework, capable of bridging physics with computer science and logic (where "observers" could be seen as contexts or processes, and one uses categorical semantics to relate them). The price is a high level of abstraction, but the payoff is unifying disparate ideas (symmetry, relativity principle, reference frames, information transfer) under one mathematical roof.

4.10.3.6 Differential-geometric observer space structures

Differential geometry provides another powerful framework for formalizing an intrinsic space of observers. Here, one treats the collection of observers as a manifold or fiber bundle and endows it with geometric structures. A clear example is again the set O of all future-directed unit timelike vectors in a given spacetime (assuming a Lorentzian manifold M). O can be thought of as the unit timelike tangent bundle of M—it is a fiber bundle over spacetime M, where each fiber (at a point $x \in M$) is the hyperboloid of unit timelike vectors (the possible 4-velocity directions for an observer at x). For 4-dimensional spacetime, O is 7-dimensional (4 for position + 3 for the velocity direction). We can call O the observer manifold. Now, any field on O that is appropriately invariant can represent a physical quantity measured by observers. For instance, a function on O could assign to each possible observer a value (like "the temperature that observer measures")—different values for different states of motion if the effect is observer-dependent (as with Unruh radiation). More powerfully, one can formulate dynamics directly on O. Gielen and Wise demonstrated that one can reformulate general relativity in terms of a Cartan geometry on observer space [62]. In their formulation, the usual Einstein field equations can be derived from curvature conditions and fields on O, and spacetime itself emerges as a derived concept (as an equivalence class or quotient of observer trajectories). The geometric idea is that O comes naturally equipped with two distributions (roughly: horizontal directions correspond to moving an observer in spacetime and vertical directions correspond to changing an observer's velocity at the same spacetime point). A Cartan connection on O can encode both the spacetime curvature and how local reference frames rotate or accelerate. The equivalence principle (local Lorentz symmetry) is built in by the fact that the structure group on O is the Lorentz group, which acts on the fibers (different velocity directions at one point are related by Lorentz transformations). By using Cartan geometry, one ensures that at each point of O the geometry looks like a "model geometry" (Minkowski space for the horizontal part, and a velocity-space model for the vertical part).

One major benefit of a differential-geometric observer space formalism is that it can handle observer-dependent effects in a smooth, quantitative way. Notions like "spatial vs temporal direction" become geometric: at each observer-state in O, one can identify the subspace of directions that correspond to that observer's spatial axes (those directions perpendicular to the observer's 4-velocity) and the time direction (along the 4-velocity). In the Cartan setup, these come from splitting the tangent space of O using the observer's velocity field. Geometric structures on O can encode things like an observer's proper time (a natural time coordinate along the observer's worldline in O), gravitational fields (which might appear as curvature or torsion in the connection on O), and fictitious forces in non-inertial frames (which appear as real geometric effects in an accelerating observer's bundle). In this picture, a specific observer is represented as a curve in O (e.g., an observer moving through spacetime traces out a path through different points of O, since their position and velocity may change). The physics along that curve is that observer's experience. But because we have the whole manifold O that includes all other observers, we can relate different observers by geometric relations in O. For instance, two observers might come into contact (literally meet at an event): this is represented by two curves in O intersecting at a point (meaning they share the same x and u at that instant). Or an observer accelerating is moving vertically in O (changing u while staying at roughly the same x, in an instantaneous sense). The geometry of O might tell us, for example, how an accelerating observer's spatial slice tilts and how their notion of simultaneity shifts—information encoded in the connection on O.

Importantly, this approach yields insight into background independence. If O is fundamental, we do not assume spacetime M exists as a separate stage; it can be constructed from O by identifying all observer-states that share the same event regardless of velocity (that quotient gives back spacetime M if it exists). The conditions for this to work are essentially that certain fields on O (like an observer congruence field) are integrable. When those conditions are met, the standard spacetime picture emerges smoothly from the observer picture. When they are not, it suggests something like a "foamy" situation where different observers' views cannot be stitched into a single manifold—possibly an avenue to understand quantum gravitational scenarios where classical spacetime breaks down. But even aside from that extreme, working on O has practical advantages. In O, one can separate "absolute" properties from observer-relative ones. For example, a tensor field on spacetime, when pulled back to O, can be split into parts seen by a given observer (like splitting an electromagnetic field into electric and magnetic fields depends on the observer's velocity). On O, that splitting is just evaluating the field with the additional data of u. Maxwell's equations can be pulled back to O and yield a set of equations that explicitly show how different observers see electric/magnetic fields mix. In gravitational dynamics, the Hamiltonian (canonical) formulation chooses an observer foliation; working on O lets one derive the Hamiltonian constraints without ever explicitly choosing a foliation—instead, an "observer field" (a choice of a representative observer at each spacetime point) can be considered a gauge fixing that O allows us to handle flexibly [62].

Another differential-geometric insight is how measurement invariants appear in observer space. In general relativity, an "observable" must be invariant under diffeomorphisms (since coordinates are arbitrary). In observer-space terms, an observable might be a function on O that is invariant under the local Lorentz transformation on the fibers (because changing the inertial axes of a given observer shouldn't change a scalar physical quantity). Thus true invariants live on O but do not depend on the u aspect—only on the spacetime event (like proper scalar curvature at a point). Those correspond to usual scalar invariants in spacetime. However, one can also consider observer-dependent observables—quantities that *do* depend on u, i.e. on the observer's state of motion. These are not invariants of the full diffeomorphism + Lorentz gauge, but they are well-defined as functions on O. An example is the energy density of a field as measured by an observer with 4-velocity u. This is not an invariant scalar on spacetime (it depends on the observer), but it is a well-defined scalar on O. By working on the observer manifold, we can talk about such quantities legitimately and track how they change as one moves in O (i.e. as the observer changes). This can be useful in relativistic statistical mechanics or black hole thermodynamics, where one wants to compare what different families of observers see.

In summary, the differential-geometric approach builds an intrinsic coordinate system on the space of observers itself. It treats observer transformations as fundamental symmetries (a kind of extended gauge symmetry that includes reference-frame changes). This allows us to express laws of physics in a manifestly observer-covariant way. Ultimately, such formalisms impact how we think of space, time, and measurement: they blur the line between what is "physical (spacetime) geometry" and what is "perspective." All frames live in one big space, and a given frame's coordinates are just one patch on this observer manifold. This is an explicit way of enforcing no preferred frame—the geometry doesn't care which observer you label as origin because any point in O is just as good as any other for describing physics. It also provides new tools to analyze physical problems by lifting them to O, solving symmetrically, and then projecting results back down to particular observers.

4.10.3.7 Observer-dependent physics and relational measurements

Physical theories increasingly recognize that what is measured or observed can depend on the state of the observer. We have already seen examples in relativity (time dilation, length contraction, simultaneity shifts, particle detection differences in the Unruh effect) where different observers experience different values or even different qualitative phenomena. By formalizing the space of observers, we get a handle on how to transform measurements from one observer to another and what structures are invariant versus what are observer-dependent. In classical physics, these transformations are given by kinematic symmetry groups (Galilean or Lorentz transformations). In modern physics, we encounter observer-dependence in broader contexts, for example:

Quantum measurements: In quantum mechanics, the result of a measurement can depend on the "context"—which is often tied to the observer's experimental setup or frame. Different observers (especially in thought experiments like Wigner's friend) might not even agree on what has been measured or the state of a system. Relational quantum mechanics (Rovelli) posits that the quantum state is not absolute but is relative to each observing system. This is analogous to how in relativity an event's time coordinate is observer-dependent; here, the outcome (or state assignment) is observer-dependent, and only when two observers exchange information and correlate their records do they find a consistent story. An observer-space for quantum contexts could formalize this by letting each observer have their own space of possible quantum states of the world, and "bridging maps" when observers interact and compare notes. Such ideas are under development, often using category theory or extended Hilbert space formalisms.

Thermodynamics and horizons: As mentioned, an accelerated observer perceives a horizon; for example, an observer free-falling into a black hole and one hovering just outside it register drastically different phenomena, even though they are both describing the same underlying physics—namely, the covariant, horizon-free field equations. Temperature and entropy can be observer-dependent. The entropy associated with a horizon (like black hole entropy or de Sitter horizon entropy) might be seen as an observer-dependent count of inaccessible information. In an observer-space picture, one might label points not just by location and velocity but also by region of spacetime accessible to that observer (horizons create a partition of what can be observed). Then laws like the second law of thermodynamics might hold in a form that depends on that partition. Cosmological observations too depend on the observer's worldline (our current observations of the universe are from one very specific vantage point). When we talk about the universe's properties, we often implicitly mean "as seen by comoving observers" or "as would be seen by an ideal inertial observer at rest with respect to the CMB." Transforming to another observer (say moving at 0.9c relative to the CMB rest frame) would complicate those properties (the CMB would be highly anisotropic, etc.). Formally including observers in the model helps make those dependencies explicit and thus clarifies which statements are invariant.

Gauge and symmetry breaking: Sometimes choosing an observer can be like choosing a gauge in field theory. For instance, in the Higgs mechanism, one typically works in unitary gauge to interpret the physics, which is analogous to working in the rest frame of the Higgs field's "observer." If one chooses a different gauge, the interpretation changes. In gravitational physics, choosing a particular time slicing (observer family) can break time-translation symmetry that might otherwise be present. Thus, observer choices can effectively break symmetries that the underlying equations have, leading to different conserved quantities or lack thereof. Only by checking invariant structures (like energy measured at infinity, etc.) can one get observer-independent conclusions.

By extending our framework to include observers, we also get a clearer picture of what an "observation" fundamentally is. In an observer-centric view, an observation is an event that involves both an object system and an observer system, resulting in a correlation between them. For example, a measurement in quantum mechanics entangles the apparatus (observer) with the measured system; in classical terms, a measurement imprints information about the system onto the observer's state (like a meter reading). In a relational view, the basic ingredients are triadic: an observer, an observed phenomenon, and the interaction linking them. If we imagine the space of all possible such interactions, that itself might be structured (one could use category theory here, too, with interactions as morphisms between observer and system). The feedback loops we discuss next build on this idea that observation is not one-way—the observer can influence the system as well.

Ultimately, making physics observer-dependent (in the formalism) doesn't mean giving up objectivity, but rather refining what objectivity means. It means that a statement is objective if it is formulated in the language of the observer space and does not actually depend on which observer-state we pick (or if it does, we know exactly how to translate between them). It's similar to how in general relativity an "objective" statement is one that is tensorial (covariant)—you can write it down in any coordinate system and it's true in all. Here, an objective statement might be one that, say, all observers agree upon when they compare (like a properly invariant scalar), or a relationship that holds between any two observers' measurements when transformed appropriately. By contrast, something like "observer O sees a particle with energy E" is not objective by itself; but "observer O sees a particle with energy E *and* observer O' (moving at X relative to O) sees it with energy E', related by the Lorentz factor" is a complete, transformable statement. Observer-space formalisms strive to encode such complete relations from the start.

4.10.3.8 Measurement and feedback cycles with embedded observers

When the observer is included as an integral part of formalism, the act of measurement is no longer a passive reading of a pre-existing value—it becomes an interactive, dynamical process. Measurement can be thought of as a mapping from the observed system's state to the observer's own state (e.g., a thermometer absorbing heat and its mercury rising, encoding the temperature). In an intrinsic observer framework, one explicitly represents this mapping. For instance, in a minimal observer model (common in cybernetics and control theory), we have an observer with an internal state space X, receiving inputs Y (sensory data) and producing outputs Z (actions or signals). The observer's update rule $f : X \times Y \rightarrow X$ takes the current internal state and a new input to produce an updated state, and an output rule $g : X \rightarrow Z$ generates an output based on its state. This quintet (X, Y, Z, f, g) defines a simple observing system. Measurement events in this model are inputs $y \in Y$ from the environment that cause state transitions $x \rightarrow f(x, y)$; the outcome of the measurement can be considered the pair of new state x' and perhaps an output $z = g(x')$ (if the observer announces or uses the information). Crucially, the observer's state is altered by acquiring information—the observer "remembers" or reflects the measurement.

This naturally leads to a feedback loop when we allow the observer to not only sense but also act. The observer's output z might influence the environment or the system being observed. In engineering terms, the observer (or controller) might then affect the next input it receives. Thus, we get a closed-loop system: environment state → sensor input to observer → observer state update → observer output action → environment changes → new sensor input, and so on. Cybernetics has long studied such loops, emphasizing that the observer (or agent) and the environment co-evolve in response to each other. The intrinsic observer concept directly incorporates this: an observer is not an abstract entity outside the system but a subsystem engaged in a feedback cycle. The field of second-order cybernetics explicitly considers the observer observing the system and itself. Key ideas from second-order cybernetics include: (1) *Observer inclusion:* the observer is part of the feedback process, not a neutral external vantage point; (2) *Self-reference:* the observer can observe and modify its own state or rules, leading to learning or adaptation; (3) *Constructivism:* what is perceived as reality emerges through the observer's interactions and interpretations, rather than being a fixed external truth.

Bringing these ideas into physics and computation, we analyze feedback loops in measurement. For a physical example, consider a robotic sensor observing a pendulum. The robot reads the pendulum angle (input), then perhaps adjusts a motor (output) to change the pendulum's motion, maybe to stabilize it. The robot is an observer with a goal (keeping the pendulum upright). Its internal state might include an estimate of the pendulum's angle and angular velocity (an internal model). It continually updates this estimate with sensor readings and outputs motor torques. This setup can be described in the observer-space framework: each state of the robot (observer) together with a state of the pendulum (system) is a point in a combined space, and the dynamics form a closed loop. To analyze it properly, one must consider both together—the combined system has no fixed external reference, it's just two interacting parts. But one can also adopt the robot's perspective: from its "intrinsic" view, it tries to measure and control the pendulum, treating itself as the reference. If we swap out the robot for a different controller with a different internal mechanism, the outcomes differ—this is essentially a different observer in the intrinsic space, and whether it can achieve the goal or make the same measurements is an observer-dependent matter.

Given a formal observer model, we can define when two observers are equivalent in their measurement and control capabilities. Earlier, we discussed an equivalence relation based on a bijective homomorphism between two observers' state-input-output structures. That theorem essentially states that if you can relabel the internal states and I/O of observer O_1 to get O_2 such that their update (f) and output (g) functions correspond exactly under that relabeling, then the two observers are behaviorally identical—they will react to inputs and produce outputs in the same way up to renaming. In terms of measurement and feedback, this means no outside entity could tell the difference between O_1 and O_2 by interacting with them. This idea is very close to the concept of bisimulation in computer science: two systems are observationally equivalent if an external observer cannot distinguish their behaviors via any sequence of tests. Here we are considering the *observers* as the systems of interest—so we are looking at equivalence from a meta-perspective. If two observers are equivalent (isomorphic), they have the same "observational power" and the same kind of feedback dynamics. They belong to the same equivalence class in observer-space and can be treated as the same point if we quotient out those symmetries.

Considering feedback cycles also raises the question of stability and adaptation. An observer with a feedback loop might reach a fixed point or a limit cycle in its state (e.g., a thermostat will reach an equilibrium temperature reading when the room stabilizes). If we change the observer (say, make the thermostat twice as sensitive), the equilibrium might shift—but there might be an invariant (like the fact that equilibrium is when room temp equals target temp, regardless of sensitivity). Understanding these feedback invariants is a part of a generalized observer theory. In this study, for example, results are derived concerning loop efficiency and adaptation speed, which depend on the observer's structure (such as its response time) but not on arbitrary labeling. This illustrates that within the space of observers, one can define metrics or partial orders: some observers are "faster," "more complex," or "more capable" than others, in ways that are invariant under relabeling (so they are intrinsic properties of the equivalence class). For instance, an invariant might be the number of internal states or the presence of a certain feedback sub-loop. These invariants help classify observers beyond simple equivalence. In physics, one could imagine classifying observers by their acceleration (which distinguishes an inertial vs. non-inertial class) or by their field of view (horizon or no horizon), etc.

Finally, embedding the observer clarifies the measurement uncertainty and disturbance. In quantum mechanics, this is usually discussed via the uncertainty principle and back-reaction. In a fully observer-space approach, the measuring apparatus is just another physical system (another "observer") interacting with the system of interest. So one can, in principle, track how the joint system's state evolves under interaction and see the trade-off—information gained by the apparatus corresponds to something (like entanglement or disturbance) in the system. In classical terms, including the observer's dynamics can show how measurement noise and delays affect results. A laboratory measurement often involves a chain of observers: for example, a particle's position influences a detector (observer 1) that converts it to an electrical signal, which is read by a computer (observer 2), and interpreted by a scientist (observer 3). Each link is an observer relative to the previous stage. Only by considering them together can we understand the full measurement record. The intrinsic observer framework encourages thinking in this compositional way—observers observing observers, etc., which category theory handles well (via composition of morphisms).

In computational contexts, these feedback considerations are concrete. In software or AI, an agent observing an environment and adjusting to it can be modeled by the same kind of state-machine observer described above. The concept of observational equivalence in computer science (two programs are equivalent if no test can distinguish them) is directly analogous to the observer-isomorphism idea. In fact, one can think of an algorithm as an observer of the input data: two algorithms are observationally equivalent if for every input (stimulus) they produce the same output (response)—this is essentially the idea of two functions being extensionally equal, or two state machines being bisimilar. The formalism we discussed thus bridges to computer science: an observer is basically an abstract machine processing inputs to outputs. The earlier theorem establishing an equivalence relation on observers by homomorphism is very much a computing notion (an isomorphism of state machines) cast in our generalized observer language. This underscores that the space of observers can include not just physical observers (people, particles, detectors) but also computational observers (algorithms, robots, AI agents), and the same formal ideas apply.

By treating observers and their interactions as first-class entities, we gain a unified perspective on measurement, feedback, and the relational nature of observation across disciplines. Physics gains a language to incorporate the agent who is observing, and computer science gains physical insight (e.g., any computation can be seen as an interaction in some physical substrate observed by some entity). Observer-space formalisms thus impact how we understand knowledge and information: knowledge is no longer an abstract absolute; it is something held by an observer, and information is what is communicated from one observer to another. The feedback loop viewpoint also emphasizes learning and adaptation, which are crucial in fields like robotics and even in evolutionary contexts (organisms as observers of their environment, adapting via feedback). All these perspectives indicate that defining an intrinsic space of observers and insisting on background-independent, equivalence-respecting structures provides a powerful framework. It forces us to carefully distinguish what is observer-specific from what is truly universal, and it provides mathematical tools (from group theory, category theory, and differential geometry) to navigate between perspectives. In doing so, it enriches our understanding of measurement processes, ensures consistency across different viewpoints, and potentially helps reconcile differences between how computations/observations occur in different domains (quantum vs. classical, physical vs. virtual). Ultimately, it highlights that observation is an active, context-dependent process—one that can be formalized and studied on its own terms, rather than always being externalized or ignored.

4.10.4 Mathematical formalization and theorems

Having presented the conceptual framework of *intrinsic observer space*, category-theoretic perspectives, and background independence in the previous sections, we now turn to a precise algebraic formalism. Our goal is to define how one observer may be mapped to another in a manner preserving the structure of state transitions and outputs, thereby showing a formal equivalence that mirrors the isomorphisms discussed earlier. We also introduce quantitative measures (complexity and adaptation speed) that capture how observers process and respond to inputs in a feedback loop.

4.10.4.1 Observer equivalence and invariants

Remark 4.1—**On commutativity and "homomorphisms":** Strictly speaking, the mappings introduced below capture the requirement that the relevant transition diagrams *commute*: for any internal state x and input y, the map ϕ_X must intertwine with f to preserve transitions, and ϕ_Z must likewise intertwine with g to preserve outputs. In a category-theoretic sense, we are demanding *compatibility of compositions* in a commutative diagram, rather than using the term "homomorphism" in a strictly group-theoretic sense.

Definition 4.2 (observer homomorphism): Let

$$O_1 = (X_1, Y_1, Z_1, f_1, g_1, \mathcal{B}_1) \quad \text{and} \quad O_2 = (X_2, Y_2, Z_2, f_2, g_2, \mathcal{B}_2)$$

be two observers (as in Definition 4.1). A *homomorphism* from O_1 to O_2 is a triple of functions (ϕ_X, ϕ_Y, ϕ_Z) such that:

$$\begin{aligned} \phi_X &: X_1 \to X_2, \\ \phi_Y &: Y_1 \to Y_2, \\ \phi_Z &: Z_1 \to Z_2, \end{aligned}$$

and for all $x \in X_1$ and $y \in Y_1$, the following commutation conditions hold:

$$\begin{aligned} \phi_X\big(f_1(x, y)\big) &= f_2\big(\phi_X(x),\, \phi_Y(y)\big), \\ \phi_Z\big(g_1(x)\big) &= g_2\big(\phi_X(x)\big). \end{aligned}$$

Intuitively, (ϕ_X, ϕ_Y, ϕ_Z) ensures that the transition function f_1 and output function g_1 in O_1 map in a structure-preserving way to f_2 and g_2 in O_2. Equivalently, we have two commutative diagrams (see Figs. 4.10 and 4.11):

Theorem 2 (equivalence relation): *Let O_1 and O_2 be two observers. Define $O_1 \sim O_2$ if and only if there exists a* bijective *homomorphism (ϕ_X, ϕ_Y, ϕ_Z) between O_1 and O_2. Then $\sim$ is an equivalence relation on observers.*

Reflexivity: Take $\phi_X = \mathrm{id}_{X_1}$, $\phi_Y = \mathrm{id}_{Y_1}$, $\phi_Z = \mathrm{id}_{Z_1}$. These trivially satisfy the commutation conditions in both diagrams.

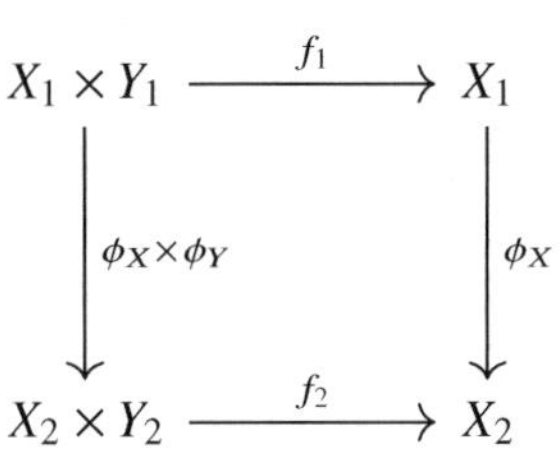

Figure 4.10 Commutative diagram for the transition function f.

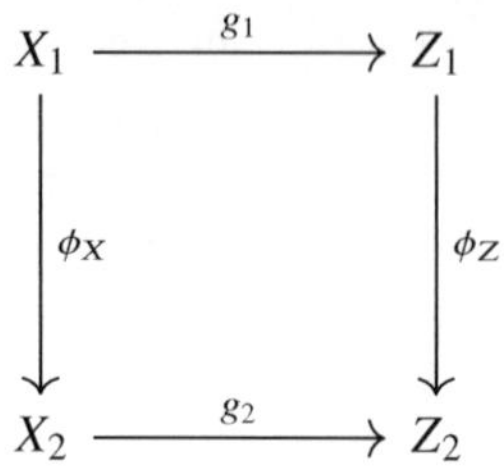

Figure 4.11 Commutative diagram for the output function g.

Symmetry: If (ϕ_X, ϕ_Y, ϕ_Z) is a bijection from O_1 to O_2, then $(\phi_X^{-1}, \phi_Y^{-1}, \phi_Z^{-1})$ is a bijection from O_2 to O_1, satisfying the same commutative properties in reverse.
Transitivity: If (ϕ_X, ϕ_Y, ϕ_Z) is a bijection from O_1 to O_2 and (ψ_X, ψ_Y, ψ_Z) is a bijection from O_2 to O_3, then composing them yields a bijection from O_1 to O_3 preserving all commutation requirements.

Thus, $\sim$ is an equivalence relation.

Two observers O_1 and O_2 are said to be *equivalent* if they differ only by a bijective, diagram-commuting relabeling. In category-theoretic language, this corresponds to an *isomorphism* between observer objects. Any property of an observer that remains invariant under such isomorphisms (e.g., minimal cardinalities of X, Y, Z, the presence of certain feedback cycles, or stable attractors) qualifies as an *invariant* of the equivalence class.

4.10.4.2 *Complexity metrics and results*

OBSERVATIONAL COMPLEXITY MEASURE

To quantify the "size" or "sophistication" of an observer $O = (X, Y, Z, f, g, \mathcal{B})$, define

$$C(O) = \log(|X| \times |Y| \times |Z|) - \Lambda(O),$$

where $\log(|X| \times |Y| \times |Z|)$ captures the combinatorial capacity of internal states, inputs, and outputs, and $\Lambda(O)$ accounts for redundancies or symmetries in f and g. (If multiple $x \in X$ respond identically and produce identical outputs, they do not increase genuine complexity.)

Proposition 4.1 (bounds on observational complexity): For a minimal observer with $|X| > 1,\ |Y| \geq 1,\ |Z| \geq 1$, we have

$$C(O) \geq \log(2).$$

Furthermore, if O can adapt or learn (increasing $|X|$ or altering f, g), then $C(O)$ can grow arbitrarily large.

Proof. A minimal observer requires $|X|\times|Y|\times|Z| \geq 2$, so $\log(|X|\times|Y|\times|Z|) \geq \log(2)$. While $\Lambda(O)$ subtracts redundancies, it cannot push $C(O)$ below zero for a structurally minimal system. As $|X|, |Y|, |Z| \to \infty$, or as f, g become more varied, $C(O)$ can increase without bound.

LOOP EFFICIENCY AND ADAPTATION SPEED

Lastly, consider how quickly an observer O "adapts" to a given environment. Define an *adaptation function*

$$\alpha_O : X \times Y^* \to \mathbb{N},$$

which, for a sequence of inputs $(y_1, y_2, \ldots) \in Y^*$, returns the time or number of state transitions required for O to reach a stable configuration or fulfill a measurement/control goal. In finite-state systems, one often proves boundedness of α_O using Markov chain hitting-times; in continuous domains, Lyapunov methods or approximate dynamic programming may be invoked.

Optimizing α_O across all allowable designs of (f, g) is akin to an *optimal control* or *reinforcement-learning* problem, seeking minimal expected adaptation time. Notably, two observers belonging to the same equivalence class via Theorem 2 (i.e. isomorphic) must exhibit the same adaptation profile, complexity measure, and other invariants, up to a relabeling of states, inputs, and outputs. This highlights the relational consistency emphasized previously and ensures that observer structure, rather than mere notation, dictates the system's dynamics.

4.10.5 Addressing counterarguments

4.10.5.1 Reductionism critique

Some worry that labeling simple devices (thermostats) as "observers" trivializes the concept. We rebut that minimal observers are foundational building blocks: layering, second-order loops, or enriched predictive mechanisms h can yield the complexity of conscious or social systems. The presence of feedback and boundary definition $\mathcal{B}$ is the *sine qua non* of observation, whether in a simple or advanced entity.

4.10.5.2 Infinite regress in self-reference

Second-order or multi-layer observers can appear to regress infinitely: who observes the observer's observer, etc.? We propose hierarchical encapsulation: each observer only references or modifies a finite subset of its own states. Formally, we forbid cycles of observation that do not converge or yield stable references. This ensures a well-founded partial order in the lattice of meta-observation relations.

4.11 Conclusions and outlook

We have presented a comprehensive, rigorously formalized theory of minimal observers that unifies concepts from cybernetics, quantum measurement, digital physics, and philosophical discussions of realism, meaning, and consciousness. Our model establishes the *minimal* criteria (sensing, action, state transitions, boundary) that constitute an observer, demonstrates how fundamental notions (measurement outcomes, reference frames, hierarchical organization) hinge on the presence of such observers, engages with Kantian, Husserlian, and Wittgensteinian perspectives, situating our feedback-based approach within classical philosophical discourse, offers rigorous mathematical results on observer equivalences, complexity metrics, and loop adaptation speeds, and addresses key critiques (reductionism, infinite regress) by highlighting scalability, hierarchical encapsulation, and boundary reconfiguration as essential elements. Through explicit *diagrams* illustrating core feedback loops and boundary definitions, we have shown how the observer concept can be visualized in contexts ranging from simple thermostats to multi-layer experimental apparatus in physics. In bridging computational, physical, and philosophical dimensions, this framework aspires to be a definitive, self-contained theory of observation.

In sum, recognizing *observation* as a fundamental, feedback-driven process—constrained by boundary definitions, state transitions, and sensor-actuator loops—offers a powerful lens for explaining measurement, emergent complexity, and the construction of meaning. We hope this work will inspire further efforts across disciplines to refine and adopt the minimal observer framework, illuminating the deep interweave of cognition, physics, computation, and philosophy in shaping our understanding of reality.

Acknowledgments

We thank Stephen Wolfram for engaging discussions on observers. We also like to thank John DeBrota for useful suggestions and comments on QBism. This work was supported by the following grants: The Foundational Questions Institute and Fetzer-Franklin Fund, a donor advised fund of Silicon Valley Community Foundation [FQXi-RFP-1817]; the John Templeton Foundation [Grant ID 62106]; and the Australian Research Council [Grant DP210100919].

Notes

1. Notably, from the perspective of Giulio Tononi's *Integrated Information Theory* (IIT), even such a minimal feedback system could be considered to have an extremely rudimentary form of consciousness, as it integrates information (the sensed temperature) and produces a differentiated response (heating action)—albeit at a very low level of complexity or Φ, IIT's measure of consciousness [20].
2. A natural extension of this minimal observer into the quantum domain could leverage epistemic interpretations of quantum mechanics, such as QBism. In QBism, probabilities associated with quantum states represent an observer's subjective degrees of belief or betting odds about measurement outcomes. Thus, a quantum minimal observer might be formalized as an entity whose internal states correspond to evolving belief states updated via quantum Bayesian inference rules, driven by the outcomes of quantum measurements relative to their actions on the world.

References

[1] Heisenberg, W. (1927). Über den anschaulichen Inhalt der quantentheoretischen Kinematik und Mechanik. *Zeitschrift für Physik*, 43(3–4), 172–198.

[2] Einstein, A. (1905). Zur Elektrodynamik bewegter Körper. *Annalen der Physik*, 17(10), 891–921.

[3] Heisenberg, W. (1958). *Physics and Philosophy: The Revolution in Modern Science*. New York: Harper & Row.

[4] Von Neumann, J. (1955). *Mathematical Foundations of Quantum Mechanics*. Princeton: Princeton University Press. (Original work published in 1932).

[5] Wheeler, J. A. (1989). Information, physics, quantum: The search for links. In *Proceedings of the 3rd International Symposium Foundations of Quantum Mechanics* (pp. 354–368). Tokyo: Physical Society of Japan.

[6] Wiener, N. (1948). *Cybernetics: Or Control and Communication in the Animal and the Machine*. Cambridge, MA: MIT Press.

[7] Ashby, W. R. (1956). *An Introduction to Cybernetics*. London: Chapman & Hall.

[8] Maturana, H. R., & Varela, F. J. (1980). *Autopoiesis and Cognition: The Realization of the Living*. Dordrecht: D. Reidel Publishing.

[9] Newell, A., & Simon, H. A. (1976). Computer science as empirical inquiry: Symbols and search. *Communications of the ACM*, 19(3), 113–126.

[10] Turing, A. M. (1936). On computable numbers, with an application to the Entscheidungsproblem. *Proceedings of the London Mathematical Society*, 2(42), 230–265.

[11] Smolin, L. (2019). *Einstein's Unfinished Revolution: The Search for What Lies Beyond the Quantum*. New York: Penguin Press.

[12] Wolfram, S. (2002). *A New Kind of Science*. Champaign, IL: Wolfram Media.

[13] Zurek, W. H. (2003). Decoherence, einselection, and the quantum origins of the classical. *Reviews of Modern Physics*, 75(3), 715–775.

[14] Varela, F. J., Thompson, E., & Rosch, E. (1999). *The Embodied Mind: Cognitive Science and Human Experience*. Cambridge, MA: MIT Press.

[15] Von Foerster, H. (1974). Cybernetics of cybernetics. In K. Krippendorff (Ed.), *Communication and Control in Society* (pp. 5–8). New York: Gordon and Breach.

[16] Varela, F. J. (1979). *Principles of Biological Autonomy*. Amsterdam: Elsevier.

[17] Braitenberg, V. (1984). *Vehicles: Experiments in Synthetic Psychology*. Cambridge, MA: MIT Press.

[18] Arsiwalla, X. D., Herreros, I., & Verschure, P. (2016). On three categories of conscious machines. In *Biomimetic and Biohybrid Systems: 5th International Conference, Living Machines 2016, Proceedings 5* (pp. 389–392). Cham: Springer International Publishing.

[19] Arsiwalla, X. D. (2024). Qualia and the formal structure of meaning. *arXiv preprint* arXiv:2405.01148.

[20] Tononi, G. (2012). Integrated information theory of consciousness: An updated account. *Archives Italiennes de Biologie*, 150(2–3), 290–326.

[21] Arsiwalla, X. D., & Verschure, P. F. (2013). Integrated information for large complex networks. In *The 2013 International Joint Conference on Neural Networks (IJCNN)* (pp. 1–7). IEEE.

[22] Arsiwalla, X. D., & Verschure, P. (2016). Computing information integration in brain networks. In *Advances in Network Science: 12th International Conference and School, NetSci-X 2016, Proceedings 12* (pp. 136–146). Cham: Springer International Publishing.

[23] Arsiwalla, X. D., Herreros, I., Moulin-Frier, C., Sanchez, M., & Verschure, P. (2016). Is consciousness a control process? In *Artificial Intelligence Research and Development* (pp. 233–238). IOS Press.

[24] Arsiwalla, X. D., Herreros, I., Moulin-Frier, C., & Verschure, P. (2017). Consciousness as an evolutionary game-theoretic strategy. In *Biomimetic and Biohybrid Systems: 6th International Conference, Living Machines 2017, Proceedings 6* (pp. 509–514). Cham: Springer International Publishing.

[25] Arsiwalla, X. D., & Verschure, P. F. (2016). High integrated information in complex networks near criticality. In *Artificial Neural Networks and Machine Learning–ICANN 2016: 25th International Conference on Artificial Neural Networks, Proceedings, Part I 25* (pp. 184–191). Cham: Springer International Publishing.

[26] Arsiwalla, X. D., & Verschure, P. F. (2016). The global dynamical complexity of the human brain network. *Applied Network Science*, 1(1), 16.

[27] Arsiwalla, X. D., & Verschure, P. (2018). Measuring the complexity of consciousness. *Frontiers in Neuroscience*, 12, 424.

[28] Rovelli, C. (1996). Relational quantum mechanics. *International Journal of Theoretical Physics*, 35(8), 1637–1678.

[29] Luhmann, N. (1995). *Social Systems*. Stanford: Stanford University Press.

[30] Wheeler, J. A. (1990). Information, physics, quantum: The search for links. In *Proceedings of the Santa Fe Institute, Vol. VIII*, pp. 309–336. Reading, MA: Addison-Wesley.

[31] Fredkin, E. (2003). Digital philosophy. *International Journal of Theoretical Physics*, 42(2), 189–247.

[32] Rickles, D., Elshatlawy, H., & Arsiwalla, X. D. (2023). Ruliology: Linking computation, observers and physical law. *arXiv preprint* arXiv:2308.16068.

[33] Arsiwalla, X. D., Elshatlawy, H., & Rickles, D. (2023). Pregeometry, formal language and constructivist foundations of physics. *arXiv preprint* arXiv:2311.03973.

[34] Tavanfar, A. (2018). HoloQuantum HyperNetwork Theory. *arXiv preprint* arXiv:1801.05286.

[35] Arsiwalla, X. D. (2020). Homotopic foundations of Wolfram models. *Wolfram Community*. `https://community.wolfram.com/groups/-/m/t/2032113`

[36] Arsiwalla, X. D., & Gorard, J. (2024). Pregeometric spaces from Wolfram model rewriting systems as homotopy types. *International Journal of Theoretical Physics*, 63, 83.

[37] Arsiwalla, X. D., Gorard, J., & Elshatlawy, H. (2021). Homotopies in Multiway (Non-Deterministic) Rewriting Systems as n-Fold Categories. *arXiv preprint* arXiv:2105.10822.

[38] Arsiwalla, X. D., Chester, D., & Kauffman, L. H. (2024). On the operator origins of classical and quantum wave functions. *Quantum Studies: Mathematics and Foundations*, 11(2), 193–215.

[39] Chester, D., Arsiwalla, X. D., Kauffman, L. H., Planat, M., & Irwin, K. (2024). Quantization of a new canonical, covariant, and symplectic Hamiltonian density. *Symmetry*, 16(3), 316.

[40] Arsiwalla, X. D. (2010). *Supersymmetric black holes as probes of quantum gravity* (PhD Thesis, University of Amsterdam). `https://pure.uva.nl/ws/files/871677/75420_thesis.pdf`

[41] Du Plessis, J. F., & Arsiwalla, X. D. (2023). A cosine rule-based discrete sectional curvature for graphs. *Journal of Complex Networks*, 11(4), cnad022.

[42] Everett, H. (1973). *The Theory of the Universal Wave Function*. Princeton: Princeton University Press.

[43] Hartle, J. B. (2005). The physics of now. *American Journal of Physics*, 73(2), 101–109.

[44] Wittgenstein, L. (1953). *Philosophical Investigations*. Edited by G. E. M. Anscombe. New York, NY: Wiley-Blackwell.

[45] Bacciagaluppi, G. (2016). Everett's servomechanisms and decoherence. *Philosophy of Science Archive*. `https://philsci-archive.pitt.edu/12693/`

[46] Fields, C. (2014). On the Ollivier-Poulin-Zurek definition of objectivity. *Axiomathes*, 24(1), 137–156.

[47] Nowakowski, M. (2023). Towards physics of internal observers: Exploring the roles of external and internal observers. *arXiv preprint* arXiv:2304.01677.

[48] Bhaskar, R. (1975). *A Realist Theory of Science*. Leeds: Leeds Books Ltd.

[49] Arsiwalla, X. D., Freire, I. T., Vouloutsi, V., & Verschure, P. (2019). Latent morality in algorithms and machines. In *Biomimetic and Biohybrid Systems: 8th International Conference, Living Machines 2019, Proceedings 8* (pp. 309–315). Springer International Publishing.

[50] Freire, I. T., Urikh, D., Arsiwalla, X. D., & Verschure, P. F. (2020). Machine morality: From harm-avoidance to human-robot cooperation. In *Biomimetic and Biohybrid Systems: 9th International Conference, Living Machines 2020, Proceedings 9* (pp. 116–127). Springer International Publishing.

[51] Signorelli, C. M., & Arsiwalla, X. D. (2019). Moral dilemmas for artificial intelligence: A position paper on an application of compositional quantum cognition. In *Quantum Interaction: 11th International Conference, QI 2018, Revised Selected Papers 11* (pp. 123–138). Springer International Publishing.

[52] Gardner, R., Ostrom, E., & Walker, J. M. (1990). The nature of common-pool resource problems. *Rationality and Society*, 2(3), 335–358.

[53] Freire, I. T., Moulin-Frier, C., Sanchez-Fibla, M., Arsiwalla, X. D., & Verschure, P. F. (2020). Modeling the formation of social conventions from embodied real-time interactions. *PLOS ONE*, 15(6), e0234434.

[54] Freire, I. T., Arsiwalla, X. D., Puigbò, J. Y., & Verschure, P. (2023). Modeling theory of mind in dyadic games using adaptive feedback control. *Information*, 14(8), 441.

[55] Moulin-Frier, C., Puigbò, J. Y., Arsiwalla, X. D., Sanchez-Fibla, M., & Verschure, P. F. (2017). Embodied artificial intelligence through distributed adaptive control: An integrated framework. In *2017 Joint IEEE International Conference on Development and Learning and Epigenetic Robotics (ICDL-EpiRob)* (pp. 324–330). IEEE.

[56] von Glasersfeld, E. (1984). An introduction to radical constructivism. In P. Watzlawick (Ed.), *The Invented Reality* (pp. 17–40). New York: Norton. Available at `https://app.nova.edu/toolbox/instructionalproducts/itde8005/weeklys/1984-vonGlaserfeld_RadicalConstructivism.pdf`.

[57] Ladyman, J., & Ross, D. (2007). *Every Thing Must Go: Metaphysics Naturalized*. Oxford: Oxford University Press.

[58] Rickles, D., & Bloom, J. (2016). Things ain't what they used to be: Physics without objects. In T. Bigaj & C. Wüthrich (Eds.), *Metaphysics in*

Contemporary Physics, *Poznań Studies in the Philosophy of the Sciences and the Humanities*, Vol. 104, 59–77. Leiden, The Netherlands:Rodopi Brill.

[59] Merleau-Ponty, M. (1962). *Phenomenology of Perception*. London: Routledge.

[60] Amodei, D., Olah, C., Steinhardt, J., Christiano, P., Schulman, J., & Mané, D. (2016). Concrete problems in AI safety. *arXiv preprint arXiv:1606.06565*.

[61] Friston, K. (2010). The free-energy principle: A unified brain theory? *Nature Reviews Neuroscience*, 11(2), 127–138.

[62] Gielen, S., & Wise, D. K. (2013). Lifting general relativity to observer space. *Journal of Mathematical Physics*, 54(5), 052501.

[63] Rickles, D. (2017). All possible perspectives: A (partial) defence of Eddington's physics. In I. T. Durham & D. Rickles (Eds.), *Information and Interaction: Eddington, Wheeler, and the Limits of Knowledge* (pp. 59–72). Cham:Springer.

[64] Bamonti, N. (2025). Re(l)ality: The view from nowhere vs. the view from everywhere. *arXiv preprint arXiv:2412.20520*.

[65] Oziewicz, Z. (2011). Ternary relative velocity; Astonishing conflict of the Lorentz group with relativity. *arXiv preprint arXiv:1104.0682*.

[66] Lachièze-Rey, M. (2001). Space and observers in cosmology. *Astronomy & Astrophysics*, 376(1), 17–27.

[67] Crane, L. (1993). Categorical physics. *arXiv preprint hep-th/9301061*.

[68] Fuchs, Christopher A. (2023). QBism, where next? *arXiv preprint*. https://arxiv.org/abs/2303.01446.

[69] DeBrota, John B., & Stacey, Blake C. (2018). FAQBism. *arXiv preprint*. https://arxiv.org/abs/1810.13401.

[70] DeBrota, John B., Fuchs, Christopher A., & Schack, Rüdiger. (2024). QBism's account of quantum dynamics and decoherence. *Physical Review A*, 110(5), 052205. https://doi.org/10.1103/PhysRevA.110.052205.

Part 3

Computational Languages for Physics

5 Topological QBits in Flux-Quantized Super-gravity

Hisham Sati and Urs Schreiber

5.1 An open problem

While the hopes associated with *quantum computation* [28] are hard to overstate, it is a public secret that fundamental new methods are needed for realizing useful quantum computers *at scale*. Plausibly, these methods will inevitably need to involve topological stabilization, notably via *anyonic* quantum states (e.g. [56, 78], i.e., via solitons whose states pick up purely topological quantum phases when moved around each other).

At the same time, despite the resulting attention that the idea of **topological quantum computation** [44, 65] (see [64] for a survey) has thus received, the microscopic understanding of anyonic topological order has arguably remained sketchy, due to the general lack of first-principles understanding of strongly-coupled/correlated quantum systems – which may also explain the dearth of experimental realizations of *topological q bits* to date: Better fundamental theory may be needed to understand how anyonic quantum states can actually arise in quantum materials.

5.2 Quantum gravity...

M-branes: Remarkably, a potential solution – to this host of theoretical problems arguably impeding practical progress – has emerged from the study of *quantum gravity* (e.g. [75]). In its locally super-symmetric enhancement, super gravity (SuGra) shows hints of having a completion to a general theory of strongly-coupled interactions, where the dynamics of strongly-correlated

DOI: 10.4324/9781003546139-8

quantum systems may usefully be mapped onto the fluctuations of *m*embranes ([8, §2], whence the working title: "M-theory" [7, 8]) and higher dimensional *5-branes* [8, §3][24, 25] inside an auxiliary higher dimensional spacetime (11D SuGra [8, §1][23]), a phenomenon famous as *holographic duality* [77].

For example, the phase transitions of quantum-critical superconductors, not amenable to traditional weak-coupling ("perturbative") analysis, have been understood at least qualitatively by these gravitational M-theoretic methods [6, 20, 21, 29, 30] (review in [30, 46, 48, 77]). More precise quantitative results cannot be expected without an actual formulation of M-theory/holography beyond the usual but unrealistic *large-N limit* of a macroscopic number of coincident such branes.

Progress on developing M-theory any further had stagnated, but we may notice that a fundamental non-perturbative phenomenon already in classical super-gravity has received little to no attention in this context, namely the issue of "flux-quantization". We find this to be crucial:

Flux-quantization: While the non-perturbative quantization of gravity famously remains a fundamental open problem of theoretical physics, we may observe that the higher (categorical symmetry) gauge fields that appear in the graviton super-multiplet call for a non-perturbative completion already at the classical level, namely by a *flux-quantization* law [63], which determines the topologically stabilized solitonic field configurations. This is classical for ordinary electromagnetism (where *Dirac charge quantization* in ordinary cohomology stabilizes the Abrikosov vortices observed in type II super conductors, cf. [63, §2.1]) and it is famous for the RR-field in 10D supergravity (which a popular conjecture sees flux-quantized in topological K-cohomology, stabilizing certain non-supersymmetric D-branes, cf. [63, §4.1]), but it had received little attention for the C-field in the pivotal case of 11D SuGra (where flux-quantization will stabilize non-supersymmetric M-branes and solitons on M5-brane worldvolumes, cf. [63, §4.2]).

Non-abelian cohomology: In fact, due to the *non-linear* electric Gauss law,

$$\mathrm{d}G_7 = \tfrac{1}{2}G_4 G_4, \quad \text{where } \mathrm{d}G_4 = 0, \quad G_7 = \star G_4 \tag{5.1}$$

on the flux densities sourced by M-branes in 11D super gravity (cf. [41, §3.1.3][23, Thm. 3.1]), none of the familiar *Whitehead-generalized cohomology* theories may serve here as flux-quantization laws (since they are intrinsically linear, or: abelian); what is needed are instead [63, §3] generalized *non-abelian cohomology* theories ([18, §2], generalizing the ordinary non-abelian cohomology of Chern-Weil theory which classifies gauge and gravitational instantons) that are receiving attention only more recently (such as in the study of non-abelian Poincaré duality [40]).

Classifying spaces and characters: The idea behind this powerful concept of non-abelian cohomology becomes quite simple once one realizes that all reasonable cohomology theories have been characterized by *classifying spaces* $\mathcal{A}$ (cf. [63, p. 19]), so that the cohomology classes on a given space X are just the homotopy classes of maps from X to $\mathcal{A}$, denoted by

$$H^1(X; \Omega\mathcal{A}) = \mathrm{Maps}(X, \mathcal{A})_{/\mathrm{hmtpy}} . \tag{5.2}$$

A simple but profound rule (cf. [63, Prop. 3.7], using the fundamental theorem of dg-algebraic rational homotopy theory, cf. [18, §5]) determines the **admissible flux quantization laws** $H^1(-; \Omega\mathcal{A})$ for given flux densities $(F^i)_{i\in I}$ satisfying Bianchi identities $\mathrm{d}F^i = P^i(\vec{F})$: The flux species F^i of degree $\deg_i$ must span the real $\deg_i$-homotopy groups of the classifying space $\mathcal{A}$, and the cohomology of the Bianchi identities on the free graded algebra generated by the flux densities must coincide with the real cohomology of $\mathcal{A}$.

For example (see also [63, p. 21]), vacuum electromagnetism with $\mathrm{d}\,F_2 = 0$ requires a classifying space whose real-homotopy and real-cohomology both are generated by a single element in degree 2, such as the universal first Chern class on infinite-projective space $\mathcal{A} \equiv \mathbb{C}P^\infty \simeq BU(1)$ which classifies the ordinary 2-cohomology known from Dirac charge quantization; while the NS B-field with $\mathrm{d}\,H_3 = 0$ may similarly be flux-quantized by the next such *Eilenberg-MacLane space* $\mathcal{A} \equiv B^2\mathrm{U}(1)$ which classifies "bundle gerbes"; and the RR-fields with $\mathrm{d}F_{2k} = H_3\,F_{2k-2}$ require a classifying space with such a generator in every even degree – such as $\mathrm{KU}_0 \equiv \varinjlim_n BU(n) \times \mathbb{Z}$ with its higher universal Chern classes – twisted to incorporate the H_3-generator, such as the Borel-construction space $\mathrm{KU}_0 /\!\!/ BU(1)$ that classifies 3-twisted topological K-theory (cf. [63, §4.1]):

$$\begin{array}{ccc} \mathrm{KU}_0 \longrightarrow \mathrm{KU}_0 /\!\!/ BU(1) & & \mathrm{d}\,F_{2\bullet} = H_3\,F_{2\bullet-2} \\ \downarrow & \overset{\text{Chern character}}{\rightsquigarrow} & \\ B^2\mathrm{U}(1) & & \mathrm{d}\,H_3 = 0\,. \end{array}$$

The construction of such *characters* generalizes [18] to non-abelian cohomology theories:

M-brane charge in cohomotopy: Among the admissible flux-quantization laws for the C-field sourced by M-branes, there is a theory that turns out to be the most fundamental and most ancient non-abelian cohomology theory, known as (unstable) "co-homotopy" (since its classifying spaces are nothing but spheres), introduced by Pontrjagin in the 1930s (and later baptized by Spanier).

Concretely, the real-homotopy groups of S^4 have a generator in degree 4 (the identity map) and in degree 7 (the quaternionic Hopf fibration $S^7 \xrightarrow{h_{\mathbb{H}}} S^4$), while the real-cohomology only has a generator G_4 in degree 4. Hence G_4 must be closed while the other generator G_7 must be a coboundary for the otherwise induced cohomology class of $G_4\,G_4$. This way, the character map on

4-Cohomotopy reproduces the equation of motion [Eq. (5.1)] of the 11D SuGra C-field [50, 61]:

$$S^4 \quad \overset{\text{character}}{\rightsquigarrow} \quad \begin{aligned} \mathrm{d}\, G_7 &= \tfrac{1}{2} G_4\, G_4 \\ \mathrm{d}\, G_4 &= 0\,. \end{aligned} \tag{5.3}$$

Careful analysis shows that assuming ("Hypothesis H") 11D supergravity to be globally completed by demanding the C-field flux densities to be quantized in (tangentially twisted) 4-Cohomotopy provably implies various subtle topological effects that are expected in M-theory [14, 15, 17, 53], notably the condition that the sum of G_4 with $1/4$th of the first Pontrjagin form of the spin-connection is integral [14, Prop. 3.13].

3-Form flux on M5-branes: Moreover, given an M5-brane $\Sigma^{1,5} \overset{\phi}{\hookrightarrow} X^{1,10}$ probing the bulk spacetime $X^{1,10}$, its worldvolume $\Sigma^{1,5}$ famously (but quite [24, 32]) carries itself a non-linearly self-dual 3-flux density H_3 (sourced by string-like solitons inside the M5), satisfying the Bianchi identity

$$\mathrm{d}\, H_3 = \phi^* G_4\,, \tag{5.4}$$

which, while nominally linear, inherits the non-linearity [Eq. (5.1)] of the source term G_4 on the right. An admissible non-abelian flux quantization law for the combination of Eq. (5.4) and Eq. (5.1) turns out to be (tangentially twisted) 7-Cohomotopy *relative to* the bulk 4-Cohomotopy (where Eq. (5.4) reflects the vanishing of the class of the volume form of S^4 upon pullback to S^7). This means that where the latter has as classifying space the 4-sphere, the former has as classifying space the 3-sphere *fibers* of the quaternionic Hopf fibration $\phi_{\mathbb{H}}$ [14, §3.7]:

$$\begin{array}{ccccc} S^3 \longrightarrow S^7 & \simeq & S(\mathbb{H}^2) & & \\ \quad\;\; \big\downarrow \phi_{\mathbb{H}} & & \big\downarrow \mathrm{mod}\, \mathbb{H}^\times & \overset{\text{character}}{\rightsquigarrow} & \begin{aligned} \mathrm{d}\, H_3 &= \phi^* G_4 \\ \mathrm{d}\, G_7 &= \tfrac{1}{2} G_4\, G_4 \\ \mathrm{d}\, G_4 &= 0\,. \end{aligned} \\ \quad\;\; S^4 & \simeq & \mathbb{H}P^1 & & \end{array} \tag{5.5}$$

Gauge field on A_1-singularities: More generally, for an M5-brane probing its would be black brane horizon, namely probing an A_1-type orbi-singularity of spacetime (i.e., locally the fixed locus of the $\mathbb{Z}_2 \subset \mathrm{Sp}(1)$-action on a patch $X^7 \times \mathbb{H} \subset X^{11}$) a further flux density F_2 appears (e.g. [71, p. 92], cf. [52]) and modifies Eq. (5.4) to

$$\mathrm{d}\, H_3 = \phi^* G_4 + F_2\, F_2\,. \tag{5.6}$$

For vanishing $\phi^* G_4$ this relation is of the same form as Eq. (5.1) and hence readily seen to be flux-quantized by the 2-sphere. Further inspection [16] shows that in general Eq. (5.6) is flux-quantized by the 2-sphere fibration over the 4-sphere that is also known as the *twistor fibration*, whose total space is $\mathbb{C}P^3$:

$$
\begin{array}{ccc}
S^2 \longrightarrow \mathbb{C}P^3 & \simeq & S(\mathbb{H}^2)/S(\mathbb{C}) \\
\big\downarrow \phi_{\mathbb{C}} & & \big\downarrow \mathrm{mod}\,\mathbb{H}^\times \\
S^4 & \simeq & \mathbb{H}P^1
\end{array}
\quad \xrightarrow{\text{character}} \quad
\begin{array}{rcl}
\mathrm{d}H_3 & = & \phi^* G_4 + F_2 F_2 \\
\mathrm{d}G_7 & = & \tfrac{1}{2} G_4 G_4 \\
\mathrm{d}G_4 & = & 0\,.
\end{array}
\tag{5.7}
$$

Anyonic solitons in 2-Cohomotopy: Now something remarkable happens: A deep theorem by Segal ([70], cf. [38, §4.1]) shows that the moduli space of codimension = 2 solitons[1] [63, §2.2] sourcing flux that is quantized in the 2-Cohomotopy [Eq. (5.7)] have moduli space the pointed mapping space

$$
\mathrm{Maps}\big(\mathbb{R}^2_{\mathrm{cpt}}, S^2\big) \;\simeq\; \mathbb{G}\mathrm{Conf}(\mathbb{R}^2)\,, \tag{5.8}
$$

equivalent to the "group completion" $\mathbb{G}$ of the *configuration space* Conf of points in the plane $\mathbb{R}^2$ (i.e. in the transverse space to the codim = 2 solitons, in which they appear as points, but (e.g. [27, 76])

$$
\mathrm{Conf}(\mathbb{R}^2) \;\simeq\; \coprod_{n \in \mathbb{N}} B\mathrm{Br}(n) \tag{5.9}
$$

is the classifying space for the *braid groups* $\mathrm{Br}(n)$ of motion of n anyons in the plane.

On this, the "group completion" $\mathbb{G}$ says essentially ([62, p. 6]) that, besides the solitons that appear as points, there may also be *anti-solitons* that appear as points carrying a negative unit charge. This means that loops $\ell \in \Omega\,\mathbb{G}\mathrm{Conf}(\mathbb{R}^2)$ describe just the kind of processes traditionally envisioned in discussion of topological quantum computation, where anyon/anti-anyon pairs are created out of the vacuum, then moved around each other, to eventually pair-annihilate again into the vacuum – whereby their worldlines form knots and generally links.

In fact, careful analysis [62, §6] shows that these loop processes in Eq. (5.8) are *framed* links (e.g. [47, p. 15]) and that homotopy classes of these processes are the cobordism classes $[L]$ of these framed links, and that these are classified by their total linking number $\#L$, including the framing number (cf. Fig. 5.1):

$$
\pi_1\big(\mathbb{G}\,\mathrm{Conf}(\mathbb{R}^2)\big) \;\simeq\; \big\{\text{Framed links}\big\}_{/\text{cobordism}} \xrightarrow[\sim]{\;\#\;} \mathbb{Z}\,. \tag{5.10}
$$

Quantum observables on quantized fluxes: Once supergravity is completed by a flux-quantization law $\mathcal{A}$ this way, then [57] for every spacetime – or world-volume – domain, its mapping space into $\mathcal{A}$ constitutes the moduli space of

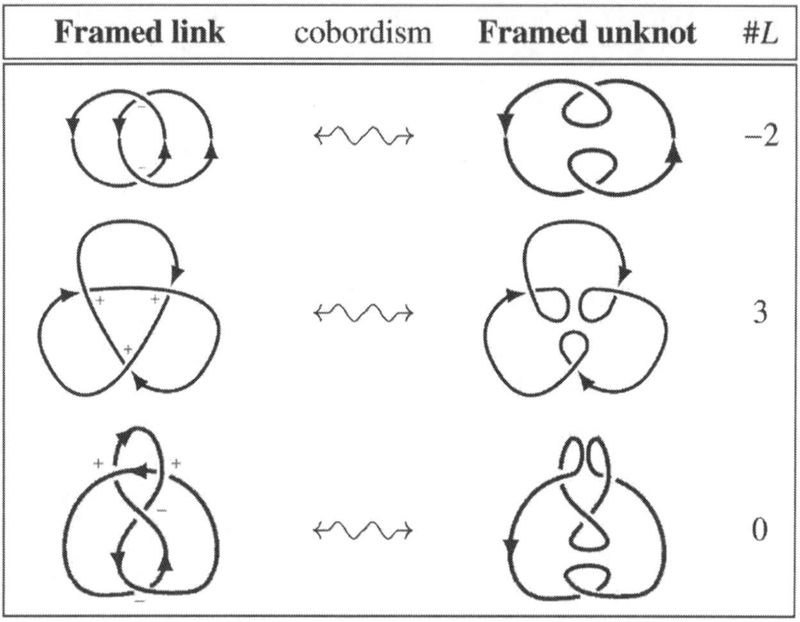

Figure 5.1 Some framed links L with the framed unknots that they are cobordant to. The number $\#L$ in Eq. (5.10) is the sum of the linking- and self-linking (framing) number.

topological sectors of solitonic higher gauge field configurations, being the higher analog of configurations of Abrikosov vortices in electromagnetism. When the domain is a principal bundle $\Sigma^{1,5} \to \Sigma^{1,4}$ of circle fibers (such as the "M-theory circle"), then the quantum observables on the topological sectors of such flux-quantized fields form the Ponrjagin homology algebra (cf. [57, §3]) of the loop space of this moduli space:

$$\underset{\substack{\text{Topological quantum observables} \\ \text{of } \mathcal{A}\text{-quantized fluxes} \\ \text{on M5-worldvolume } \Sigma^{1,4} \times S^1}}{\mathrm{Obs}_\bullet} \;=\; H_\bullet\big(\Omega\,\mathrm{Maps}(\Sigma^{1,4},\,\mathcal{A});\,\mathbb{C}\big)\,. \tag{5.11}$$

This is the non-perturbative quantization of a small but crucial fragment of super-gravity, rather complementary to the traditional focus of interest: Instead of local quantum effects such as of graviton scattering visible on any coordinate chart, here we deal with the global topological quantum effects.

Anyonic quantum observables on M5-branes: Concretely, consider a world-volume domain on which to measure the charge of 3-brane solitons inside the

M5-brane (cf. [75, §14.6.1][55, p. 26]), wrapped over the M-theory circle,

$$\overbrace{\Sigma^{1,5}}^{\text{5-brane worldvolume}} = \underbrace{\mathbb{R}^{1,2} \times S^1}_{\text{3-brane worldvolume}} \times \overbrace{\mathbb{R}^2_{\text{cpt}}}^{\text{transverse space}} \tag{5.12}$$

in, for simplicity, a background with vanishing C-field. Then, for any choice of admissible flux quantization law $\mathcal{A}$, the topological quantum observables on these 3-brane solitons is given by Eq. (5.13), which for the choice [Eq. (5.7)] is [62, §4] the group algebra of cobordism classes of framed links, under their connected sum:

$$\begin{aligned}
\mathrm{Obs}_0 &\equiv H_0\Big(\Omega\,\mathrm{Maps}\big(\mathbb{R}^{1,2} \times \mathbb{R}^2_{\text{cpt}},\, S^2\big);\, \mathbb{C}\Big) && \text{by Eq. (5.13)} \\
&\equiv H_0\Big(\Omega\,\mathrm{Maps}\big(\mathbb{R}^2_{\text{cpt}},\, S^2\big);\, \mathbb{C}\Big) && \text{since } \mathbb{R}^{1,2} \text{ is cntrctbl} \\
&\equiv H_0\Big(\Omega\,\mathbb{G}\,\mathrm{Conf}(\mathbb{R}^2);\, \mathbb{C}\Big) && \text{by Eq. (5.8)} \\
&\equiv \mathbb{C}\Big[\pi_1\big(\mathbb{G}\mathrm{Conf}(\mathbb{R}^2)\big)\Big] && \text{by 0-Hurewicz} \\
&\equiv \mathbb{C}\Big[\big\{\text{Framed links}\big\}_{/\text{cbrdsm}}\Big] && \text{by Eq. (5.10)}\,.
\end{aligned} \tag{5.13}$$

Anyonic quantum states on M5-branes: This implies, by the rules of algebraic quantum theory, that [62, Cor. 3.3] the corresponding *pure* quantum states $|\psi\rangle$ are, via the expectation values that they induce on the observables [Eq. (5.13)], the algebra homomorphisms

$$\begin{array}{ccc}
\mathbb{C}\Big[\big\{\text{Framed links}\big\}_{/\text{cbrdsm}}\Big] & \xrightarrow{\langle k|-|k\rangle} & \mathbb{C} \\
{[L]} & \longmapsto & \exp\big(\tfrac{\pi \mathrm{i}}{k}\#L\big)\,,
\end{array} \tag{5.14}$$

which are generated by states $|k\rangle$ for $k \in \mathbb{Z}$, as shown. These are exactly the traditional[2] quantum observables of U(1) Chern-Simons theory as expected for abelian anyons. It is believed that such quantum states have been observed [45] in fractional quantum Hall (FQH) systems [26, 73].

However, as may often be overlooked, anyonic states in this form are not yet useful for quantum computation: While the anyonic braiding statistics is visible in the phase factor [Eq. (5.14)], there is no control yet over the movement of these anyons around each other in order to implement topological quantum gate operations (cf. [44, §3]).

We next see how this control arises in our holographic theory.

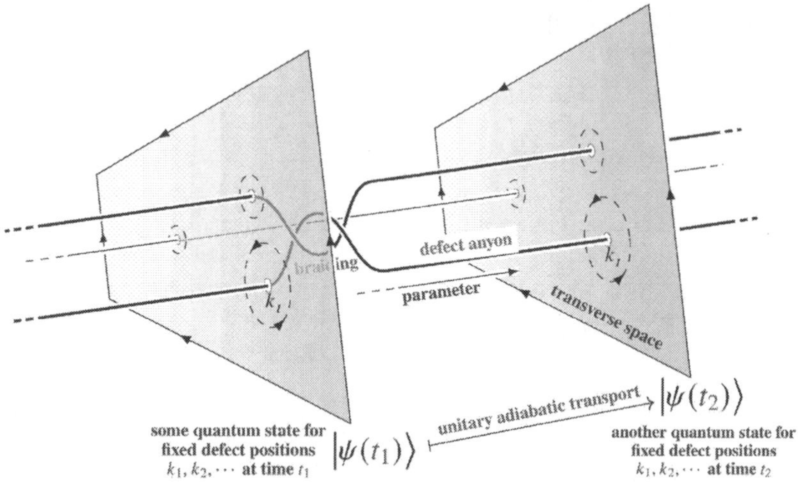

Figure 5.2 Topological quantum gate based on braiding of anyon worldlines.

Topological quantum gates: Namely, to model classically controllable anyonic *defects* in addition to the above anyon "virtual particles", consider deleting a subset $\mathbf{n} \subset \mathbb{R}^2$ of $n \in \mathbb{N}$ defect points from the transverse plane (just as one deletes the singular locus of a black hole or black brane defect from the spacetime domain, cf. [63, §2.2]) and take the transverse space of the 3-brane soliton inside the M5-brane now to be the **n**-punctured plane, generalizing Eq. (5.12) to

$$\overbrace{\Sigma^{1,5}}^{\text{5-brane worldvolume}} = \underbrace{\mathbb{R}^{1,2} \times S^1}_{\text{3-brane worldvolume}} \times \overbrace{\left(\mathbb{R}^2 \setminus \mathbf{n}\right)_{\mathrm{cpt}}}^{\text{transverse space}} . \tag{5.15}$$

The topological symmetries of the worldvolume domain fixing the 3-brane locus is the mapping class group of $\mathbb{R}^2 \setminus \mathbf{n}$ fixing the original point at infinity, which in turn is the braid group Br_n (quotiented by its center, cf. [27, §1.4]) which as such canonically acts on the resulting quantum observables, formed as in Eq. (5.13) (Fig. 5.2)

$$\mathrm{Obs}_\bullet \;:=\; H_\bullet\Big(\Omega\,\mathrm{Maps}\big((\mathbb{R}^2 \setminus \mathbf{n})_{\mathrm{cpt}},\, S^2\big);\, \mathbb{C}\Big).$$

But this in turn gives an action of Br_n on the corresponding Hilbert space of quantum states. This is what counts as a set of topological quantum gates, where an adiabatic braid-motion of anyon defects around each other acts by quantum phases on the system's Hilbert space.

An analogous analysis for a more sophisticated situation of *intersecting* M5-branes and resulting in *non-abelian* anyons was given in [55].

In summary so far, this shows that the fundamentals of topological quantum logic gates, acting by adiabatic braiding of worldlines of anyonic defects, arise quite naturally from the non-perturbative quantization of the topological sector of solitons on single M5-branes in 11D supergravity, *if* flux-quantization is taken into account, of the bulk C-field and of the self-dual tensor field on the worldvolume, whose non-linear Gauß law [Eq. (5.7)] is seen to reflect anyonic soliton charges in *non-abelian* generalized cohomology (concretely, in unstable Cohomotopy).

5.3 ... and computation

Formulation in homotopically typed programming language: To bring out this relation between flux-quantized supergravity and (quantum) computation more manifestly, we may observe [44] that the elementary algebro-topological/homotopy-theoretic nature [57, 63] of quantum observables on flux-quantized fields – as exhibited, for example, in Eq. (5.13) – lends itself (exposition in [42]) to formalized expression in novel *homotopically-typed* programming languages (cf. §5.4 and [74], such as `Agda` [9] or `cubicalAgda` [43]) and better yet [67][58–60] in languages with *linear homotopy types*, of which a prototype design has recently been described [49].

To wit, the core mechanism of topological holonomic quantum gates [80, 82], parallel-transporting the quantum state of a system along paths of classical parameters (such as anyon defect positions) is, strikingly, *native* to such languages ("type transport", cf. [44, p. 39][43, §2.5]) (Fig. 5.3):

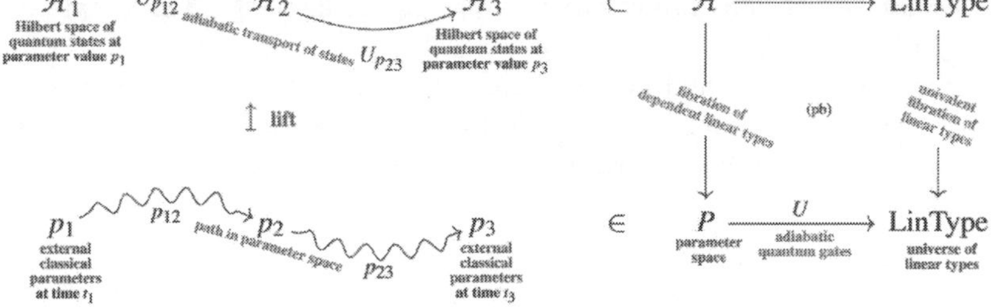

Figure 5.3 The physics notion of adiabatic transport of quantum states is neatly encoded in the type-theoretic notion of transport of (linear) types.

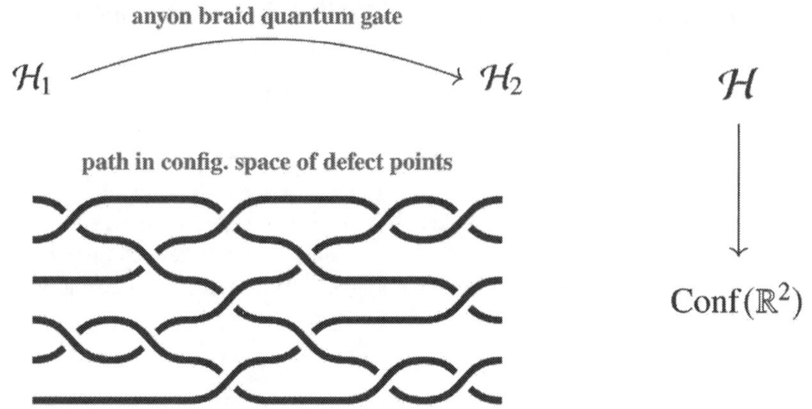

Figure 5.4 A braid quantum gate enacts parallel transport in a bundle of Hilbert spaces over a configuration space of points.

Also native to homotopically-typed languages is the declaration of classifying spaces, such as for the braid group in Eq. (5.9), which means that their general knowledge of transport readily specializes in the case of braid gates:

Thereby, the encoding of topological quantum gates in a homotopically-typed programming language becomes essentially a 1-liner [44, Thm. 6.8].

This is remarkable: When topological quantum computers become a reality (or their quantum simulation becomes refined enough, cf. recent progress in [36]) their hardware-level quantum gates will be completely different from the idealized gates familiar from traditional qbit-based quantum circuits (Hadamard, CNOT, etc.) and efficient (hardware-aware) topological quantum programming languages will need to reflect this (cf. [59, p. 3]).

In final conclusion this means that the embedding ("geometric engineering") of topological qbits into quantized topological sectors of flux-quantized supergravity with M-brane probes illuminates the quantum-physical and the quantum-information theoretic nature of anyons, without relying on the unrealistic large-N limit of existing holographic descriptions of quantum materials.

5.4 Vista

In reaction to and amplification of some of the thoughts of our editors expressed in [1], we close with more meta-physical remarks on the relevance of (cohesive) homotopy type theory in the foundations not just of (quantum-)computation but of fundamental physics and potentially of M-theory.

Computation and physical process: In our age of electro-mechanical computers, and at the plausible dawn of a new age of quantum-mechanical computers, it is a truism that any *computation is a physical process*, possibly a very fundamental physical process (say, if we think of photonic quantum computation). The reverse of this truism, that possibly all *physical processes are computations*, hence that the history of the universe is the unfolding of an ancient primordial algorithm ([79, 81, 83]), is thought provoking – all the more since it is less clear what it would actually mean.

Computation and mathematical proof: On this issue we highlight that the field of mathematical logic has long developed an analogous relation: In the view of *constructive mathematics* (cf. [81, §8], essentially what was originally called "intuitionism") the proof of a theorem must consist of the actual *construction* of a *witness* of its truth – notably existential statements, such as that "every surjection has some section" (the *axiom of choice*), are not regarded as constructively true unless the existence of at least one instance is concretely established. In its modern guise as *intuitionistic type theory* (short for: *data type* theory!, cf. [4][44, §5.1]), this paradigm of constructive mathematics means (cf. [44, p. 42]) that *proofs are algorithms* (hence are physical processes, when executed on a mechanical computer): With given assumptions as input, their output constructively witnesses the existence of data of their specified output type.

Hence if physical processes are (or were) algorithms also, then *physical processes are proofs* – echoing Wittgenstein's identification of "the world" with "all facts".

Generalizing sets to types: But to make sense of this, we highlight another lesson of type theory – famous among specialists (cf. [4, §2.1]) but otherwise underappreciated: As the name indicates, type theory is a foundation of mathematics whose fundamental elementary objects are not necessarily just *sets* of isolated *elements*. Instead, types:

(i) may carry extra *structure* (cf. [44, p. 53]),

(ii) need not be determined by its elements (now called *terms*), and

(iii) have their (higher categorical) symmetries built-in (cf. [44, pp. 40]).

Another way to say this is: Where set theory is realized (only) by the ordinary category of sets, intuitionistic (homotopy) type theory is realized ("modeled" by "semantics", cf. [37]) more generally by categories called (higher) *toposes* – from $\tau \acute{o} \pi o \varsigma$ for "place": Already according to [39]; (cohesive) toposes are where *physics may take place* (exposition in [68], more details in [22]).

Space is a cohesive type:

- As an example for (i): In (homotopy) *cohesive type theories* [5, 69, 72] to be realized in (higher) *cohesive toposes* [51, §3.1], the real line, hence **the continuum**

$$\mathbb{R}$$

as understood not just in classical physics but notably in quantum physics (where $\mathbb{C} \simeq \mathbb{R} \times \mathrm{i}\mathbb{R}$), exists, including its smooth- and ring-structure, on the same fundamental level as any plain set. This suggests that the notorious trouble that set-based approaches to algorithmic physics have with "the continuum limit" may be an artifact of not considering non-discrete cohesive types.

- As an example for (ii): In *super-cohesive* toposes [51, §3.1.3] also the **super-point**

$$\mathbb{R}^{0|1}$$

(having a single element/term 0, but equipped with a "fermionic infinitesimal halo", cf. [19, Fig. 4][2, §3.1]) exists on the same fundamental level as any set – in fact including its (abelian) super-Lie algebra structure.

- As an example for (iii): The homotopy type of the circle, namely the **classifying space** of the integers (having a single element, but equipped with $\mathbb{Z}$-symmetry):

$$\int S^1 \simeq \mathbf{B}\mathbb{Z}$$

exists on the same fundamental level as plain sets (cf. [44, (147)]), as do all its "higher deloopings" $\mathbf{B}^n\mathbb{Z}$ (having a single element, but equipped with n-categorical higher $\mathbb{Z}$-symmetry, cf. [44, (192)]).

- As a combined example: Every homotopy Lie algebra (L_∞-algebra) exists (cf. [66, §4.5.1]) as a cohesive homotopy type with a single element but equipped with any infinitesimal higher symmetry. In particular, for every classifying space $\mathcal{A}$ as in Eq. (5.2), there exists its *Whitehead L_∞-algebra* ([18, Prop. 5.11])

$$\mathfrak{l}\mathcal{A}$$

such that flat $\mathfrak{l}\mathcal{A}$-valued differential forms [18, Def. 6.1] are precisely flux densities for which $\mathcal{A}$ is an admissible flux-quantization law, as in §5.2 (cf. [63, §3]).

Super-spacetime emerges: Like a mustard seed, the super-point $\mathbb{R}^{0|1}$ is tiny and yet carries seminal internal structure. Homotopy types detect this inner structure via a non-trivial 2-cocycle, namely a non-null map of super-L_∞ algebras

$$\mathbb{R}^{0|1} \xrightarrow{\mathrm{d}\theta \wedge \mathrm{d}\theta} \mathfrak{l}\mathbf{B}^2\mathbb{Z}\,. \tag{5.16}$$

An equivalent incarnation of cocycles are the *extensions* that they classify, which in turn are equivalently the *homotopy fibers* (cf. [18, Def. 1.14][44, p. 41]) of their classifying map. But for Eq. (5.16) this turns out to be [35, p. 18] the real *super-line*, (or "super-continuum")

$$
\begin{array}{ccc}
\mathbb{R}^{1|\mathbf{1}} & & \\
{\scriptstyle \mathrm{hofib}}\big\downarrow & & \\
\mathbb{R}^{0|\mathbf{1}} & \xrightarrow{\mathrm{d}\theta\wedge\mathrm{d}\theta} & \mathfrak{l}B^2\mathbb{Z}
\end{array}
$$

equipped with its super-translation structure, hence the $D = 1$, $\mathcal{N} = 1$ supersymmetry algebra.

Yet more remarkably, the doubled superpoint

$$
\mathbb{R}^{0\,|\,\mathbf{1}\oplus\mathbf{1}} \simeq \mathbb{R}^{0\,|\,1} \underset{\mathbb{R}^0}{\sqcup} \mathbb{R}^{0\,|\,1}
$$

carries 3 independent 2-cocycles whose corresponding extension is [35, Prop. 9] nothing but $D = 3$, $\mathcal{N} = 1$ super-spacetime

$$
\begin{array}{ccc}
\mathbb{R}^{1,2\,|\,\mathbf{2}} & & \\
{\scriptstyle \mathrm{hofib}}\big\downarrow & & \\
\mathbb{R}^{0\,|\,\mathbf{1}\oplus\mathbf{1}} & \xrightarrow{\mathrm{d}\theta^{(i}\wedge\mathrm{d}\theta^{j)}} & \mathfrak{l}B^2\mathbb{Z}^3
\end{array}
$$

with its metric structure encoded in its external automorphism algebra [35, Prop. 6].

Proceeding in this manner by doubling the fermions on this super-space, its maximal $\mathrm{Spin}(1,2)$-equivariant extension next is [35, Thm. 14] nothing but $D = 4$, $\mathcal{N} = 1$ super-spacetime

$$
\begin{array}{ccc}
\mathbb{R}^{1,3\,|\,\mathbf{4}} & & \\
{\scriptstyle \mathrm{hofib}}\big\downarrow & & \\
\mathbb{R}^{1,2\,|\,\mathbf{2}\oplus\mathbf{2}} & \longrightarrow & \mathfrak{l}B^2\mathbb{Z}
\end{array}
$$

again with its metric structure encoded in its external automorphisms.

This progression continues [35, Thm. 14] and discovers next $D = 6$, then $D = 10$, and finally $D = 11$ super-spacetime, see Fig. 5.5. We have hence a kind of *emergence of spacetime* from pure computational logic ("It from Bit"), rather different from traditional set-based approaches and right away recovering the continuum structure of spacetime together with its local (super-)metric structure.[3]

Super-branes emerge: When seen in the higher super-cohesive topos, these super-spacetimes sprout a whole bouquet of further invariant *higher extensions*

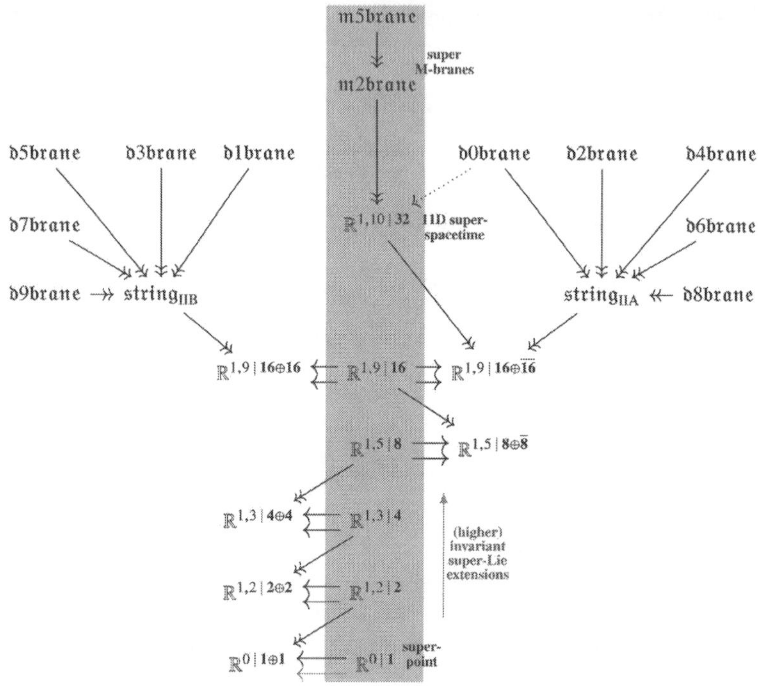

Figure 5.5 **The Brane Bouquet.** In cohesive homotopy theory, there emerges, from the super-point, a bouquet of (invariant central) *higher extensions* which first [35] grows the super-spacetimes in the critical dimensions of string theory, then [10] sprouts the corresponding super-brane species, and eventually blossoms into the M-brane species on 11D super-space classified in rational 4-Cohomotopy [11, 12] (animated exposition in [84], more background in [13, Fig. 1][34, Fig. 3]).

[13] which may be understood [3] as (higher super-spacetimes extended by charges of) brane species: The *brane bouquet* [10][34, p. 14] shown in Fig. 5.5. Notably, 11D super-space carries an invariant 4-cocycle which is the WZW term of the M2-brane sigma model, whence the higher central extension it classifies is known as $\mathfrak{m2brane}$ ([33, §3.1.3][10, Def. 4.2][34, p. 13] to be read as: "11D super-space extended by M2-brane charges"):

$$\begin{array}{l} \mathfrak{m2brane} \\ \big\downarrow {\scriptstyle \text{hofib}} \\ \mathbb{R}^{1,10\,|\,\mathbf{32}} \xrightarrow{\quad G_4^0 := \frac{1}{2}(\overline{\psi}\,\Gamma_{a_1 a_2}\,\psi) e^{a_1} e^{a_2} \quad} \mathfrak{l}B^4\mathbb{Z}\,. \end{array}$$

This, in turn, carries yet one more invariant 7-cocycle, being the WZW term of the M5-brane sigma model:

$$
\begin{array}{lcl}
\mathfrak{m5brane} & & \\
\big\downarrow{\scriptstyle \mathrm{hofib}} & & \\
\mathfrak{m2brane} & \xrightarrow{\quad \widetilde{G}_7 := \frac{1}{5!}\left(\overline{\psi}\,\Gamma_{a_1 \cdots a_5}\,\psi\right) e^{a_1} \cdots e^{a_5} - \frac{1}{2} c_3\, G_4 \quad} & \int B^7 \mathbb{Z}\,.
\end{array}
$$

This is a "structural" or "synthetic" emergence of spacetime, which is possible in topos/type theory, quite distinct in nature from attempts to see spacetime emerge via point-set or graph models.

The C-field emerges. Finally, the abelian M2- and M5-brane cocycles unify [11, §3][13, (57)] into a single non-abelian cocycle in 4-Cohomotopy [12, Cor. 2.3], via *homotopy pullback* (cf. [18, Ex. 1.12]) along the Hopf fibration [Eq. (5.5)]:

$$
\begin{array}{ccc}
\mathfrak{m5brane} & \longrightarrow & * \\
\big\downarrow & \swArrow{\scriptstyle (\mathrm{hpb})} & \big\downarrow \\
\mathfrak{m2brane} & \xrightarrow{\quad \widetilde{G}_7 \quad} & \int S^7 \\
\big\downarrow & \swArrow{\scriptstyle (\mathrm{hpb})} & \big\downarrow{\scriptstyle \int h_{\mathbb{H}}} \\
\mathbb{R}^{1,10\,|\,\mathbf{32}} & \xrightarrow{\quad (G_4,\, G_7) \quad} & \int S^4
\end{array}
$$

But this 4-Cohomotopy cocycle in 11D – which thus emerges from the superpoint – is the avatar (in a precise sense, [23, Thm. 3.1]) of the C-field that we started the discussion with in Eq. (5.3).

This may be seen to close a grand circle, where (super-)gravitational spacetime emerges from homotopical logic (§5.4), as such holographically exhibits topological qbits (§5.2), which in turn are naturally described in homotopy-typed language (§5.3).

Acknowledgments

The authors acknowledge the support by *Tamkeen* under the *NYU Abu Dhabi Research Institute grant* CG008.

Notes

1. The subscript $(-)_{\mathrm{cpt}}$ on a worldvolume domain – as in Eqs. (5.8) and (5.15) – denotes its *one-point compactification*, reflecting the characteristic condition that solitonic charges *vanish at infinity*, cf. [54, pp. 7, 14, 43][63, §2.2].
2. In traditional discussion of these observables, the framing on the links and the inclusion of the self-linking number are introduced in an *ad hoc* manner in order to work around an

otherwise ill-defined term obtained by path-integral heuristics. In contrast, in our derivation above these features emerge by rigorous analysis of quantum observables of the flux-quantized self-dual higher gauge field.

3. More precisely, what emerges here are the Kleinian local model spaces of higher-dimensional supergravities; but from these, curved supergravity follows as the super-Cartan geometric extension (cf. [23]).

References

[1] Arsiwalla, X. D., Elshatlawy, H., & Rickles, D. (2023). Pregeometry, formal language and constructivist foundations of physics. `https://arxiv.org/abs/2311.03973`.

[2] Cherubini, F. (2018). Cartan geometry in modal homotopy type theory. `https://arxiv.org/abs/1806.05966`.

[3] Chryssomalakos, C., de Azcárraga, J. A., Izquierdo, J. M., & Pérez Bueno, C. (2000). The geometry of branes and extended superspaces. *Nuclear Physics B*, *567*, 293–330.

[4] Constable, R. (2011). The triumph of types: Creating a logic of computational reality. Lecture presented at *Types, Semantics and Verification*, Oregon. `https://ncatlab.org/nlab/files/ConstableTriumphOfTypes.pdf`.

[5] Corfield, D. (2020). *Modal Homotopy Type Theory: The Prospect of a New Logic for Philosophy*. Oxford University Press.

[6] Donos, A., Gauntlett, J. P., & Pantelidou, C. (2013). Semi-local quantum criticality in string/M-theory. *Journal of High Energy Physics*, *2013*(3), 103.

[7] Duff, M. J. (1996). M-theory (the theory formerly known as strings). *International Journal of Modern Physics A*, *11*, 5623–5642.

[8] Duff, M. J. (1999). *The World in Eleven Dimensions: Supergravity, Supermembranes and M-theory*. IoP Publishing.

[9] Escardó, M. H. (2019). Introduction to univalent foundations of mathematics with Agda. *arXiv preprint arXiv:1911.00580*.

[10] Fiorenza, D., Sati, H., & Schreiber, U. (2015). Super Lie n-algebra extensions, higher WZW models and super p-branes with tensor multiplet fields. *International Journal of Geometric Methods in Modern Physics*, *12*(02), 1550018.

[11] Fiorenza, D., Sati, H., & Schreiber, U. (2015). The WZW term of the M5-brane and differential cohomotopy. *Journal of Mathematical Physics*, *56*, 102301.

[12] Fiorenza, D., Sati, H., & Schreiber, U. (2017). Rational sphere valued supercocycles in M-theory and type IIA string theory. *Journal of Geometry and Physics*, *114*, 91–108.

[13] Fiorenza, D., Sati, H., & Schreiber, U. (2019). The rational higher structure of M-theory. In *Proceedings of the LMS-EPSRC Durham*

Symposium: Higher Structures in M-Theory 2018. Fortschritte der Physik, *67*(8–9), 1903.02834.

[14] Fiorenza, D., Sati, H., & Schreiber, U. (2020). Twisted cohomotopy implies M-theory anomaly cancellation on 8-manifolds. *Communications in Mathematical Physics*, *377*, 1961–2025.

[15] Fiorenza, D., Sati, H., & Schreiber, U. (2021). Twisted cohomotopy implies M5 WZ term level quantization. *Communications in Mathematical Physics*, *384*, 403–432.

[16] Fiorenza, D., Sati, H., & Schreiber, U. (2022). Twistorial cohomotopy implies Green–Schwarz anomaly cancellation. *Reviews in Mathematical Physics*, *34*(05), 2250013.

[17] Fiorenza, D., Sati, H., & Schreiber, U. (2021). Twisted cohomotopy implies twisted String structure on M5-branes. *Journal of Mathematical Physics*, *62*, 042301.

[18] Fiorenza, D., Sati, H., & Schreiber, U. (2023). *The Character Map in Nonabelian Cohomology: Twisted, Differential and Generalized*. World Scientific.

[19] Freed, D. (1999). *Five Lectures on Supersymmetry*. American Mathematical Society. `https://bookstore.ams.org/FLS`.

[20] Gauntlett, J. P., Sonner, J., & Wiseman, T. (2009). Holographic superconductivity in M-Theory. *Physical Review Letters*, *103*(15), 151601.

[21] Gauntlett, J. P., Sonner, J., & Wiseman, T. (2010). Quantum criticality and holographic superconductors in M-theory. *Journal of High Energy Physics*, *2010*(2), 060.

[22] Giotopoulos, G., & Sati, H. (2023). Field theory via higher geometry I: Smooth sets of fields. `https://arxiv.org/abs/2312.16301`.

[23] Giotopoulos, G., Sati, H., & Schreiber, U. (2024). Flux quantization on 11d superspace. *Journal of High Energy Physics*, *2024*(7), 82.

[24] Giotopoulos, G., Sati, H., & Schreiber, U. (2024). Flux quantization on M5-branes. *Journal of High Energy Physics*, *2024*(10), 140.

[25] Giotopoulos, G., Sati, H., & Schreiber, U. (2024). Holographic M-brane super-embeddings. `https://arxiv.org/abs/2408.09921`.

[26] Girvin, S. M. (2004). Introduction to the fractional quantum Hall effect. *Séminaire Poincaré*, *2*, 53–74. Reprinted in: *The Quantum Hall Effect*, Progress in Mathematical Physics, *45*, Birkhäuser (2005).

[27] González-Meneses, J. (2011). Basic results on braid groups. *Annales Mathématiques Blaise Pascal*, *18*(1), 15–59.

[28] Grumblin, E., & Horowitz, M. (Eds.). (2019). *Quantum computing: Progress and prospects*. The National Academies Press.

[29] Gubser, S. S., Pufu, S. S., & Rocha, F. D. (2010). Quantum critical superconductors in string theory and M-theory. *Physics Letters B*, *683*, 201–204.

[30] Hartnoll, S., Lucas, A., & Sachdev, S. (2018). *Holographic Quantum Matter*. MIT Press. `https://arxiv.org/abs/1612.07324`.

[31] Herzog, C. P., Kovtun, P., Sachdev, S., & Son, D. T. (2007). Quantum critical transport, duality, and M-theory. *Physical Review D*, *75*, 085020.

[32] Howe, P. S., Sezgin, E., & West, P. C. (1997). The six-dimensional self-dual tensor. *Physics Letters B*, *400*, 255–259.

[33] Huerta, J. (2012). Division algebras and supersymmetry III. *Advances in Theoretical and Mathematical Physics*, *16*, 1485–1589.

[34] Huerta, J., Sati, H., & Schreiber, U. (2019). Real ADE-equivariant (co)homotopy and super M-branes. *Communications in Mathematical Physics*, *371*, 425–524.

[35] Huerta, J., & Schreiber, U. (2018). M-theory from the superpoint. *Letters in Mathematical Physics*, *108*, 2695–2727.

[36] Iqbal, M., Tantivasadakarn, N., Verresen, R., et al. (2024). Non-Abelian topological order and anyons on a trapped-ion processor. *Nature*, *626*, 505–511.

[37] Jacobs, B. (1998). *Categorical Logic and Type Theory*. Studies in Logic and the Foundations of Mathematics, *141*. Elsevier.

[38] Kallel, S. (2024). Configuration spaces of points: A user's guide. *Encyclopedia of Mathematical Physics* (2nd ed.), *4*, 98–135.

[39] Lawvere, F. W. (1997). *Toposes of Laws of Motion*. Talk presented in Montreal. `https://ncatlab.org/nlab/files/Lawvere-ToposesOfLawsOfMotion.pdf`.

[40] Lurie, J. (2014). *Nonabelian Poincaré Duality*. Lecture 8 in Tamagawa Numbers via Nonabelian Poincaré Duality. `http://www.math.harvard.edu/~lurie/282ynotes/LectureVIII-Poincare.pdf`.

[41] Miemiec, A., & Schnakenburg, I. (2006). Basics of M-theory. *Fortschritte der Physik*, *54*, 5–72.

[42] Myers, D. J. (2024). Topological Quantum Gates. Talk at *Homotopy Type Theory and Computing – Classical and Quantum*, CQTS @ NYU Abu Dhabi. `https://cdnapisec.kaltura.com/html5/html5lib/v2.73.2/mwEmbedFrame.php/p/1674401/uiconf_id/23435151/entry_id/1_k89ytmg4?wid=_1674401&iframeembed=true&playerId=kaltura_player&entry_id=1_k89ytmg4`.

[43] Myers, D. J., & Riley, M. (2023). *Introduction to Cubical Agda*. Course notes at CQTS, NYUAD. `https://github.com/CQTS/introduction-to-cubical`.

[44] Myers, D. J., Sati, H., & Schreiber, U. (2024). Topological quantum gates in homotopy type theory. *Communications in Mathematical Physics*, *405*, 172.

[45] Nakamura, J., Liang, S., Gardner, G. C., & Manfra, M. J. (2020). Direct observation of anyonic braiding statistics. *Nature Physics*, *16*, 931–936.

[46] Nastase, H. (2017). *String Theory Methods for Condensed Matter Physics*. Cambridge University Press.

[47] Ohtsuki, T. (2001). *Quantum Invariants: A Study of Knots, 3-Manifolds, and Their Sets*. World Scientific.

[48] Pires, A. (2014). *AdS/CFT Correspondence in Condensed Matter*. Morgan & Claypool.

[49] Riley, M. (2022). *A Bunched Homotopy Type Theory for Synthetic Stable Homotopy Theory*. PhD thesis.

[50] Sati, H. (2018). Framed M-branes, corners, and topological invariants. *Journal of Mathematical Physics, 59*(6), 062304.

[51] Sati, H., & Schreiber, U. (2020). Proper Orbifold Cohomology. `https://arxiv.org/abs/2008.01101`.

[52] Sati, H., & Schreiber, U. (2020). The Character Map in Equivariant Twistorial Cohomotopy Implies the Green–Schwarz Mechanism with Heterotic M5-Branes. `https://arxiv.org/abs/2011.06533`.

[53] Sati, H., & Schreiber, U. (2021). Twisted cohomotopy implies M5-brane anomaly cancellation. *Letters in Mathematical Physics*, *111*, 120.

[54] Sati, H., & Schreiber, U. (2023). M/F-theory as Mf-theory. *Reviews in Mathematical Physics*, *35*(4), 2350028.

[55] Sati, H., & Schreiber, U. (2023). Anyonic defect branes and conformal blocks in twisted equivariant differential K-theory. *Reviews in Mathematical Physics*, 35(6), 2350009.

[56] Sati, H., & Schreiber, U. (2023). Anyonic topological order in TED K-theory. *Reviews in Mathematical Physics*, *35*(3), 2350001.

[57] Sati, H., & Schreiber, U. (2024). Quantum observables of quantized fluxes. *Annales Henri Poincaré*, in print. `https://arxiv.org/abs/2312.13037`.

[58] Sati, H., & Schreiber, U. (2023). Entanglement of Sections. `https://arxiv.org/abs/2309.07245`.

[59] Sati, H., & Schreiber, U. (2023). The Quantum Monadology. `https://arxiv.org/abs/2310.15735`.

[60] Sati, H., & Schreiber, U. (2023). Quantum and Reality. `https://arxiv.org/abs/2311.11035`.

[61] Sati, H., & Schreiber, U. (2024). Flux Quantization on Phase Space. *Annales Henri Poincaré* **26**: 815-919.

[62] Sati, H., & Schreiber, U. (2024). Cohomotopy, Framed Links, and Abelian Anyons. `https://arxiv.org/abs/2408.11896`.

[63] Sati, H., & Schreiber, U. (2025). Flux Quantization. In *Encyclopedia of Mathematical Physics* (2nd ed., Vol. 4, pp. 281–324).

[64] Sati, H., & Valera, S. J. (2024). Topological Quantum Computing. In *Encyclopedia of Mathematical Physics* (2nd ed., Vol. 4, pp. 325–345).

[65] Sau, J. (2017). A roadmap for a scalable topological quantum computer. *Physics*, 10, 68.

[66] Schreiber, U. (2013). Differential Cohomology in a Cohesive ∞-Topos. `https://arxiv.org/abs/1310.7930`.

[67] Schreiber, U. (2014). Quantization via Linear Homotopy Types. Talks at Paris Diderot and ESI Vienna. *arXiv preprint arXiv:1402.7041*.

[68] Schreiber, U. (2025). Higher Topos Theory in Physics. In *Encyclopedia of Mathematical Physics* (2nd ed., Vol. 4, pp. 62–76).

[69] Schreiber, U., & Shulman, M. (2014). Quantum gauge field theory in cohesive homotopy type theory. *Electronic Proceedings in Theoretical Computer Science*, *158*, 109–126.

[70] Segal, G. (1973). Configuration-spaces and iterated loop-spaces. *Inventiones Mathematicae*, *21*, 213–221.

[71] Shimizu, H. (2018). *Aspects of Anomalies in 6D Superconformal Field Theories* (Doctoral dissertation, University of Tokyo).

[72] Shulman, M. (2018). Brouwer's fixed-point theorem in real-cohesive homotopy type theory. *Mathematical Structures in Computer Science*, *28*(6), 856–941.

[73] Stormer, H. L. (1999). Nobel lecture: The fractional quantum Hall effect. *Reviews of Modern Physics*, *71*(4), 875–889.

[74] Univalent Foundations Project. (2013). *Homotopy Type Theory – Univalent Foundations of Mathematics*. Institute for Advanced Study, Princeton. `https://homotopytypetheory.org/book`.

[75] West, P. (2012). *Introduction to Strings and Branes*. Cambridge University Press.

[76] Williams, L. (2020). Configuration spaces for the working undergraduate. *Rose-Hulman Undergraduate Mathematics Journal*, *21*(1), Article 8.

[77] Zaanen, J., Liu, Y., Sun, Y.-W., & Schalm, K. (2015). *Holographic Duality in Condensed Matter Physics*. Cambridge University Press.

[78] Zeng, B., Chen, X., Zhou, D.-L., & Wen, X.-G. (2019). *Quantum Information Meets Quantum Matter – From Quantum Entanglement to Topological Phases of Many-Body Systems*. Springer.

[79] Wolfram, S. (2002). *A New Kind of Science*. Wolfram Media.

[80] Zanardi, P., & Rasetti, M. (1999). Holonomic quantum computation. *Physics Letters A*, 264(2–3), 94–99.

[81] Zenil, H. (Ed.). (2012). *A Computable Universe – Understanding and Exploring Nature as Computation*. World Scientific.

[82] Zhang, J., et al. (2023). Geometric and holonomic quantum computation. *Physics Reports*, 1027, 1–53.

[83] Zuse, K. (1969). *Rechnender Raum*. Viehweg & Sohn. English trans.: (1970). *Calculating Space*, MIT Technical Translation AZT-70-164-GEMIT.

[84] *Animated Exposition of the Brane Bouquet*. `https://ncatlab.org/schreiber/files/BraneBouquetAnimated.gif`.

6 Linear Homotopy Type Theory: A Computational Language for Quantum Physics

David Corfield

6.1 Fivefold way

In recent years, a number of authors have observed that various subsets of the quintuple of *category theory*, *physics*, *topology*, *logic* and *computation* share considerable common ground. The logic-computation connection, the oldest, may be dated back to at least the 17th century with Leibniz's *Calculus Ratiocinator*. Interconnections multiply increasingly rapidly in the past century. To mention just a few key moments in this five-way convergence, topological semantics for intuitionistic logic was investigated by Tarski and McKinsey from the 1930s. Category theory arose in the 1940s to capture widespread patterns in mathematics, in particular in the field of algebraic topology. It was extended to the treatment of logic in the 1960s. Then in the 1970s it was discovered that the gauge fields of the physicists may be formulated as connections on fibre bundles, concepts from topology. Also, from around this time, a great deal of work related computer science to category theory. A recent development of this line of research comes in the form of *homotopy type theory*, closely related to the category-theoretic concept of an ∞-topos, and dubbed by Michael Shulman 'The logic of space' [1]. Physics and category theory have been brought together via the concept of monoidal categories in the work of Bob Coecke and colleagues, with connections to quantum computing [2].

The occurrence of such elaborate relations are now used as important pointers in theory development. Rather like the navigational idea of triangulation, it, being observed that a construction independently has a clear meaning in a

DOI: 10.4324/9781003546139-9

number of different domains, is taken as a sign that the investigation is on track. Some have spoken explicitly about multiple such connections. Robert Harper jokingly named the doctrine that good ideas in computation, category theory and logic coincide, *Computational Trinitarianism*, three manifestations of a single notion [3]. Meanwhile John Baez and Mike Stay were perhaps first to name the whole set, although they privileged one of them by depicting category theory's ability to represent the common core of the other four components via the analogy of the *Rosetta Stone* [4], where constructions common to all fields are set out in the rows of a table.

This five fold convergence[1] reaches a pinnacle of development in the recent work of Sati, Schreiber and coauthors [5]. Here we see that *linear homotopy type theory*, itself the 'internal logic' of a certain kind of pair of ∞-categories, provides a certification language for quantum computation, both in terms of hardware in the form of verification of topologically-protected quantum gates, and in terms of software in the form of a quantum programming language. But they take linear homotopy type theory to be more than this, as a computational language for modern quantum physics in general and for a 'synthetic' treatment of an important branch of mathematics known as stable homotopy theory. Let's now build up the ingredients of this calculus.

6.2 Dependent type theory

Here I mean in particular the kind of type theory associated with the Swedish logician, Per Martin-Löf. This is a logical calculus that has built into it a constructivist, and even a computational, attitude. From this perspective, there is considered a strict parallel between a proof of a theorem and a program, meeting a specification. For instance, one might require a program, which takes as input a finite list of natural numbers and converts it into an ordered list of natural numbers. *Quicksort* is the name of one program me which can do this. Similarly mathematicians can construct a function from $List(\mathbb{N})$ to $List(\mathbb{N})$ with the requirement that the target list is ordered and the underlying multisets are identical. To have a constructive proof of this result is to have a programme which meets the specification and *vice versa*.

- Program = Constructive proof

That one should opt for a *typed* logical calculus is something widely observed by computer scientists, but less commonly by philosophers. Computer scientists realize that type-checking in a programming language saves many errors, and makes it easier to see that a programme acts on desired kinds of entity to produce required outputs. Much of ordinary language philosophy was a complaint against typing errors, known as 'category mistakes'. Gilbert Ryle himself spoke of 'type-trespassing'. But these philosophers often also carried the conviction that formalisms of any kind would not be conducive to capturing natural language with all of its elasticity of meaning. An argument for the countervailing claim that dependent types ought to be deployed in philosophy, as regarding both language and metaphysics, is put forward in Chapter 2 of *Modal Homotopy Type Theory* [6]. For instance, via its framing, metaphysics may keep properly separate kinds of being: objects, events, properties, states of affairs, etc.

Let us now sketch what a pure dependent type theory gives us in terms of type formation:

- Empty type **0**, unit type **1**, sum type $A + B$, product type $A \times B$, function type $[A, B]$.
- A type of types (indeed an infinite series) $Type_i$.
- Types depending on other types: $x : A \vdash B(x) : Type$
- Two consequent type formations: dependent sum (pair/coproduct), $\sum_{x:A} B(x)$ and dependent product (function), $\prod_{x:A} B(x)$.
- Identity types: $A : Type, a, b : A \vdash Id_A(a, b) : Type$

Along with the formation of types, such as that for any two already formed types, A and B, there is a product type $A \times B$; there are also rules concerning the terms of the type. In the case of a product, given we have already established $a : A$ and $b : B$, we may pair these to *introduce* the element $(a, b) : A \times B$. Similarly, given $t : A \times B$, we may *eliminate* t to give its two components, $\pi_1(t) : A$ and $\pi_2(t) : B$. This is in complete agreement with the category-theoretic notion of product.

An important part of Martin-Löf type theory is the notion of a *dependent* type, denoted as $x : A \vdash B(x) : Type$. Here the type $B(x)$ *depends* on an element of A, as in

- $m : Month \vdash Days(m) : Type$
- $t : Team \vdash Players(t) : Type$

Generally, we may think of these dependent types as setting one type *fibred* above another. For instance, in the football case, one may imagine the collection of league players lined up in fibres above their team name. Then two central constructions we can apply to these types are *dependent sum* and *dependent product*.[2] The *dependent sum* is the total type of all players, its elements being pairs of a team and a player of that team. Likewise, an element of the *dependent product* is a choice of a player from each team, such as $captain(t)$ or $top\_scorer(t)$.

Considering types at this stage as either sets or propositions, in a sense that will be made precise in the next chapter, we have Table 6.1:

The final construction mentioned in the list above, namely, identity types, will now be considered in the broader context of *homotopy* type theory.

Table 6.1 A comparison of dependent types

Dependent sum	Dependent product
$\sum_{x:A} B(x)$ is the collection of pairs (a, b) with $a : A$ and $b : B(a)$	$\prod_{x:A} B(x)$, is the collection of functions, f, such that $f(a) : B(a)$
When A is a set and $B(x)$ is a constant set B: The product of the sets.	When A is a set and $B(x)$ is a constant set B: The set of functions from A to B.
When A is a proposition and $B(x)$ is a constant proposition, B: The conjunction of A and B.	When A is a proposition and $B(x)$ is a constant proposition, B: The implication $A \to B$.
When A is a set and $B(x)$ is a varying proposition: Existential quantification.	When A is a set and $B(x)$ is a varying proposition: Universal quantification.

6.3 Homotopy type theory

A central choice for a mathematical foundation is what to consider as the basic shape of mathematical entities. One time-honoured choice is:

- Set: a bag of dots which are different and yet indistinguishable.

Irrespective of the way one chooses to describe sets formally, 'materially' or 'structurally', it is an astonishing idea that mathematics could rely on such a conception. From the perspective of dependent type theory, the most important feature of the structure of a set, A, is that to ask of any two of its elements, $x, y : A$, whether or not they are the same, is to wonder whether or not a certain proposition is true. We have that

- $x, y : A$, then $(x =_A y)$ is a proposition.

When the type theorist specifies something as a proposition, they are understanding that thing as a type. Indeed, $(x =_A y)$ is a type. Its truth depends on whether or not it is inhabited. For it to be inhabited, we would have some element $p : (x =_A y)$, which would act as a witness or proof of the equality of x and y. Note that in this framework we are only ever to form an identity type for elements of the same type. If we have formed $x : A$ and $y : B$, then we are unable to form $(x = y)$.

So we take a proposition: if it contains two elements, then they must be equal. In other words, a proposition is a subsingleton. In the *extensional* form of dependent type theory, to construct an identity type of any type is to ask *whether* two of its elements are the same, not *how* they are the same. This corresponds to the bag of dots image we have of a set. However, arising from the needs of current geometry and current physics, we find that relying solely on such a basic shape is a restriction. We need to know the type of ways that two entities are the same – the *how* of identity. And further, we need to know how these *hows* are related. Besides sets, we need:

- *Homotopy types* or *n-groupoids*: points, reversible paths between points, reversible paths between paths, ...

We arrive there by iteration of identity type formation:

- Where we have a type A and $x, y : A$, we form the type $x =_A y$.
- Then from $x =_A y$, and $p, q : x =_A y$, we form $p =_{(x=_A y)} q$.

Some type theories stop here and insist that any such p and q must be the same. However, we may decide to forgo the 'Uniqueness of Identity proofs' in the sense that we need not insist that any two proofs of identical entities are themselves the same. We reject the axiom that claims this is the case, or in other words, we don't insist that the following type is necessarily uniquely inhabited:

$$p =_{(x=_A y)} q \,.$$

This iteration of identity types allows us to speak of a hierarchy of *homotopy* types.

Now we have a hierarchy of kinds of types to be treated uniformly, where the level corresponds to how many iterations of identity type formation bottom out in triviality:

...	...
2	2-groupoid
1	groupoid
0	set
−1	mere proposition
−2	contractible type

We may have types of any and indeed infinitely many levels. These correspond to ∞-groupoids.

These n-types may seem complicated, and from a set-theoretic perspective, they are more complicated. But from the perspective of *intensional* dependent type theory, they appear as the basic entities, and sets will have to be picked out from them by some specification as 0-types via the characteristic that their identity types are propositions.

To the above we need to add one more condition, this time an axiom, one that gives its name to *Univalent Foundations*.

- Univalence Axiom: $Equiv(A, B) \simeq A =_U B$

This axiom is dictating that whenever we have two equivalent types, A and B, which essentially means there are structure preserving maps between them, then their corresponding elements in a relevant universe of types, U, are equal, now in the sense that whatever we can establish in the calculus about A may be transferred to B and *vice versa*.[3] HoTT is a structural theory *par excellence*.

So HoTT with univalence is an intensional dependent type theory, where types are characterized by their formation rules and the corresponding introduction and elimination of their terms, specifying what is it to form a term of that type and how to use one. The structure of types is thus given intrinsically. We may then contrast this *internal* view of the language with the *external* view which corresponds to the *interpretation* of the type theory, what its models look like.

Now mathematicians had noticed that

- Gathering together all sets and functions results in a collection or category which behaves nicely: a *topos*.
- Gathering together all homotopy types/∞-groupoids and ∞-functors results in a collection or ∞-category which behaves *extremely* nicely: an ∞-*topos*.

We may tell a justificatory story internal to mathematics, running at least from Grothendieck to Lurie, which explains the reason for this formulation. The connection to physics comes from seeing ∞-toposes as a particularly suitable environment to understand *cohomology*, where cohomology itself is precisely the mathematical concept which captures gauge fields in fundamental physics of all kinds from Yang-Mills to gravity.

'Homotopy type theory' may be parsed as both *(homotopy type) theory* **and** *homotopy (type theory)*.

- *Homotopy type theory* as *(homotopy type) theory* is a synthetic theory of homotopy types or ∞-groupoids. A structurally invariant theory of ∞-groupoids, where the structure emerges from iterated identity types. It is modelled by spaces (but also by lots of other things).
- *Homotopy type theory* as *homotopy (type theory)* is the internal language of ∞-toposes. It is a type theory in the logical sense, and may be implemented on a computer. It allows a synthetic treatment of abstract spatial structure – homotopy types.

In terms of more familiar logical calculi, homotopy type theory for the lower levels of the hierarchy encapsulates:

- Propositional logic

- (Typed) predicate logic
- Structural set theory

Considering the full type theory, with *higher inductive types* included, the line between logic and mathematics is profoundly blurred – constructions such as the homotopy groups of the spheres, group actions and invariants, may now be seen as purely logical.

As has been stressed earlier, an *intensional* dependent type theory is very much tied to a notion of computation. We're seeing this played out in Kevin Buzzard's *Xena* programme with Lean used as a proof assistant [7], one with the ambitions to take on the most advanced mathematics, such as Wiles's proof of Fermat's Last Theorem, where there need to be type-theoretic constructions to capture the concepts of automorphic forms and representations, Galois representations, the arithmetic of varieties, class field theory, arithmetic duality theorems, Shimura varieties and much more. This is unthinkable with any theorem prover based on ZFC set theory.

For a brief glimpse of its representational capacity, consider the following piece of typical mathematical text:

> Let k be a field, V a finite-dimensional vector space over k, and f an endomorphism of V. Then define $E(V, k, f)$, the eventual image of f, as the vector space which is the intersection of all $f^n(V)$. Show that $f(E) = E$.

We can begin to parse the construction of a relevant type, and display what it would be to prove the theorem:

- $k : Field, V : FinVect(k), f : Endo(V, k) \vdash E(V, k, f) : FinVect(k)$,

Then we need to construct an element in the following type:

- $k : Field, V : FinVect(k), f : Endo(V, k) \vdash g : (f(E) = E)$

At the present time, Lean relies on the uniqueness of identity proofs, so it has no higher-level types.

6.4 Modal homotopy type theory

Philosophers and computer scientists have sought *modal* variants of propositional and predicate logic and of type theory. It was natural then to expect a *modal* HoTT. From the perspective of category theory, modalities are kinds of *monad* and *comonad*, operators arising from adjunctions, used in computer science to treat *effects* and *context dependence*. To illustrate the first of these briefly: a computer does more than compute – it also affects the world, e.g., by sending out an error signal or a command to print. Monads provide a way to express such effects while staying in the domain of functional programming languages. Hence the title of the paper *The Quantum Monadology* [5], indicating the extension of the monadic treatment of effects and contexts to quantum computing.

Modalities may also be used in mathematics to allow us to go beyond the merely structural, combinatorial ∞-groupoids by capturing further important mathematical structure synthetically, such as topological cohesion and smoothness, but also supergeometry (for fermions), equivariance (a form of invariance under group actions) and orbifold structure (singularities). From the *external* point of view, a variety of modal HoTT is the internal language of a system of ∞-toposes.[4]

There is a family of pairs of native modalities already given for any ∞-topos, which arises from morphisms between its *slices*. The slice of a category over an object collects together arrows into that object and so allows the expression of objects varying over some fixed object. One natural pair of examples to consider includes the modalities arising from dependent sum and dependent product.

- $w : World \vdash A(w) : Prop$
- $\prod_{w:World} A(w)$: 'For all worlds, A holds'.
- $\sum_{w:World} A(w)$ 'The worlds where A holds', may be *truncated* to 'In some world, A holds'.
- From these we may derive the operators act on world-dependent propositions to act as necessity and possibility.
- E.g., $w : World \vdash A(w) : Prop$, then $w : World \vdash \Box A(w) : Prop$.[5]

But notice that these constructions are purely structural. We need not take the type of variation to be a collection of worlds, and we need not take the dependent types as mere propositions. In the footballing case above, we have variation over the type of teams. There, the effect of the possibility operator is to place a copy of the type of all footballers above each team name. Likewise, the effect of the necessity operator would be to place above each team name a copy of the type of all sections, that is, all choices of a player per team. We can now construct morphisms $\Box A(w) \rightarrow A(w) \rightarrow \diamond A(w)$ via the so-called *counit* and *unit* of the adjunctions.

Then the corresponding implications

$$necessity \rightarrow actuality \rightarrow possibility$$

above a team, c, correspond to (i) taking a section, such as $goalkeeper(t)$, and evaluating it at c to give the goalkeeper of that team, $goalkeeper(c)$; and (ii) then inserting this player $goalkeeper(c)$ into the collection of all players over c.

Aside from these intrinsic modalities, which may be defined for any object W, and indeed any morphism $f : W \to V$, one may also specify modal operators for various mathematical purposes, e.g., cohesive, smooth, singular, and linear structure. It is the latter, linear modality, that we shall need for quantum computing.

Summing up, we have

1. HoTT: a synthetic language to describe structure.
2. Modal HoTT:
   a. *Cohesive* HoTT, etc.: synthetic languages for topological, differential, singular-orbifold, and supergeometric structure, differential cohomology of (higher) supersymmetric gauge theory.
   b. *Linear* HoTT: a synthetic language for stable homotopy theory, for 'linear' structure (infinitesimal, tangent, abelian, stable, etc.), quantum information.

In terms of category-theoretic semantics, these correspond to:

1. HoTT: ∞-topos
2. Modal HoTT: systems of ∞-toposes and geometric morphisms
   a. *Cohesive, differential, supergeometric, singular* HoTT: interrelated adjoint quadruples between pairs of ∞-toposes
   b. *Linear* HoTT: bireflective inclusion of one ∞-topos inside another.

Let's turn briefly to this *linear* variant.

6.5 Linear homotopy type theory

The '*linear*' in this version of HoTT echoes its use in 'linear logic'. The latter is a resource sensitive logic by means of which it is possible to represent the idea that we may only be able to use a proposition once. Here, for example, in an inference from A and B to produce some conjunction, $A \otimes B$, we may only do this once because we have consumed the original A and B in the inference. In logic we typically allow *weakening* and *contraction*; in linear logic we do not. So, where we have

- $\vdash B$, therefore $A \vdash B$
- $A, A \vdash B$, therefore $A \vdash B$

in classical or intuitionistic logic, these are not allowed in linear logic.

It was recognized early that there is a relationship between linear logic and quantum mechanics, which can be seen through the semantics of each in something like vector spaces. A feature of the category of vector spaces compared to the category of sets is that product defined in the former is not *cartesian*. This means we cannot merely project out from some joint vector space $A \otimes B$ to give an A-component and a B-component. The entanglement of quantum systems reflects this – the state of an entangled system is not given merely by states of each system. Nor can we clone states, since we have no duplication map $A \longrightarrow A \otimes A$.

Now linear HoTT is designed to add this linear feature to the intensional dependent type theory that is HoTT. In this calculus we have a construction which allows us to map a type to a purely nonlinear type. We also have a means to map a type to a purely quantum type. What then appears is that we may consider any type as a linear type depending on a nonlinear base type. In terms of semantics, linear HoTT is represented as concerned with parameterized linear spaces, in its general sense to include the *spectra* of algebraic topology, the source of values for abelian cohomology. A particular kind of such parameterized linear spaces, the 0-truncated $\mathbb{C}$-linear sector, concern finite-dimensional complex vector spaces indexed by a finite set, which are what we need to represent the states relevant to quantum computing. We are dealing here with complex vector spaces indexed over the values of measurement outcomes. Other sectors are relevant for broader purposes, such as certifying quantum gates.

Picking up on the earlier treatment of the modalities of necessity and possibility, in the case of finitely-indexed vector spaces, it is the case that linear dependent sum and linear dependent product coincide. Thus, the linear equivalent of the implications

$$necessity \longrightarrow actuality \longrightarrow possibility$$

compose into a self-map, one which corresponds to a projection onto the subspace corresponding to the measurement outcome. The first map represents collapse of the wave function, the second quantum state preparation. We see here a rapprochement between the possible worlds of classical modal logic and the many worlds of the interpretation of quantum mechanics.

Linear HoTT then shows itself supremely capable of representing: measurement and preparation, the deferred measurement principle, the equivalence of the Copenhagen and Everettian pictures, density matrices, the Born rule, dynamic lifting and much more. And, it is claimed, the broader setting which allows for dependency on types with path structure, not just finite sets, and which allows a wider range of spectra, is suitable for quantum physics more broadly.

As a type theory, linear HoTT relies on what was a surprising discovery. There is a result in category theory which shows that toposes and abelian categories are about as far apart as possible. In the context of all categories that share their common features, so-called AT-categories, any one of these may be factorized uniquely into its topos component and its abelian category component. In a sense, this sharply divides the nonlinear world from the linear world. The strange finding was that when lifted to the world of ∞-categories, there is a construction which acts like that of forming a tangent space to a manifold, known as forming the tangent ∞-category, and when this is carried out on an ∞-topos, the result is still an ∞-topos. Somehow this blending of the linear with the non-linear has resulted in an overall non-linear structure. This means that when devising a type theory for such a situation, it is possible to start out from plain HoTT to describe even dependent linear types. We have two ∞-toposes, one being like an infinitesimal thickening of the other, $\mathbf{H} \hookrightarrow \mathbf{H}_{th}$.

- There is a map $\mathbf{H}_{th} \to \mathbf{H}$ which forgets the thickening, projecting to the underlying parameter space.
- The inclusion of a space as parameterized 0-spectra over that space, $\mathbf{H} \hookrightarrow \mathbf{H}_{th}$, is left and right adjoint to projection to the indexing base.
- We add to HoTT the self-adjoint modality for round trips, $\natural$.
- We also freely add linear connectives $\otimes$ and $\multimap$.

Now the claim is that linear HoTT is a universal quantum certification language, embedded in which it is possible to construct a quantum computing language – *QS* [8]. The larger Sati-Schreiber programme looks to combine the linear modalities with the cohesive modalities for a logic of quantum physics in general. From this perspective, we may see all physical processes as a form of computation.

6.6 Conclusion

There is a great deal to take in from the brief survey covered in this paper, which in turn is intended to provide an alternative entry point to the Sati-Schreiber article, and the substantial literature their programme has generated. I consider their work to be of the greatest philosophical interest. We are seeing emerge before us a simultaneous revolutionary shift in the foundations of logic, mathematics and physics. In [9], I discuss this simultaneous revolution in relation to the philosophical perspective of Michael Friedman. Something resembling his description of a shift of constitutional languages is happening here. We see emerging a new logic allowing the expression of a new mathematics for a new physics. This new logic, *cohesive linear HoTT*, is to provide a logic for quantum physics through its encoding of orbi-singular, supergeometric, differential cohomology. In a certain sector, cohesive linear HoTT provides a language for certified quantum programming of classically controlled quantum circuits compiled from topological quantum gates in physically realistic quantum materials. We may consider then the extent to which all of quantum physics is a form of computation.

Notes

1. An argument could be made that we might include *algebra* as a sixth.
2. Sometimes these are called *dependent pair* and *dependent function*, respectively.
3. The Univalence Axiom has a non-computational flavour. Alternatives to HoTT with UA are provided by amongst others *cubical* HoTT, and now Higher Observational Type Theory.
4. Interested readers should look to understand how monads and comonads arise from adjunctions between categories.
5. Cf. Chapter 4 of my Modal HoTT book for discussion.

References

[1] Coecke, B., & Gogioso, S. (2023). *Quantum in Pictures*. Quantinuum Publications.

[2] Shulman, M. (2021). Homotopy Type Theory: The Logic of Space. In M. Anel & G. Catren (Eds.), *New Spaces in Mathematics: Formal and Conceptual Reflections* (pp. 322–404). Cambridge University Press. https://arxiv.org/abs/1703.03007

[3] Harper, R. (2011). The Holy Trinity. https://existentialtype.wordpress.com/2011/03/27/the-holy-trinity/

[4] Baez, J., & Stay, M. (2011). Physics, Topology, Logic and Computation: A Rosetta Stone. In B. Coecke (Ed.), *New Structures for Physics* (pp. 95–174), Lecture Notes in Physics, 813. Springer.

[5] Sati, H., & Schreiber, U. (2023). The Quantum Monadology. https://ncatlab.org/schreiber/show/The+Quantum+Monadology

[6] Corfield, D. (2020). *Modal Homotopy Type Theory: The Prospect of a New Logic for Philosophy*. Oxford University Press.

[7] Buzzard, K. (n.d.). What is the Xena Project? https://xenaproject.wordpress.com/what-is-the-xena-project/

[8] Myers, D. J., Sati, H., & Schreiber, U. (2023). QS: Quantum Programming via Linear Homotopy Types. https://ncatlab.org/schreiber/show/Quantum+Certification+via+Linear+Homotopy+Types

[9] Corfield, D. (2024). Thomas Kuhn, Modern Mathematics and the Dynamics of Reason. In Y. Shan (Ed.), *Rethinking Thomas Kuhn's Legacy* (pp. 51–76). Springer.

7 Pregeometry, Formal Language, and Constructivist Foundations of Physics

Xerxes D. Arsiwalla, Hatem Elshatlawy, and Dean Rickles

7.1 Introduction

Ever since the inception of general relativity and quantum field theories early and mid-20th century, an outstanding open question in theoretical physics has concerned the quantum nature of gravity, or equivalently, the quantum geometry of space and time at the Planck scale. Contemporary approaches to quantum gravity today have thrown up a rather wide range of proposals on the question of what the underlying building blocks of quantum geometry may be: From discretizing topology [41, 72], to triangulated spacetime foam [60], to geometric operators [66, 77], to extended objects as quantum gravitational fluctuations [23], to holographic duals [3]. What perhaps unites these ostensibly diverse theories is the recognition of a pregeometry at the foundations of spacetime, which appears at energies close to the Planck scale.

How then does one undertake a comparative investigation of pregeometric structures? More specifically, is there a universal structure underlying pregeometric building blocks of physics? This work is an attempt at examining the metaphysics of pregeometric structures. That necessitates a conceptual analysis of the structure of structures upon which existing notions of quantum (or at the least, non-classical) geometry can be founded. Based on tools from formal language theory, this work is a philosophical attempt at addressing a meta-theory of structures, such that different formulations of quantum/non-classical geometries may be investigated within a common theoretical framework. It is hoped that the synthesis of ideas presented in this work might pave the way toward a

DOI: 10.4324/9781003546139-10

mathematical "theory of pregeometry", which may serve as a formal unifying framework for conceptualizing and analyzing precise definitions of quantum and classical spaces.

The term "pregeometry" was first coined by John Wheeler as an approach to the foundations of physics that ought to encompass any underlying explanation of spacetime or quantum gravity (as per Wheeler, this would also include an explanation of elementary particles) [64, 89]. One may argue that this term merely functions as a placeholder for whatever more elementary structure is eventually found to serve its intended function. Wheeler treated the problem as a kind of exercise in structure-substitution. That is, test every known structure, "from crystal lattices to standing waves and from Borel sets to the calculus of propositions" [9, pp. 17–18]. As an historic anecdote, it was also the inadequacy of most seemingly plausible structures that led Wheeler to his ideas encapsulated in the phrase "It from Bit". For our purposes here, by pregeometry we will refer to symbolic/linguistic structures which do not come endowed with any pre-assigned geometric attributes. Instead, geometric (and also non-trivial topological) structures should be derived properties of abstract building blocks (within suitable limits, of course). As we will discuss here, higher homotopical constructions in formal languages, expressed using higher categories, turn out to operationalize such a framework of pregeometry. This exercise may be seen as a modern-day incarnation of Wheeler's original intuition.

We will argue that a meta-theory of structures, or for that matter any analysis concerning the structure of structures, would be incompatible with a metaphysics based on material realism. That would lead to the well-known infinite regress problem. Even a "Foundationalist" stance with its "universal self-evident truths" may not provide a satisfactory resolution to the problem. Instead, we posit that a "Coherentist" philosophy of physics based on mathematical constructivism provides the appropriate foundations for the kind of pregeometric structures that Wheeler had in mind. Typed languages, and in particular, computational languages, are inherently constructivist. We will argue that what we refer to as "structureless structures", are in fact syntactic entities (or types) that realize programs (or proofs). The study of "pre-physics" is then presented as a constructivist paradigm, where spaces and algebras relevant to physical theories are modeled as computational routines built from compositional rules of the underlying formal language.

Apart from Wheeler, language-theoretic approaches to the foundations of physics have also been proposed in several earlier studies pioneered by Isham and collaborators, within the context of topos theory [36–40, 52, 53]. More recently, the Wolfram Model [13, 17, 92], Constructor Theory [34], Categorical Quantum Mechanics, [1, 29], Quantum ZX Calculus [28, 47–49], Mechanics from Intuitionistic Mathematics [46], Operator Mechanics [16], and Assembly Theory [82] can all be seen to be instances of different syntactic formalisms, possibly expressible within a broader language-theoretic framework. Also, some of the modern conceptions in theories of quantum gravity (those mentioned above) such as spin networks, causal sets, group field theories, simplicial calculus, C^*–algebras, tensor networks, matrix models, and so on, being purely algebraic, can potentially be constructible within the context of a formal language internal to an appropriate (higher) topos.[1] The key point is that the framework of formal language subsumes many of the above-mentioned theories and models, and thus provides the appropriate foundation for discussing various notions of pregeometry and geometry proposed in theories of quantum gravity.

The outline of this chapter is as follows: To set the motivation for pregeometry as "structureless structure" expressed within a formal language, in Section 7.2 we begin with a classification of four types of investigations surrounding the foundations of theoretical physics. Based on that, Section 7.3 leads to a discussion on foundationalism versus coherentism as philosophies of physics. Following the latter, in Section 7.4 we build the case for pregeometry as structureless structure. Section 7.5 presents key ideas from Leibniz's Monadology [57], as well as Pauli-Jung's monism [19], both of which, one may argue, allude to modern-day ideas of pregeometry. Then in Section 7.6, we elaborate how formal languages express structureless structures. Finally, in Section 7.7 we conclude with closing remarks and future directions.

7.2 Types of theories of fundamental physics

By and large, investigations probing the modern-day foundations of theoretical physics[2] can broadly be categorized into four main classes (though not completely unrelated to each other):

(i) Those involving **interpretation**s of existing physical concepts and structures;

(ii) Those involving new physical mechanisms or new **phenomenological models** (often relying on existing theoretical frameworks);

(iii) Those involving **new physical structures**, generalizing existing physical notions of space, time or matter in the search for new physics; and

(iv) Those investigations seeking **radical new conceptualizations** of existing physics, in order to address questions related to ontological and metaphysical origins of structure and the role of observers.

Efforts involving class (i) concern issues such as wave function realism [5], quantum measurement and contextuality [50], the nature of wave-particle duality [8], the problem of time in quantum gravity [7], and many others.
Examples that fall within class (ii) involve prospective mechanisms to explain the free parameters, such as the masses and couplings of the standard model [42], the cosmological constant problem [88], the dark matter/modified gravity puzzle [33, 78], etc. While investigations of this kind seek to extend known

models of particle theory or cosmology, they do so largely within or with limited modifications to the existing structural framework of quantum field theories and general relativity.

On the other hand, present-day efforts involving class (iii) typically concern new proposals for quantum gravity [59, 76], non-perturbative completions of gauge theories [12, 14, 35], noncommutative geometry [30], physics of higher dimensions [70], emergence of spacetime from holography [86], standard model symmetries from division algebras [43], among many others. Investigations of this third type often propose resolutions for outstanding foundational problems by invoking new physical principles and/or new mathematical structures that seek to generalize or extend the scope of existing frameworks of theoretical physics.

And finally, efforts involving class (iv) investigations involve questions regarding the origins of space, time, and matter itself. Rather than generalizing existing physical structures, these investigations seek new ontological origins such that existing physical structures may be seen to be emergent or derivable from an underlying "pre-physical" framework. The classic example of this is Wheeler's pregeometry [64, 89].

Historically, several of the then new ideas and developments in quantum theory and relativity originated as new mathematical formalisms of class (iii) and eventually became amenable to investigations of class (ii) and subsequently class (i). This was true for even what were considered abstract mathematical structures back in the day. Examples include Clifford algebras and spinors introduced in quantum theory (which eventually made their way into fermionic quantum field theories) by Dirac or non-Euclidean geometries of spacetime by Einstein and Grossmann. Falsifiable theories, at the very least, are marked by a transition from class (iv) or class (iii) to class (ii). Based on this premise, the issue isn't whether or not new mathematics is necessary to inform the foundations of physics, but rather, whether or not new mathematical formalisms will eventually become amenable to investigations of class (ii) and class (i). In particular, investigations seeking to address the origins of established theoretical frameworks such as quantum field theories or cosmological models, will inevitably require the introduction of new mathematical structures, very likely, radical new ones too. The challenge then being: How does one effectively filter out choices that do not refer to the observable universe?

Consider for instance, the longstanding problem of reconciling quantum field theory with general relativity. It is widely believed that such a reconciliation between these founding pillars of modern physics will markedly alter our understanding of major open problems in theoretical physics today, including the origin of our universe, the origin of matter and the fundamental forces, the quantum mechanics of black holes, and the nature of space and time, among others. A key question underlying many of these issues is the following: What are the underlying building blocks of space, time and quantum fields? Any theory attempting to bring together quantum field theory with general relativity will, at the very least, have to take a definitive stance on the nature of these building blocks. While it seems there is widespread consensus on the view that space, time, and matter are fundamentally discrete [45, 52], specific proposals concerning the nature of this discretization (often described in terms of "quanta" of space, time, or matter) and consequently the underlying mathematical structure one needs to start with, differ quite a bit. Notable examples include: (a) Theories of quantum gravity, including loop quantum gravity (LQG) [77], string theory and its proposed non-perturbative completion, M-theory [23], group field theory [66], the causal sets program [41], causal dynamical triangulation (CDT) [60]; (b) approaches seeking a unification of the fundamental forces, either within the context of supersymmetric gauge theories [6], or within the framework of F-theory [22], or based on the representation theory of exceptional Lie algebras such as those associated to the group E_8 [58]; (c) models of emergent spacetime from entanglement entropy [42], from energetic causal sets [32], from emergent gravity [87], the Anti-de Sitter/conformal field theory (AdS/CFT) correspondence [3], and other realizations of black hole holography [10, 11, 15]; and (d) pregeometric models as building blocks of spacetime, as those initiated by Wheeler [64, 89], and recent approaches based on homotopy type theory [13, 17, 18].

Extracting empirical verifiability from the multitude of above-mentioned mathematical proposals and filtering out physically redundant ones have been and continue to be rather daunting challenges. Recent collaborative projects in quantum gravity phenomenology seek in part to address this problem (see [2] for a status review of the field). Nonetheless, the point remains that each of these competing proposals of quantum gravity finds itself having to introduce into the foundations of physics new abstract mathematical structures in order to attempt extensions beyond or reconciliations between our existing notions of space, time, and matter. This is the essence of class (iii) and class (iv) investigations. Even in a future scenario, where one or more of the above proposals turns out to be an empirically adequate (or at least falsifiable) description of quantum gravity, questions about the origins of that new theory will potentially remain open and require new structures and extensions beyond its existing framework.

7.3 The epistemic regress problem: Foundationalism vs. coherentism

Any classification of theories, as the one presented above, into more and more fundamental ones, confronts us with an obvious philosophical problem. For instance, consider the principle of "inferential justification" in the context of epistemology [85]. This states that: "To be justified in believing A on the basis of B one must be, (1) justified in believing B, and (2) justified in believing that B makes probable A" [51]. Hence, any approach to the foundations of physics based exclusively on inferential justification, in the sense of seeking explanations of explanations recursively, leads to an infinite regress problem. This is particularly relevant if one were to insist on a fundamentally materialist ontology.

Of course, issues such as the one stated above, have extensively been discussed in the philosophy of physics, and more generally, in theories of epistemic justification [85]. Typical resolutions of this problem adopt either a stance of "foundationalism" or "coherentism".

- Foundationalism assumes the existence of certain self-evident truths that can, in principle, halt the regress [51];
- Coherentism requires that statements within that system self-cohere, forming an inter-dependent web of mutual justification [65].

In the context of physics, the search for realizations of foundationalism could manifest in terms of what is sometimes referred to as the "final theory".[3] On the other hand, proposals grounded in coherentism emphasize the role of relations (rather than objects) and compositionality, such that attributes of a system are described in relational terms.

More specifically, one may ask, what kind of relational frameworks should one examine as plausible candidates for investigating how structural properties of space, time, and matter might emerge?, Our answer to this is to consider formal languages as the framework for pregeometric foundations of physics.

- A **formal language** $\mathcal{L}$ refers to a collection of "well-formed" strings Σ^* over an alphabet Σ (usually a set of finitely many letters), where the superscript "$*$" denotes the Kleene product over Σ (the free monoid over the alphabet). The moniker "well-formed" refers to syntactic constraints that can be specified either by a generative grammar or a set of $n-ary$ relations based on a universal algebra. These relations recursively specify how the letters of Σ can be composed to form strings in $\mathcal{L}$.
- For use in theorem proving or automated reasoning protocols, it is often useful to additionally equip the language $\mathcal{L}$ with a deductive system.
- A **deductive system** (also called **proof system**) within a language $\mathcal{L}$ consists of an axiom schema and/or a collection of inference rules, which can be used for theorem-proving within $\mathcal{L}$.
- We will call a formal language equipped with a deductive system a **computational language**. This is sometimes also referred to as a formal system.

As alluded to above, we shall mainly be interested in formal languages that are computational languages. These are typically typed languages equipped with a deductive system, using which one may construct proofs and programs. Computational languages are inherently constructivist in the sense their proof systems only allow constructivist mathematical proofs (those in which proofs by contradiction are not permitted; consequently, a proof that a given property holds, requires constructing an algorithm that realizes an instance of that property). These systems are based on intuitionistic logic, which as we shall see, allows for expressing theories of physics that are not bound to an *a priori* use of continuum notions.

In fact, this brand of constructivism founded in formal language is not exclusive to physics. It turns out that many type theories originating from a constructivist paradigm of mathematics are formal systems that are distinct from Zermelo-Fraenkel Choice (ZFC) set theory (in that, they do not enforce the law of the excluded middle or the axiom of choice). Rather, many of these type theories and their associated toposes are based on the univalence axiom of homotopy type theory [4, 68]. In particular, this suggests that there does not exist merely one preferred axiomatization to describe all of mathematics, as proponents of the Hilbert school of thought may have hoped for. Rather, there are multiple universes (or toposes) where mathematics can be formulated. These universes may be founded on very distinct axiomatization schemes, which nonetheless can be transformed from one to another [83, 84].

How does formal language relate to pregeometry? In the next few sections we will argue that our "structureless structures" are in fact syntactic entities (types) that realize programs (proofs) in a formal language. Within the modern set-up of homotopy type theory, spaces and algebras relevant to physical theories can be modeled as computational routines built from compositional rules of a formal language. Arguably, the kind of constructivism guiding the current foundations of mathematics turns out to be important for the foundations of physics too.

7.4 Pregeometry as structureless structure

What kind of structures should a constructivist coherentism entail for class (iv) investigations relevant to the foundations of physics? As alluded to above, formal language, and in particular, computational languages based on syntactic structures, their compositions, and rules of computation provide the constructivist building blocks for a realization of pregeometry that does not hinge on pre-existing notions of space, time, and matter. Such formal constructs necessarily shift away from a materialist ontology of physics.

- **Structureless structures** thus refer to symbolic and relational entities of a formal language. These symbolic structures are the building blocks of proofs, programs, and computations.

When Wheeler introduced the notion of pregeometry, he thought about it as an all-encompassing approach to the very foundations of physics, with the idea that pregeometry ought to transcend any structural explanation of space, time, matter, and even physical law [64, 89]. Pregeometry was thus intended as a broad conceptual framework from which one may seek, or upon which one may build, descriptions of quantum gravity. As mentioned, at the time, Wheeler treated the problem as a structure-substitution exercise, meaning that he tested every known structure, with the objective of seeking structural abstractions that might serve as the building blocks of the physical universe. In particular, he examined abstractions of lattices, waves, Borel sets and importantly, the calculus of propositions. Additionally, Wheeler also introduced what he referred to as "Observership". This he deemed as crucial for any physical theory [9] (see [25] for a recent discussion on Wheeler's ideas of observership and their relation to our experiences of space and time). Indeed, some of these ideas eventually led him to his now well-known "mantra": "It from Bit".

Yet another angle that inspired Wheeler to consider pregeometry as the foundation of all of physics came from a gravitational collapse argument ("the crisis of collapse"): If the universe can collapse, then it will take space, time, matter, and law with it; therefore, there needs to be something that transcends these in such a way as to not be subject to the same demise (see [24] for an insightful historical overview). Moreover, this something (the pregeometry) needs to be such that it can forge a way for the universe to come into being. It must provide building blocks. Hence, it serves cosmogonical and cosmological functions in Wheeler's thinking. This crisis of collapse curtails the kind of approach Einstein was attempting (which Wheeler also initially pursued in his earlier geometrodynamical investigations), in which space itself is the primordial substance from which all else is constructed. Despite the impressive topological gymnastics involved in constructing mass, charge, and (with extreme difficulty) spin from space alone, contemporary theory is simply inadequate to the task when the collapse problem is faced.

The sum-over-histories ideas were part of an initial simple attempt to quantize gravity using what would now be recognized as integrating over moduli spaces of geometries and fields. It was largely Wheeler's PhD student Charles Misner who did this early work on quantum gravity [64]. The aim was to try and generate more structure in space by allowing for quantum fluctuations of the geometrical (and topological) properties, in order to produce multiply-connected wormholes that could be used to thread fields and explain how point-charges can appear to emerge in a purely continuous field theory. However, in addition to the collapse problem, Wheeler was also motivated by the a desire for the deeper constructibility of the world of physics. As he put it using an analogy:

> Glass comes out of the rolling mill looking like a beautifully transparent and homogeneous elastic substance. Yet we know that elasticity is not the correct description of reality at the microscopic level. Riemannian geometry likewise provides a beautiful vision of reality; but ... is inadequate to serve as primordial building material.
>
> [67, p. 544]

Central to this new approach to physics, in which one seeks the deepest level of structure, is the idea that one ought not to start from the upper levels in order to figure things out. In other words, part and parcel of Wheeler's approach was a quite radical constructivism. It is no good, from this point of view, to consider conventional quantization approaches in which one *begins* with the classical system and then applies a procedure to it. This corresponds to our artificial methodology, rather than nature's own technique for creating the world which is, after all, already quantum. What pregeometry and constructivism share, and what we also share, is the belief that our aim, in foundational work, must be to find the methods and materials that nature herself uses to build the world.

It is worth remarking that many of the contemporary theories of quantum gravity, in fact, are set up with a fair share of *a priori* geometric structures (see examples (a) – (c) mentioned earlier). On the other hand, a truly pregeometric description of the kind Wheeler had in mind, ought to be one from which all geometric features of the physical universe should be derived ([63] discusses this point at length). Hence, it is the precursors of geometry (and one may argue, even topology) that make up the genuinely pregeometric building blocks of the universe. Now, given the common expectation that blending general relativity and quantum mechanics would permit "foam-like" realizations of geometry, at energies close to those that existed in the early universe, this implies, at the very least, that a set of more fundamental rules regarding connectivity of spacetime that are independent of topology and dimensionality, are required (as emphasized by Wheeler). Formulations of theoretical physics based on pregeometric structures then allow one to work with deeper underlying rules that are not dependent on classical structural assumptions about the properties of space and time. As we shall see, such an approach capitalizes upon deep connections between theoretical physics, computation, proof theory, and homotopy. It is such pregeometric entities that we will hereon in refer to as "Structureless Structures".[4]

7.5 Pregeometry in metaphysics

Of course, the idea that "something" has to transcend space, time, and matter has long since been part of philosophical discourse, in particular, metaphysics [75]. One of the early proponents of relationalism in the metaphysics of space and time, was none other than Gottfried Wilhelm Leibniz (in contrast to his contemporary at the time, Isaac Newton). Leibniz, however, was thinking of something beyond merely material or physical relationalism. Those views led him to the concept of "monads" [57]. Leibniz's *Monadology* was an attempt to codify an entire philosophical system. For all its encompassing majesty, it was notable for its extreme brevity.[5] The monadology can in all likelihood also be viewed as the first example of a pre-space theory. Leibniz argued in favor of a set of features of space from principles applying to a set of relations that are not spatial themselves. That is, the relations between monads are used to set up a correspondence to phenomenal space, with its characteristic features such as extension[6]. In this sense, his monadological theory of space is more primitive than the usual kind of physical relationalism, in which relations are simply thought to involve the objects of physics, with spatial relations being secondary to those objects (i.e. supervenient) in an ontological sense.[7] In monadology, a different kind of object ends up being primary (i.e. the monad is fundamental), and the relations hold between these such that space, the objects of physics, and every other thing in the manifest world emerge from this more basic layer. Moreover, monads are *simple* ("the true atoms of nature") in the sense of admitting no further reduction or decomposition into other elements. That is, they have no structural elements of their own and so provide a kind of structureless structure. And yet, from this simple foundation, according to Leibniz, we can generate all of the incredible complexity of the world.

The monads collectively provide all possible perspectives of a world, as tiny independent mirrors (or points of view). However, there is also a sense in which the monads are carrying out a pre-set program (or entelechy), coordinated with all other monads, in a pre-established and divinely choreographed dance determined to generate (i.e. construct) the best of all possible worlds. While there are the well-known principles of sufficient reason and identity of indiscernibles providing basic constraints on this construction, the principles themselves do not directly determine what is constructed. Rather, they inform the composition of monads into complex structures which is then carried out through the pre-established harmony. A major reason for the introduction of pre-established harmony was to explain the mind-body (or soul-body) correlations. For Leibniz there was no causal link and the correlation simply follows from the common cause in which both were set on their way.

In this context, it is interesting to note the philosophical parallels of Leibniz's metaphysics with Wolfram model of physics (or rather, "pre-physics") [92]. Analogous to Leibniz's monads, in the Wolfram model, abstract rewriting events comprise the "atoms of nature". These events are generated by rewriting rules that realize abstract computation. Based on local rule application, rewriting events are knitted together via causal relations. These are the causal graphs of the Wolfram model. The irreducible rewriting events and their mutual relations, taken in some appropriate limit, are hypothesized as models of spacetime geometries [13, 17, 18]. Furthermore, Wolfram model has a remarkably similar explanation for the correspondence of the world to the mind in that they both emerge from the same initial rules for construction and emerge in parallel with the mind (or observer) simply sampling the world and providing a perspective [93, 94], much like a monad, where different observers perceive the whole universe from different points of view. Likewise, one can find a similar generation of variety in the Wolfram model through this dislocation of a single, unified structure into many points of view [71].

Of course, Leibniz's theory, as it stands, cannot provide a satisfactory foundation for physics. At least, not one of much practical value in terms of showing how our present theories and phenomena can be *constructed*. Our aim in this work is to discuss some developments, including very recent ones, in this direction. Ultimately, the approach we focus on, that is, structureless structure from formal language, places the ontological weight on the very rules of construction themselves. By contrast with Leibniz's "God as architect" (as he puts it in S.89 of his *Monadology*), here the metaphor is better expressed as "Nature constructing itself", in particular, space, time, matter, and law.

Besides Leibniz, notions of pre-physical substrates of existence have also been discussed in the philosophy of mind, in particular, ideas related to monism of mind and matter [19]. In the context of the mind-body problem, the philosophy of monism seeks to resolve the metaphysical debate between physicalism and idealism by proposing a fundamentally new neutral substrate that is by itself neither physical nor mental, but instead, whose various manifestations then realize the physical and mental components of the world. A prominent example of this school of thought is dual-aspect monism proposed by the physicist Wolfgang Pauli and the psychologist Carl Jung. Dual-aspect monism posits that the physical and mental are merely complementary perspectives of an underlying neutral substrate [19]. In other words, the physical universe, including mental states of agents within it, is to be built upon a metaphysically fundamental layer of reality that is both pre-physical and pre-mental. From this perspective, monism necessitates pregeometric building blocks for our perceived reality.

7.6 Pregeometric theories from formal language

What then should constitute the essence of pregeometric structure from which space, time, and matter all emerge? For one, pregeometry being structureless structure, cannot arise from yet another unbeknownst physical substrate. A foundationalist philosophy of pregeometry can potentially admit an irreducible underlying substrate. But then, one would need to posit that the existence of such a substrate be accepted as a "universal self-evident truth", such that one cannot ask further questions about its nature or origins. This may seem a rather unsettling predicament to have to accept within a scientific theory. Furthermore, even if such a universal self-evident entity existed, how should one describe it (as opposed to explaining it) within a given theory without alluding to any spatial or temporal notions, including internal spaces (those describing internal degrees of freedom corresponding to internal symmetries, spin or gauge indices)? On the other hand, a coherentist philosophical stance places precedence on relations rather than objects and posits that structure emerges from the metaphysics of abstract relations. In a philosophy of this kind, emphasis is placed on the ontology of relations rather than the ontology of objects.

It turns out that the appropriate mathematical framework to formalize a theory of abstract relations and their properties is what is called a "Formal System" with well-defined rules of compositionality. Formal languages are precisely such systems. Formal systems lie at the heart of mathematical logic, computer science, cryptography, and several other branches of mathematics. More pertinently, formal language, and in particular, homotopy type theory and the univalent foundations program have been at the forefront of important recent advances seeking a new constructivist foundation for all mathematics [68, 83, 84]. Here, we seek to identify appropriate parallels arising from developments in mathematics to the foundations of physics. We reckon that the application of homotopy type theory and its representation in higher category theory will be extremely useful for:
(i) Exploring higher symmetries and spaces in physics, that cannot readily be captured by current methods; and for
(ii) Seeking a constructivist foundation for physics, where structures intrinsic to notions of the continuum are not fundamental, but emerge within well-defined limits.

While Wheeler himself had suggested a propositional calculus as a pregeometric framework from which the emergence of physical structures may be sought (though he ultimately had to introduce a "participator" to deal with undecidable propositions of physics), our contention here is that formal language permits the expression of generic pregeometric calculi. As mentioned, this parallels the way mathematicians discuss universes of mathematics using homotopy type theory. A formal language encompasses a system of primitive symbols (or ground types) along with relations for constructing composite types which can be used to construct clauses and sentences. The latter constitute propositions expressible within the language. A language can additionally be equipped with axioms and inference rules for a deductive logic using which one can reason about its propositions. However, propositions are only declarative statements. One can go further. Including variables and quantifiers allows one to extend a propositional system to one that includes predicates, thus expressing formulae, whose validity (truth) may subsequently be evaluated within a specified interpretation (semantic modality). Based on the logical relations and inference rules that a given language admits, one can then construct proofs relating one formula to another, that is, prove theorems within that language.

Besides Wheeler's pregeometry, the role of formal language toward conceptualizing new foundations for quantum theory and physics on discrete spaces has been extensively investigated by Chris Isham and collaborators [40, 53]. Rather than pregeometry per se, the motivations for the latter arose in seeking an axiomatization of physical theories within a common mathematical framework – that of topos theory. Toposes are categories that behave like **sets** (the category of sets). Like **sets**, toposes are equipped with the category-theoretic analog of Cartesian products, disjoint unions, a singleton set, a notion of a set of functions, and importantly, a notion of sub-sets of objects (i.e., sub-objects). Thus, toposes are formal "places" where foundations of mathematics can be formulated. Examples of toposes other than **sets** are the category of finite graphs, the category of G-sets, and the category of presheaves over a small category.

Even more generally, attempts seeking a formal axiomatic framework for physical theories, pre-date Wheeler, going all the way back to David Hilbert. In his 1900 address at the International Congress of Mathematics in Paris, Hilbert stated his famous 23 open problems of mathematics. Of these, the sixth problem referred to a universal axiomatization of physics (see [31] for a historical overview). Apart from the issue of whether such axiomatizations ought to be universal or even complete, it set the course for seeking mathematical formulations of physical theories using a common (axiomatic and/or inferential) framework. Then, toward the latter half of the 20th century, with rapid advances in category theory, William Lawvere, one of the founders of categorical logic, sought to build the foundations of mathematics in topos theory (as opposed to set theory) [55]. Lawvere was also interested in applications of topos theory toward the formalization of physics [56], which was subsequently followed up by Isham and others (as noted above).

It is worth noting that while the topos-theoretic foundations discussed here offer the elegant possibility of expressing theories of physics through a "mathematically unified" framework, they do not carry the usual baggage of grand unification of physical theories. This allows for a potentially background independent formulation of a broad class of physical theories. In particular, Isham's work proposes specifically distinct toposes for classical and quantum mechanics [36, 37]. The key objective of their program was to do away with any a priori use of continuous spatial or temporal constructs in formulating notions related to classical or quantum systems. As stated in [36], "the use of continuous properties associated to space and time would be deemed a major error if those turned out to be fundamentally incompatible with what is needed for a theory of quantum gravity". The contention there was that theories of a physical system should be formulated within a topos that depends on both, the theory-type and the system-type. In turn, any topos-theoretic approach employs formal language. This is because of a well-known result in topos theory that there exists an internal formal language associated to each topos [54]. In fact, not only does each topos generate an internal language, but, conversely, a language satisfying appropriate conditions generates a topos [54]. The goal in [40] was to find a novel structural frameworks within which new types of theory can be constructed, and in which continuum quantities play no fundamental role. These works proposed an abstract language-theoretic formulation of classical and quantum mechanics which primarily addresses questions related to kinematics of classical and quantum systems in arbitrary spaces. Going beyond this kinematical description, the question is how does one generalize topos-theoretic approaches to address pregeometric theories as well as other effective theories at high energies?

More generally, the kind of languages admissible in toposes are typed languages. Type theory provides the building blocks to formally construct such languages. Given their constructivist flavor, the logic expressed by type theories in toposes is intuitionistic logic. This means one need not enforce the law of the excluded middle or the axiom of choice in these formal systems. Furthermore, the natural extension of intuitionistic type theory is homotopy type theory, which includes homotopy n-types, up to ∞-types. The representation of homotopy types takes us beyond the realm of standard category theory to higher category theory, which includes morphisms between morphism (representing homotopies between types). This tower of higher morphisms goes all the way to ∞-categories. Formal languages expressed in homotopy type theory are internalized in higher categories, and consequently higher toposes.

How then do topological and geometric spaces relevant to physics (and mathematics) arise from type-theoretic building blocks? One of the key takeaways from the synthetic geometry and homotopy type theory program is that the notion of space arises from functorial constructions involving ∞-toposes [79, 81, 84]. Geometry is thus inherited from higher structures, and induced upon local structures by taking sections or projections of the total space [17, 18]. Homotopy type theory provides a syntactic formalism for realizing higher structures. The objects of the ∞-topos under consideration are the so-called "∞-groupoids". The latter are categories endowed with a tower of higher morphisms, up to infinity (and invertibility conditions). Via Grothendieck's hypothesis, ∞-groupoids realize models of formal topological spaces [21]. With additional "cohesivity conditions", one also obtains synthetic geometric spaces in ∞-toposes from this construction [79–81, 83, 84]. These authors also show how quantum field theories with higher gauge symmetries can be formalized in ∞-toposes [79–81]. Higher homotopical structures in formal languages expressed using higher categories thus provide us a useful formal framework for constructing pregeometric physics as well as theories of higher symmetries.

Furthermore, a computational realization of the above ∞-groupoid constructions was shown in [13, 17, 18]. This construction was based on what are called "Multiway Systems", the non-deterministic rewriting systems of the Wolfram model [91, 92]. Using a type-theoretic representation of multiway rewriting systems, the authors of [17] provide an algorithmic construction of higher homotopies on non-deterministic rewriting systems. This connection between abstract rewriting systems and higher homotopies suggests a way to realize spatial structures and geometry from purely pregeometric models such as those based on rewriting systems (mentioned above).

7.7 Outlook and discussion

In conclusion, this work serves as an initial metaphysical exploration of a plausible description of pregeometric building blocks for the physics of spacetime, based on formal language. We have put forth the proposal that syntactic structures formalized in computational languages model the kind of pregeometric structures that Wheeler had in mind concerning the foundations of physics. We described these pregeometric structures as structureless structure to emphasize the necessity to shift away from a fundamentally material ontology. Instead, these are symbolic and relational structures of a formal language.

Our approach to pregeometry takes seriously a constructivist stance on the laws and structures of the physical universe; not merely in terms of how observers may perceive the universe, but more importantly, in metaphysical terms, as to how these laws and structures might come into being. Such a perspective closely aligns with Wheeler's intuitions of pregeometry as something that transcends space, time, matter, and law. Indeed, this flavor of constructivism, not surprisingly, resembles the kind of constructivism that has been recognized in recent advances in the foundations of mathematics, particularly in the context of homotopy type theory and the univalence foundations program [68]. Speaking of the "unreasonable effectiveness of mathematics" point of view [90], it is perhaps fitting that the metaphysics of spacetime geometry directly draws from formal advances in metamathematics. A computational realization of this connection between metamathematics and physics in terms of rewriting systems can be found in [95]. All in all, a language-theoretic constructivist framing serves as an important conceptual advance for approaches that emphasize the interplay between computation and physics, such as the Wolfram model.

Given that the coherentist constructivism discussed here follows from requiring to go beyond a materialist ontology for pregeometry, this implies that primitives of pregeometry are not merely discretizations of classical structures in physics. Its about nature constructing itself from abstract computation using syntactic compositions and relations. Simply replacing classical and quantum systems on continuous spaces with their discrete counterparts is unlikely to capture the full essence of Planck scale physics.[8] The plausible emergence of space, time, and matter, and theories describing them at or below the Planck length, will likely require new mathematical formalisms that go beyond mere replacements of classical real or complex analysis with discrete geometry. Recent advances in homotopy theory, higher algebra, and topos theory offer new mathematical methods for such investigations (see [61, 73, 74] for works in this direction). Also, worth noting that the strict dichotomy between continuous versus discrete geometry may be a bit misleading given that there exist examples of geometric formalisms that are by themselves neither continuous nor discrete, such as operator algebras that realize "pointless geometry" [16, 30]. Rather, it is the representation and spectrum of these operators that may take continuous or discrete values under different conditions. This feature has been exploited in models of quantum gravity such as loop quantum gravity, noncommutative geometry, and group field theories [30, 66, 77].

As mentioned earlier, approaches advocating the use of formal language to conceptualize the foundations of physics, by themselves, are not new. However, the newly developing mathematical formalism of homotopy type theory [68], extended topological field theories [62], operator mechanics [16, 27], infinity-categories [74], infinity-toposes [79], higher-arity algebras [96–98], etc. offer new ways to investigate pregeometric structures formalized in computational languages. A recurring theme in these investigations is that of higher structures. A language-theoretic pregeometric formalism based on higher structures will likely bridge, or at the very least, help identify crucial intersections between existing constructivist and background-independent approaches to quantum gravity.

Acknowledgments

We thank Stephen Wolfram for his enthusiasm, encouragement and discussions on the manuscript. We also like to thank Alex Blum for many useful suggestions and comments on this work.

This research has been supported by the following grants: The Foundational Questions Institute and Fetzer-Franklin Fund, a donor advised fund of Silicon Valley Community Foundation [FQXi-RFP-1817]; The John Templeton Foundation [Grant ID 62106]; and the Australian Research Council [Grant DP210100919].

Notes

1. This is soon to be reported in a forthcoming publication.
2. For our purposes here, we restrict our discussion to the fundamental physics of particles, fields, and geometry. One could reasonably well build a case in favor of including principles from condensed matter theory. And in fact, that may well be the logical extension of the work discussed in this chapter.
3. See [26] for a nice historical account of Heisenberg's Weltformel (World Formula), a final theory reducing all of physics, known and unknown, to the interactions of one elementary quantum field.
4. Here we arrive at this with a focus on pregeometry. However, connections between physics, computation, and formal systems have been discussed in other contexts too. For instance, relating to undecidable dynamics and the edge of chaos in [69]; or founded on monoidal categories, quantum processes, and cobordisms in [20].
5. In fact, it was so brief that Leibniz later added annotations pointing the reader to other works for clarification.
6. See [75] for a detailed study of Leibniz's deeper philosophy of space.
7. Wheeler himself was influenced by some of Leibniz's ideas on space, time, and matter. A historical account of this intersection of ideas can be found in [44].
8. It is well-known that quantization itself isn't solely about discretizing a system or about replacing a system of equations in classical analysis with those in discrete analysis.

References

[1] Abramsky, S., & Coecke, B. (2009). Categorical Quantum Mechanics. In K. Engesser, D. M. Gabbay, & D. Lehmann (Eds.), *Handbook of Quantum Logic and Quantum Structures*, (Vol. 2, pp. 261–325). Elsevier.

[2] Addazi, A., Alvarez-Muñiz, J., Batista, R. A., Amelino-Camelia, G., Antonelli, V., Arzano, M., Asorey, M., Atteia, J. L., Bahamonde, S., Bajardi, F., et al. (2022). Quantum Gravity Phenomenology at the Dawn of the Multi-Messenger Era – A Review. *Progress in Particle and Nuclear Physics*, 125, 103948.

[3] Aharony, O., Gubser, S. S., Maldacena, J., Ooguri, H., & Oz, Y. (2000). Large N Field Theories, String Theory and Gravity. *Physics Reports*, 323(3–4), 183–386.

[4] Ahrens, B., North, P. R., Shulman, M., & Tsementzis, D. (2021). The Univalence Principle. `https://arxiv.org/abs/2102.06275`

[5] Albert, D. Z. (2013). Wave Function Realism. In A. Ney & D. Z. Albert (Eds.), *The Wave Function: Essays on the Metaphysics of Quantum Mechanics* (pp. 52–57). Oxford University Press.
[6] Altarelli, G., & Feruglio, F. (2001). SU(5) Grand Unification in Extra Dimensions and Proton Decay. *Physics Letters B*, 511(2–4), 257–264.
[7] Anderson, E. (2012). Problem of Time in Quantum Gravity. *Annalen der Physik*, 524(12), 757–786.
[8] Angelo, R. M., & Ribeiro, A. D. (2015). Wave–Particle Duality: An Information-Based Approach. *Foundations of Physics*, 45, 1407–1420.
[9] Wheeler, J. A. (1977). Genesis and Observership. In R. E. Butts & J. Hintikka (Eds.), *Foundational Problems in the Special Sciences: Part Two of the Proceedings of the Fifth International Congress of Logic, Methodology and Philosophy of Science* (pp. 3–33). Springer.
[10] Arsiwalla, X. D. (2008). More Rings to Rule Them All: Fragmentation, 4D $\leftrightarrow$ 5D and Split-Spectral Flows. *Journal of High Energy Physics*, 2008(02), 066.
[11] Arsiwalla, X. D. (2009). Entropy Functions with 5D Chern-Simons Terms. *Journal of High Energy Physics*, 2009(09), 059.
[12] Arsiwalla, X. D. (2010). Supersymmetric Black Holes as Probes of Quantum Gravity. PhD Thesis, University of Amsterdam. `https://pure.uva.nl/ws/files/871677/75420_thesis.pdf`
[13] Arsiwalla, X. D. (2020). Homotopic Foundations of Wolfram Models. *Wolfram Community*. `https://community.wolfram.com/groups/-/m/t/2032113`
[14] Arsiwalla, X. D., Boels, R., Mariño, M., & Sinkovics, A. (2006). Phase transitions in Q-deformed 2D Yang-Mills Theory and Topological Strings. *Physical Review D*, 73(2), 026005.
[15] Arsiwalla, X. D., de Boer, J., Papadodimas, K., & Verlinde, E. (2011). Degenerate Stars and Gravitational Collapse in AdS/CFT. *Journal of High Energy Physics*, 2011(1), 066.
[16] Arsiwalla, X. D., Chester, D., & Kauffman, L. H. (2022). On the Operator Origins of Classical and Quantum Wave Functions. `https://arxiv.org/abs/2211.01838`
[17] Arsiwalla, X. D., & Gorard, J. (2021). Pregeometric Spaces from Wolfram Model Rewriting Systems as Homotopy Types. `https://arxiv.org/abs/2111.03460`
[18] Arsiwalla, X. D., Gorard, J., & Elshatlawy, H. (2021). Homotopies in Multiway (Nondeterministic) Rewriting Systems as n-fold Categories. `https://arxiv.org/abs/2105.10822`
[19] Atmanspacher, H., & Rickles, D. (2022). *Dual-Aspect Monism and the Deep Structure of Meaning*. Routledge.

[20] Baez, J., & Stay, M. (2011). Physics, Topology, Logic and Computation: A Rosetta Stone. In *New Structures for Physics* (Vol. 813, pp. 95–172). Springer.

[21] Baez, J. (2007). The Homotopy Hypothesis. Fields Institute. Available at `http://math.ucr.edu/home/baez/homotopy/`

[22] Beasley, C., Heckman, J. J., & Vafa, C. (2009). GUTs and Exceptional Branes in F-theory – I. *Journal of High Energy Physics*, 2009(01), 058.

[23] Becker, K., Becker, M., & Schwarz, J. H. (2006). *String Theory and M-Theory: A Modern Introduction*. Cambridge University Press.

[24] Blum, A., & Furlan, S. (2022). How John Wheeler Lost His Faith in the Law. In *Rethinking the Concept of Law of Nature: Natural Order in the Light of Contemporary Science* (pp. 283–322). Springer.

[25] Blum, A. S., Renn, J., & Schemmel, M. (2016). *Experience and Representation in Modern Physics: The Reshaping of Space*. Max-Planck-Institut für Wissenschaftsgeschichte

[26] Blum, A. S. (2019). *Heisenberg's 1958 Weltformel and the Roots of Post-Empirical Physics*. Springer.

[27] Chester, D., Arsiwalla, X. D., Kauffman, L., Planat, M., & Irwin, K. (2023). Quantization of a New Canonical, Covariant, and Symplectic Hamiltonian Density. `https://arxiv.org/abs/2305.08864`

[28] Coecke, B., & Duncan, R. (2011). Interacting Quantum Observables: Categorical Algebra and Diagrammatics. *New Journal of Physics*, 13, 043016.

[29] Coecke, B., & Kissinger, A. (2018). Picturing Quantum Processes. In *International Conference on Theory and Application of Diagrams* (pp. 28–31). Springer.

[30] Connes, A. (1994). *Noncommutative Geometry*. Springer.

[31] Corry, L. (1997). David Hilbert and the Axiomatization of Physics (1894–1905). *Archive for History of Exact Sciences*, 51, 83–198.

[32] Cortés, M., & Smolin, L. (2014). The Universe as a Process of Unique Events. *Physical Review D*, 90(8), 084007.

[33] De Swart, J., Bertone, G., & van Dongen, J. (2017). How Dark Matter Came to Matter. *Nature Astronomy*, 1(3), 0059.

[34] Deutsch, D. (2013). Constructor theory. *Synthese*, 190(18), 4331–4359.

[35] Dijkgraaf, R., & Vafa, C. (2002). Matrix Models, Topological Strings, and Supersymmetric Gauge Theories. *Nuclear Physics B*, 644(1–2), 3–20.

[36] Döring, A., & Isham, C. J. (2008). A Topos Foundation for Theories of Physics: I. Formal Languages for Physics. *Journal of Mathematical Physics*, 49(5), 053515.

[37] Döring, A., & Isham, C. J. (2008). A Topos Foundation for Theories of Physics: II. Daseinisation and the Liberation of Quantum Theory. *Journal of Mathematical Physics*, 49(5), 053516.

[38] Döring, A., & Isham, C. J. (2008). A Topos Foundation for Theories of Physics: III. The Representation of Physical Quantities with Arrows $\delta^o(A) : \Sigma \to \mathbb{R}^{\geq}$. *Journal of Mathematical Physics*, 49(5), 053517.

[39] Döring, A., & Isham, C. J. (2008). A Topos Foundation for Theories of Physics: IV. Categories of Systems. *Journal of Mathematical Physics*, 49(5), 053518.

[40] Döring, A., & Isham, C. J. (2010). "What Is a Thing?": Topos Theory in the Foundations of Physics. In Bob Coecke (Ed.), *New Structures for Physics* (pp. 753–937). Springer.

[41] Dowker, F. (2006). Causal Sets as Discrete Spacetime. *Contemporary Physics*, 47(1), 1–9.

[42] Froggatt, C., & Nielsen, H. (2003). Trying to Understand the Standard Model Parameters. *Surveys in High Energy Physics*, 18(1–4), 55–75.

[43] Furey, C. (2016). Standard Model Physics from an Algebra? `https://arxiv.org/abs/1611.09182`

[44] Furlan, S. (2020). Merging Labyrinths. *Studia Leibnitiana*, 52(1/2), 123–155.

[45] Gibbs, P. E. (1995). The Small Scale Structure of Space-Time: A Bibliographical Review. `https://arxiv.org/abs/hep-th/9506171`

[46] Gisin, N. (2021). Indeterminism in Physics and Intuitionistic Mathematics. *Synthese*, 199(5–6), 13345–13371.

[47] Gorard, J., Namuduri, M., & Arsiwalla, X. D. (2020). ZX-Calculus and Extended Hypergraph Rewriting Systems I: A Multiway Approach to Categorical Quantum Information Theory. `https://arxiv.org/abs/2010.02752`

[48] Gorard, J., Namuduri, M., & Arsiwalla, X. D. (2021). Fast Automated Reasoning over String Diagrams Using Multiway Causal Structure. `https://arxiv.org/abs/2105.04057`

[49] Gorard, J., Namuduri, M., & Arsiwalla, X. D. (2021). ZX-Calculus and Extended Wolfram Model Systems II: Fast Diagrammatic Reasoning with an Application to Quantum Circuit Simplification. *arXiv preprint arXiv:2103.15820*. `https://arxiv.org/abs/2103.15820`

[50] Griffiths, R. B. (2019). Quantum Measurements and Contextuality. *Philosophical Transactions of the Royal Society A*, 377(2157), 20190033.

[51] Hasan, A., & Fumerton, R. (2022). Foundationalist Theories of Epistemic Justification. In E. N. Zalta & U. Nodelman (Eds.), *The Stanford Encyclopedia of Philosophy* (Fall 2022 Edition). Metaphysics Research Lab, Stanford University. `https://plato.stanford.edu/entries/justep-foundational/`

[52] Isham, C. (1995). Structural Issues in Quantum Gravity. `https://arxiv.org/abs/gr-qc/9510063`

[53] Isham, C. J., & Butterfield, J. (2000). Some Possible Roles for Topos Theory in Quantum Theory and Quantum Gravity. *Foundations of Physics*, 30(10), 1707–1735.

[54] Johnstone, P. T. (2014). *Topos Theory*. Courier Corporation.

[55] Lawvere, F. W. (1966). The Category of Categories as a Foundation for Mathematics. In S. Eilenberg, D. K. Harrison, S. MacLane, & H. Röhrl (Eds.), *Proceedings of the Conference on Categorical Algebra: La Jolla 1965* (pp. 1–20). Springer.

[56] Lawvere, F. W. (1997). Toposes of Laws of Motion. American Mathematical Society, Transcript from Video, Montreal, September 27.

[57] Leibniz, G. W. (1989). *The Monadology: 1714 (Reprinted Edition)*. Springer.

[58] Lisi, A. G., Smolin, L., & Speziale, S. (2010). Unification of Gravity, Gauge Fields and Higgs Bosons. *Journal of Physics A: Mathematical and Theoretical*, 43(44), 445401.

[59] Loll, R., Fabiano, G., Frattulillo, D., & Wagner, F. (2022). Quantum Gravity in 30 Questions. `https://arxiv.org/abs/2206.06762`

[60] Loll, R. (2019). Quantum Gravity from Causal Dynamical Triangulations: A Review. *Classical and Quantum Gravity*, 37(1), 013002.

[61] Lurie, J. (2009). *Higher Topos Theory*. Princeton University Press.

[62] Lurie, J. (2009). On the Classification of Topological Field Theories. In David Jerison, Barry Mazur, Tomasz Mrowka, Wilfried Schmid, Richard P. Stanley, & Shing-Tung Yau (Eds.), *Current Developments in Mathematics, 2008* (pp. 129–280). International Press of Boston.

[63] Meschini, D., Lehto, M., & Piilonen, J. (2005). Geometry, Pregeometry and Beyond. *Studies in History and Philosophy of Modern Physics*, 36(3), 435–464.

[64] Misner, C. W., Thorne, K. S., & Wheeler, J. A. (1973). *Gravitation*. Macmillan.

[65] Olsson, E. (2021). Coherentist Theories of Epistemic Justification. In E. N. Zalta (Ed.), *The Stanford Encyclopedia of Philosophy* (Fall 2021 Edition). Metaphysics Research Lab, Stanford University. `https://plato.stanford.edu/entries/justep-coherence/`

[66] Oriti, D. (2006). The Group Field Theory Approach to Quantum Gravity. `https://arxiv.org/abs/gr-qc/0607032`

[67] Patton, C. M., & Wheeler, J. A. (1975). Is Physics Legislated by Cosmogony? In *Quantum Gravity* (pp. 538–605). Academic Press.

[68] Univalent Foundations Program. (2013). *Homotopy Type Theory: Univalent Foundations of Mathematics*. Institute for Advanced Study. `https://homotopytypetheory.org/book/`

[69] Prokopenko, M., Harré, M., Lizier, J., Boschetti, F., Peppas, P., & Kauffman, S. (2019). Self-Referential Basis of Undecidable Dynamics: From the Liar Paradox and the Halting Problem to the Edge of Chaos. *Physics of Life Reviews*, 31, 134–156.

[70] Randall, L. (2006). *Warped Passages: Unravelling the Universe's Hidden Dimensions*. Penguin UK.

[71] Rickles, D., Elshatlawy, H., & Arsiwalla, X. D. (2025). Ruliology: Linking Computation, Observers and Physical Law. `https://arxiv.org/abs/2308.16068`

[72] Rideout, D., & Wallden, P. (2009). Emergence of Spatial Structure from Causal Sets. In *Journal of Physics: Conference Series*, 174, 012017. IOP Publishing.

[73] Riehl, E., & Shulman, M. (2017). A Type Theory for Synthetic ∞-Categories. *Higher Structures*, 1(1), 147–224.

[74] Riehl, E., & Verity, D. (2018). Elements of ∞-Category Theory. Preprint available at `http://www.math.jhu.edu/~eriehl/elements.pdf`

[75] Risi, V. (2007) *Geometry and Monadology: Leibniz's Analysis Situs and Philosophy of Space*. Springer.

[76] Rovelli, C. (1998). Strings, Loops and Others: A Critical Survey of the Present Approaches to Quantum Gravity. *arXiv preprint* gr-qc/9803024

[77] Rovelli, C. (2008). Loop Quantum Gravity. *Living Reviews in Relativity*, 11(1), 1–69.

[78] Sanders, R. H., & McGaugh, S. S. (2002). Modified Newtonian Dynamics as an Alternative to Dark Matter. *Annual Review of Astronomy and Astrophysics*, 40(1), 263–317.

[79] Schreiber, U. (2013). Differential Cohomology in a Cohesive Infinity-Topos. `https://arxiv.org/abs/1310.7930`

[80] Schreiber, U. (2016). Higher Prequantum Geometry. New Spaces in Physics. Formal and Conceptual Reflections, Cambridge University Press, 202–278.

[81] Schreiber, U., & Shulman, M. (2012). Quantum Gauge Field Theory in Cohesive Homotopy Type Theory. *EPTCS*, 158, 109–126.

[82] Sharma, A., Czegel, D., Lachmann, M., Kempes, C. P., Walker, S. I., & Cronin, L. (2023). Assembly Theory Explains and Quantifies Selection and Evolution. *Nature*, 629, 518–525.

[83] Shulman, M. (2016). Homotopy Type Theory: A Synthetic Approach to Higher Equalities. In *Homotopy Type Theory and Univalent Foundations* (pp. 183–210). Oxford University Press.

[84] Shulman, M. (2017). Homotopy Type Theory: The Logic of Space. In M. Anel & G. Catren (Eds.), *New Spaces in Mathematics: Formal and Conceptual Reflections* (pp. 322–404). Cambridge University Press.

[85] Steup, M., & Neta, R. (2020). Epistemology. In E. N. Zalta (Ed.), *The Stanford Encyclopedia of Philosophy* (Fall 2020 Edition). Metaphysics Research Lab, Stanford University.

[86] Swingle, B. (2018). Spacetime from Entanglement. *Annual Review of Condensed Matter Physics*, 9, 345–358.

[87] Verlinde, E. (2017). Emergent Gravity and the Dark Universe. *SciPost Physics*, 2(3), 016.

[88] Weinberg, S. (1989). The Cosmological Constant Problem. *Reviews of Modern Physics*, 61(1), 1–23.

[89] Wheeler, J. A. (1980). Pregeometry: Motivations and Prospects. In A. R. Marlow (Ed.), *Quantum Theory and Gravitation*, edited volume. Academic Press.

[90] Wigner, E. P. (1960). The Unreasonable Effectiveness of Mathematics in the Natural Sciences. *Communications on Pure and Applied Mathematics*, XIII, 1–14.

[91] Wolfram, S. (2002). *A New Kind of Science*. Wolfram Media.

[92] Wolfram, S. (2020). A Class of Models with the Potential to Represent Fundamental Physics. *Complex Systems*, 29(2), 107–536. https://doi.org/10.25088/complexsystems.29.2.107

[93] Wolfram, S. (2021). The Concept of the Ruliad. `https://writings.stephenwolfram.com/2021/11/the-concept-of-the-ruliad/`

[94] Wolfram, S. (2021). What Is Consciousness? Some New Perspectives from Our Physics Project. `https://writings.stephenwolfram.com/2021/03/what-is-consciousness-some-new-perspectives-from-our-physics-project/`

[95] Wolfram, S. (2022). *Metamathematics: Foundations & Physicalization*. Wolfram Media.

[96] Zapata-Carratala, C., & Arsiwalla, X. D. (2022). An Invitation to Higher Arity Science. `https://arxiv.org/abs/2201.09738`

[97] Zapata-Carratala, C., Arsiwalla, X. D., & Beynon, T. (2022). Heaps of Fish: Arrays, Generalized Associativity and Heapoids.. `https://arxiv.org/abs/2205.05456`

[98] Zapata-Carratalá, C., Schich, M., Beynon, T., & Arsiwalla, X. D. (2023). Hypermatrix Algebra and Irreducible Arity in Higher-Order Systems: Concepts and Perspectives. *Advances in Complex Systems*, 26(06), 2350007.

8 Quantum Information Elements in Quantum Gravity States and Processes

Daniele Oriti

8.1 Introduction

The main goal of this contribution is to show how quantum gravity states and processes as identified by a number of quantum gravity formalisms can be recast in the language of quantum information and how entanglement or quantum correlations can then be seen, in the same formalisms, as essential in the very structure of quantum spacetime. It is not a review even of the few results we will summarize briefly, let alone of the substantial research done on entanglement and quantum information features in quantum gravity formalisms. For the latter, we refer to [1, 2] to remain limited to the results obtained in the quantum gravity context closer to our focus.

The perspective we find convenient to adopt, in order to appreciate the role of quantum information-theoretic structures in these quantum gravity formalisms, is that of emergent spacetime, i.e. of quantum gravity as a theory of 'spacetime constituents'with spacetime itself, geometry and fields as emergent entities [3–7]. This perspective is motivated by several results in semi-classical physics, for example, black hole thermodynamics and the information paradox, gravitational singularities, that all point in various ways to a breakdown of key notions on which standard continuum, geometric physics is based, and, more indirectly, the results of analogue gravity in condensed matter systems, showing how effective field theory on curved backgrounds can emerge rather generically from non-gravitational systems. It is also motivated by results in modern quantum gravity approaches, including the ones we

DOI: 10.4324/9781003546139-11

focus on in this contribution, with their combinatorial, algebraic and, indeed, (quantum) information-theoretic structures replacing geometric notions and spacetime-based quantum fields. Indeed, to give some examples, canonical loop quantum gravity replaces smooth metric geometries with piecewise-degenerate quantum twisted geometries encoded in combinatorial/algebraic data; lattice quantum gravity works with piecewise-flat quantum (often non-metric) geometries; string theory dualities also suggest that the fundamental degrees of freedom of M-theory are not spacetime-based; and AdS/CFT gives a concrete example of emergent gravity as well as a very partial 'emergent space', reconstructed from a lower-dimensional and non-gravitational CFT. This perspective also implies a shift away from the more traditional perspective that sees quantum gravity as the result of straightforwardly quantizing general relativity or some other classical gravitational and spacetime-based theory (whether perturbatively or non-perturbatively).[1] From an emergent spacetime perspective, a breakdown of spacetime notions, including locality, should be expected when moving to a more fundamental description. One of the key tasks is then to identify the hidden, possibly discrete microstructure replacing continuum spacetime fields in such more fundamental description of the universe, with such fields, including the metric, being then understood as collective entities and gravity and the rest of continuum spatiotemporal physics as an approximate effective description of collective dynamics; controlling such collective dynamics is the second key task. In other words, the universe itself is seen as a (background independent) quantum many-body system.

Tied to the notion of spacetime emergence is the further hypothesis that spacetime geometric and, possibly, topological structures can in fact emerge from the entanglement among more fundamental quantum constituents [8, 9], thus via a conjectured 'entanglement/geometry correspondence'. Support for this conjecture has been obtained mostly in a semi-classical context and in the AdS/CFT context, thus in the presence of well-defined spacetime and geometric notions, starting with the Ryu-Takanayagi entropy formula and related results [10]. However, they are suggestive of something more fundamental that calls for a concrete realization of this idea in full quantum gravity, thus in the absence of spacetime and fields as we know them. This call has been heard and partially answered, we would claim, in the quantum gravity formalisms on which the rest of this contribution will focus. Quantum correlations and, more generally, quantum information-theoretic notions acquire, in fact, a central role. These formalisms are canonical loop quantum gravity, spin foam models, group field theories and lattice quantum gravity in first order (tetrad-connection) variables. Despite several technical differences between them, they all share many basic features. We will discuss such shared features, and only occasionally point out specific differences; unless noted otherwise, we will only consider models of 4-dimensional quantum gravity and spacetime and a Lorentzian (as opposed to Euclidean/Riemannian) setting. We will first discuss the nature of quantum gravity states in these formalisms, and emphasize the role that entanglement among their constituents plays, their re-interpretation as quantum circuits and their use to define holographic maps and quantum information channels. Then, we will discuss the corresponding quantum gravity processes, indicating their possible formulation as quantum causal histories and, again, as quantum circuits, as well as the present limitations to such reformulation.

8.2 Quantum gravity states as entanglement networks and quantum circuits

In the quantum gravity formalisms, generic quantum gravity states can be represented as (superpositions of) entanglement networks of quantum geometric constituents (Fig. 8.1); more precisely, they are expressed by assigning algebraic data to a combinatorial graph, where the algebraic data are taken from the (representation) theory of (Lie) groups, notably the Lorentz group or the rotation subgroup thereof. In turn, they can be seen as composed of elementary quantum systems, with associated one-body Hilbert space, located on nodes of the graph, with the graph itself encoding a pattern of entanglement across the (sub-)systems living on the nodes.

The graphs are usually taken to be dual to 3-dimensional simplicial complexes in quantum gravity formalisms with a direct discrete geometric interpretation. In the following we restrict to such a case. Different formalisms, as

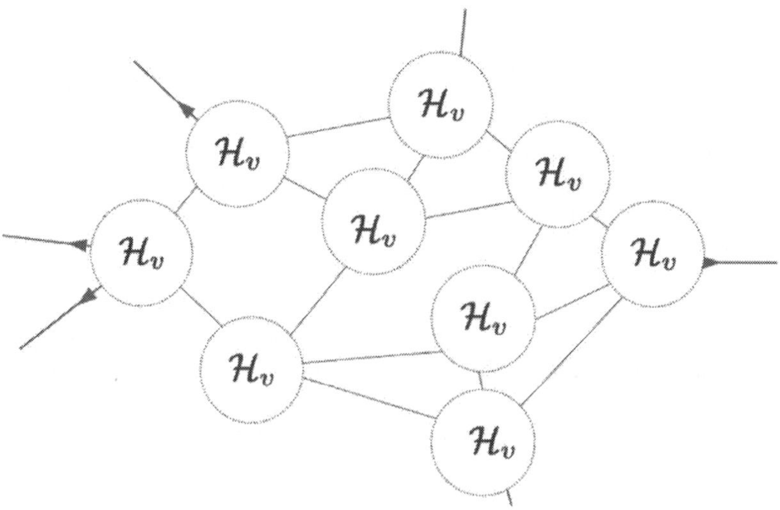

Figure 8.1 The general structure of quantum gravity states as many-body systems, with the underlying graph including both open links and links connecting pairs of nodes.

well as different models within a given formalism, will differ for the class of graphs being considered and for the choice of one-body Hilbert space associated to the nodes of graphs, as well as for the quantum dynamics assumed for such quantum gravity states. Leaving dynamical considerations aside, in this section, we now give more details on a specific (set of) proposal(s) for the kinematical structures, and illustrate their quantum information-theoretic aspect.

8.2.1 Entanglement patterns of atoms of space: Quantum space as a quantum circuit

The one-body Hilbert space can be taken to be space of states for a quantum tetrahedron, which can be constructed from (2) group data and expressed in terms of its irreducible representations:

$$\mathcal{H}_v = \bigoplus_{\vec{j}_v} \left(\otimes_{i=1}^{4} V^{j_v^i} \bigotimes \mathcal{I}^{\vec{j}_v} \right) \tag{8.1}$$

where one has a vector space $V^{j_v^i}$ for the representation label j_i (a 'spin'valued in the half-integers) for each of the four triangles of the tetrahedron, with canonical basis $|j^i, n^i\rangle$, then tensored together, and $\mathcal{I}^{\vec{j}_v} = Inv_G[V^{j_v^1} \otimes \cdots \otimes V^{j_v^4}]$ is the space of intertwiners, i.e. tensors invariant under the diagonal action of the group $G = (2)$, built from the same four representation spaces. This Hilbert space can be obtained from the direct quantization of the classical phase space of geometries of a single tetrahedron, parametrized by Lie algebra elements corresponding to normal vectors associated to its four triangles and conjugate group elements corresponding to elementary parallel transports of a (2) connection along paths dual to the same triangles. It can be depicted dually as a single vertex with four semi-links outgoing from it (each dual to one of the four triangles of the tetrahedron; see Fig. 8.2).

The discrete geometric interpretation is confirmed by the action of geometric operators encoding the tetrahedral geometries. For example, elements of the canonical basis in $V^{j_v^i}$ diagonalize the area of the corresponding triangle, while intertwiners encode information about the volume of the whole tetrahedron with the given triangle areas. For more details about this quantum geometry, see [11, 12]. Thus, generic states in the (kinematical) Hilbert space of the quantum gravity formalisms we consider here can be understood as quantum many-body states built out of this single-body Hilbert space.

In particular, an interesting class of quantum states are those that admit a natural interpretation as corresponding to quantum tetrahedra glued to one another across shared faces to form extended simplicial complexes dual to

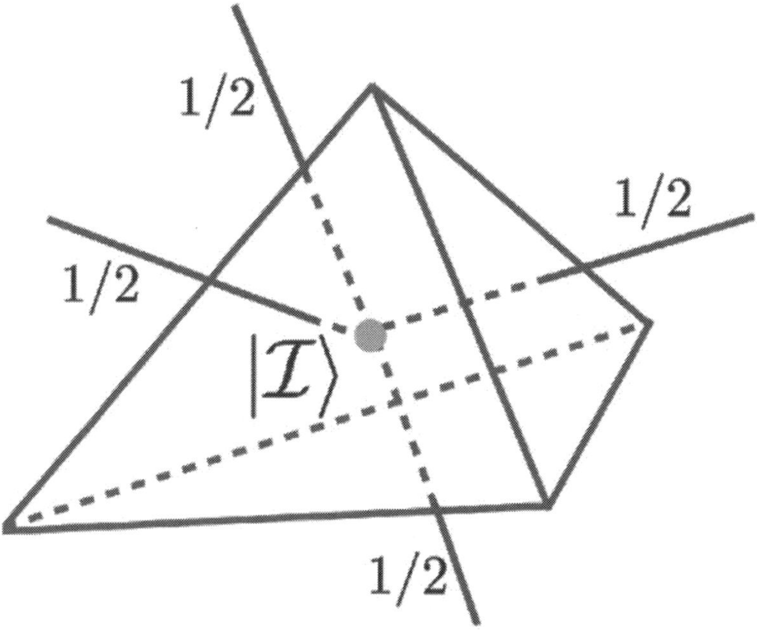

Figure 8.2 A spin network vertex, with specific choice of labels, dual to a 3-simplex (tetrahedron).

4-valent graphs (Fig. 8.3). These are (maximally) entangled states of quantum tetrahedra [13, 14], and can be obtained by imposing, on a state corresponding to a set of N disconnected quantum tetrahedra $|\psi\rangle \in \mathcal{H}_N = \bigotimes_{n=1}^{N} \mathcal{H}_v^n$, the projector $P_\gamma = \prod_{A_{xy}^i=1} P_i^{xy}$, where A_{xy}^i is the adjacency matrix of the graph γ ((x, y) label the pairs of vertices in the graph, and the additional index i runs through the possible multiple links connecting the same two vertices), and the gluing projector $P_i^{xy} : \mathcal{H}_i^x \otimes \mathcal{H}_i^y \rightarrow Inv\left(\mathcal{H}_i^x \otimes \mathcal{H}_i^y\right)$ imposes maximal entanglement along the corresponding degrees of freedom of the two semi-links one intends to connect, by tracing over the corresponding $SU(2)$ labels, and thus imposing (diagonal) $SU(2)$ invariance.

The resulting state for the graph γ will be a (linear combination of) spin network(s) living in the Hilbert space

$$\mathcal{H}_\gamma = \bigoplus_{\{j\}} \left(\bigotimes_{\{v\}} \mathcal{I}^{\vec{j}_v} \bigotimes_{\{e\} \in \partial\gamma} V^{j^e} \right) \tag{8.2}$$

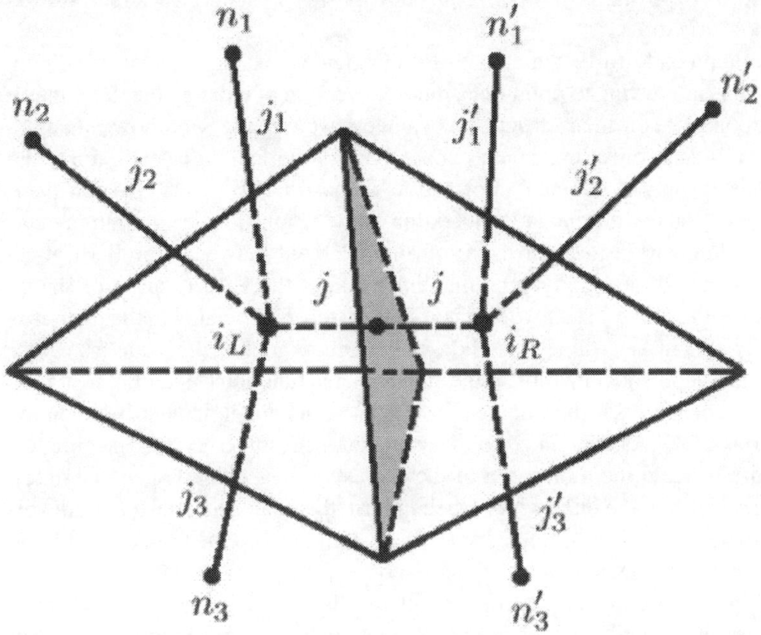

Figure 8.3 The gluing of two spin network vertices (equivalently, two quantum tetrahedra) obtained by the imposition of maximal entanglement across the shared semi-link degrees of freedom (equivalently, the quantum data associated to a triangle on their boundary).

where we considered the general case of a graph including some open (semi-)links $\{e\}$, with all graph links labelled by an irrep of $SU(2)$ and graph vertices labelled by an intertwiner between the associated irreps.

These spin network states constitute the kinematical Hilbert spaces of several quantum gravity formalisms: canonical loop quantum gravity, spin foam models, simplicial quantum gravity and tensorial group field theories (for quantum geometric models within this broader framework). The difference between these formalisms lies in how the single-graph (or before that, the single-vertex) Hilbert spaces are embedded into the full Hilbert space that includes all possible graphs and all possible numbers of vertices (as necessary to include the infinite number of degrees of freedom one would a priori expect in a full quantum theory of gravity. For example, in group field theory, this can be given by a Fock space of quantum tetrahedra. Generically, in all such formalisms generic states are thus superpositions of (open) spin network states, including a superposition of graph structures.

To appreciate further the role of entanglement in these quantum gravity states, it is interesting to point out a minimal version of entanglement/geometry correspondence in their structure [14]. We have already seen how entanglement is directly encoding graph connectivity (simplicial adjacency relations) and thus the only topological information, in fact, that is encoded in such quantum gravity states, absent any embedding of the graphs inside continuum manifolds. Moreover, a local measure of entanglement between simplices glued across a shared face is given by the dimension of the Hilbert space of shared states, i.e. the Hilbert space associated to the irrep j associated to the dual link: $D = 2j + 1$; in fact, this is also how the quantum area of the same triangular face scales, upon quantization of the classical area function ('entanglement/area correspondence'). Further, one can ask what is the entanglement between the four triangles/links associated to the same tetrahedron/vertex and the simplest measure is again the dimension of the corresponding Hilbert space of states, which scales like the intertwiner label; in turn, this scales like the quantum volume of the tetrahedron, obtained again by quantizing the corresponding volume function ('entanglement/volume' correspondence).

Before we summarize a few recent results exploiting this entanglement structure, it is also worth emphasizing that such quantum gravity states can be understood in two additional ways that make their quantum information theoretic nature manifest.

First, they can be understood as generalized tensor networks that have been central in much recent literature in quantum many-body physics, entanglement renormalization, numerical simulations of many-body systems, lattice gauge theory, neural networks, quantum computing, AdS/CFT correspondence and more [15–20]. More precisely, they are generalized projected entangled pair states (PEPS); the generalization corresponding to the fact that the link bond dimension, normally held fixed and equal in all links of the network and here corresponding to the dimension of the assigned irreps of $SU(2)$, is dynamical and assigned independently in each link, and to the fact that a generic state is actually a superposition of tensor networks with given combinatorics and bond dimensions, with the superposition affecting also the combinatorial structure (one has a superposition of different graphs).

Second, they can be reformulated as defining quantum circuits [21] (see also [22]). Consider a spin network state associated to an oriented graph with a number of open links, and consider the corresponding spin network wavefunction, depending on group elements (holonomies) associated with the bulk links, in the corresponding irreps of $SU(2)$. This wavefunction can be seen as a boundary-to-boundary map, from the Hilbert space corresponding to the tensor product of representation spaces for the incoming boundary links to the one

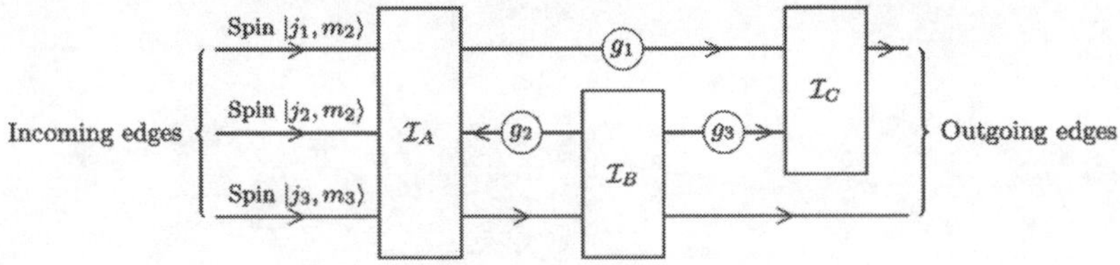

Figure 8.4 The quantum circuit corresponding to a spin network wavefunction; holonomies are one-spin unitary gates, while intertwiners are multi-spin gates.

corresponding to the tensor product of the representation spaces for the outgoing boundary links:

$$\psi(\{g_e\}_{e\in\gamma}) : \bigotimes_{e\in\partial\gamma, t(e)\in\gamma} V_{j_e} \rightarrow \bigotimes_{e\in\partial\gamma, s(e)\in\gamma} V_{j_e} \,. \tag{8.3}$$

This map defines a quantum circuit for the degrees of freedom living in the boundary Hilbert spaces, with holonomies playing the role of unitary one-spin gates and intertwiners being instead multi-spin gates (Fig. 8.4).

This reformulation is intriguing as it is potentially useful for further applications of quantum information ideas in fundamental quantum gravity.

8.2.2 *Holographic maps and quantum channels from quantum gravity states*

The correspondence between spin network states and generalized tensor networks has been exploited in a number of works, starting from [18, 23], then in [24–26] and more recently in [27–29]. Here we give a brief summary of the last set of results. For other related results with a similar formal setting and goals, although a slightly different perspective, see also [30].

The starting point is to consider quantum spin network states associated to a generically open graph (which is held fixed in the following), of the general form (Fig. 8.5):

$$|\varphi_\gamma\rangle = \bigoplus_{\{j\}} \sum_{\{n\}} \sum_{\{\iota\}} \varphi^{\{j\}}_{\{n\},\{\iota\}} P_\gamma \bigotimes_v |\{j^v\}, \{n^v\}, \iota_v\rangle \tag{8.4}$$

where we have highlighted its construction from a product basis built from the one-body Hilbert spaces, but kept generic the assignment of irreps j for each link and intertwiner labels ι for each vertex, as well as the vector indices in each representation space.

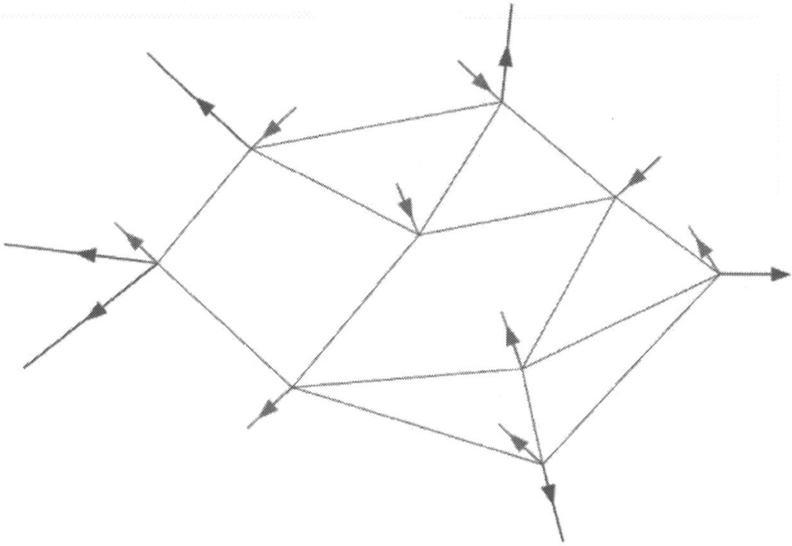

Figure 8.5 A schematic representation of a quantum state associated to an open (4-valent) graph.

Given such states, one can identify two subsets of data: the spin labels and magnetic numbers associated to the boundary links of the graph, which we may call *boundary dofs*, and the spin labels and intertwiner labels associated with the internal links and with the vertices of the graph, respectively, which we call *bulk dofs*.

In terms of this partition, we are interested in defining two types of maps for a given quantum state: bulk-boundary maps, and boundary-boundary maps, where the name indicates their domain and target. In the first case, we are interested in particular in the conditions that make such bulk-boundary maps *holographic*; in the second, we aim also to identify the conditions that would characterize such boundary-boundary maps as good quantum channels of information.

In order to apply more straightforwardly tensor network techniques, we restrict to a subclass of quantum states, whose modes factorize per vertex, i.e. $\varphi^{\{j\}}_{\{n\}\{\iota\}} = \prod_v (f_v)^{j^v}_{n^v,\iota_v}$.

Consider the first issue. To simplify the analysis, we fix the bulk spins to specific values collectively labelled J. This leads to a decoupling of boundary and remaining bulk dofs. Thus the relevant Hilbert space factorizes as: $\mathcal{H}^J = \bigotimes_v \mathcal{I}^{\vec{j}_v} \bigotimes_{e\in\partial\gamma} V^{j_e}$. Now we can define a map M_{φ_γ}, depending on the chosen quantum state, between the two bulk and boundary sub-spaces, mapping a generic bulk state $|\zeta\rangle = \sum\{\iota\}\zeta_{\{\iota\}}|\{\iota\}$ to the boundary state $|\varphi_{\partial\gamma}(\zeta)\rangle = \langle\zeta|\varphi_\gamma\rangle$, which is clearly fully specified by the chosen quantum state for bulk+ boundary.

As a proxy condition for holographic behaviour, we take the isometry of this map, i.e. the condition: $M^\dagger_{\varphi_\gamma} M_{\varphi_\gamma} = I$, where I is the identity in the bulk subspace.

One can then show that the isometry condition is satisfied if and only if the reduced bulk density matrix $\rho_{bulk} = Tr_{\partial\gamma}\left[\frac{|\varphi_\gamma\rangle\langle\varphi_\gamma|}{\prod_v D_{j^v}}\right]$ is maximally mixed, i.e. it has maximal entropy. In turn its entropy can be estimated, in terms of Renyi entropies, via standard randomization techniques applied to tensor networks [31, 32], which allow to translate the problem of maximizing the entropy of the reduced density matrix of the quantum system into that of minimizing the free energy of a dual Ising model. In our case, the same randomization method shows that, in the regime in which spins are large (naively, a semi-classical regime), the boundary-bulk map defined by our quantum gravity state is, roughly speaking, the more isometric (holographic) the more inhomogeneous is the assignment of spin labels. The precise mathematical conditions can be found in [27]. For related work, although relying on different methods, see [21]. This is interesting because it may indicate an avenue for a microscopic, quantum gravity realization (and explanation?) of holography in a non-spatiotemporal and information-theoretic context.

Consider now the second issue, i.e. transmission of information and the entanglement between two portions of the boundary, for the same quantum state and within the same restrictions (fixed bulk spins, fixed graph, factorized state). We partition boundary dofs into two complementary sets A and $\bar{A}$, and look at the reduced density $\rho_A = Tr_{\bar{A}}[\rho]$ for the region A, where ρ is the density matrix for the full quantum state φ_γ.

We are interested in the entanglement entropy between the two subregions of the boundary (Fig. 8.6). As a proxy for it, we can compute the second Renyi entropy of the reduced density matrix, using again the same random tensor network techniques, and thus the same dual Ising model. The calculation can be performed for both the homogeneous, same-spin case (all bulk spins assumed equal) and the inhomogeneous one.

From the calculation, in the case of vanishing bulk (intertwiner) entropy, one obtains an exact Ryu-Takayanagi-like formula

$$S(\rho_A)_2 \simeq K(j,\gamma)\min_{\Sigma_A}|\Sigma_A| \tag{8.5}$$

where K is a factor depending on the details of the bulk spin assignment and $|\Sigma_A|$ is the size (i.e. the number of crossing links) of the minimal surface in the bulk separating the two boundary regions (Fig. 8.7).

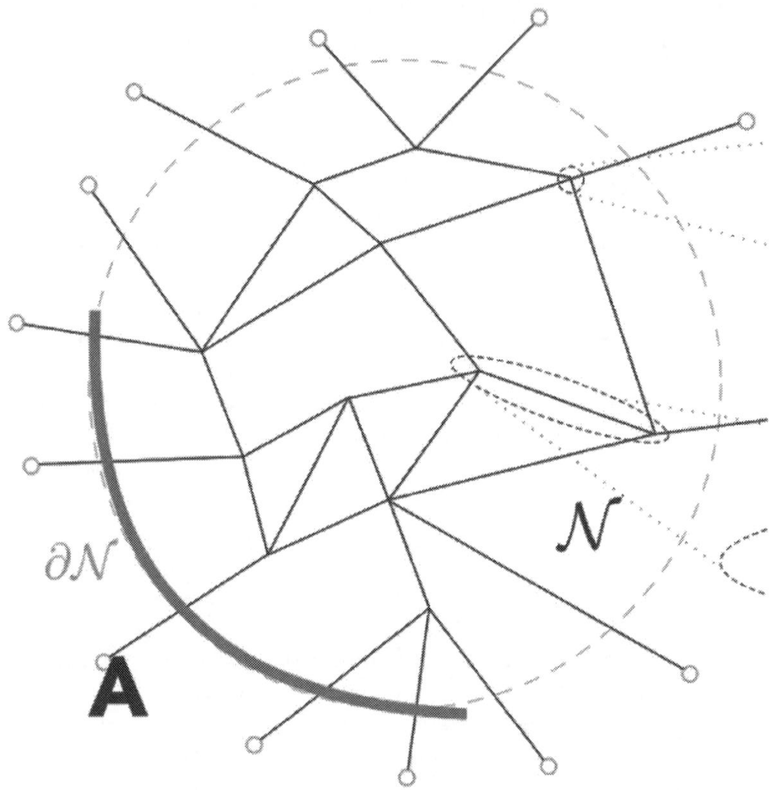

Figure 8.6 Partitioning the boundary links into two subsets.

When the bulk (intertwiner) entropy is not negligible, one finds that the RT formula acquires a correction term measuring the bulk contribution

$$S(\rho_A)_2 \simeq K(j,\gamma)\min_{\Sigma_A}|\Sigma_A| + S_{bulk} \tag{8.6}$$

which can also be computed. Interestingly, when the bulk entropy increases, a smaller and smaller portion of the RT surface enters the bulk regions, and, under the same increase, when the boundary region A tends to occupy the whole boundary, the RT surface tends to coincide with the boundary of this high-entropy bulk region. This is closely reminiscent of a black hole horizon, whose surface coincides with the RT surface, in the continuum geometry picture, and which encloses a maximal entropy bulk region. This is intriguing because it suggests a possible realization of holographic behaviour and effective

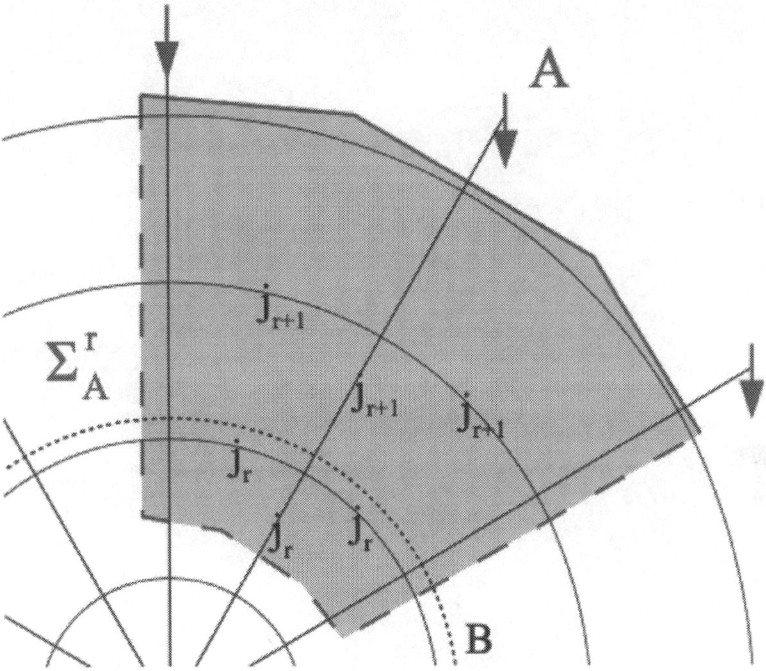

Figure 8.7 RT-like behaviour for the entanglement entropy between boundary subregions.

black hole geometries in a non-spatiotemporal and information-theoretic context (Fig. 8.8).

In trying to perform the same type of analysis for quantum gravity states involving a sum over spins, thus a sum over (discrete) quantum geometries, one has to face the issue that the corresponding Hilbert space does not present any obvious factorization between bulk and boundary dofs or between subsets of boundary ones (but it possesses a direct sum structure with respect to the possible spin assignments to the whole graph). The very notion of entanglement between any subset of dofs becomes ambiguous. If one tries to bypass this ambiguity by embedding the Hilbert space into a larger, factorized one and work with a more clearly defined notion of subspaces and entanglement there, the ambiguity presents itself again in the choice of embedding. One way to proceed is to work at the level of algebras of observables (acting on the given Hilbert space of quantum gravity states) and to identify a notion of subsystem in terms of subalgebras, rather than subspaces. Moreover, holographic behaviour is then best characterized in terms of 'transmission of information', rather than

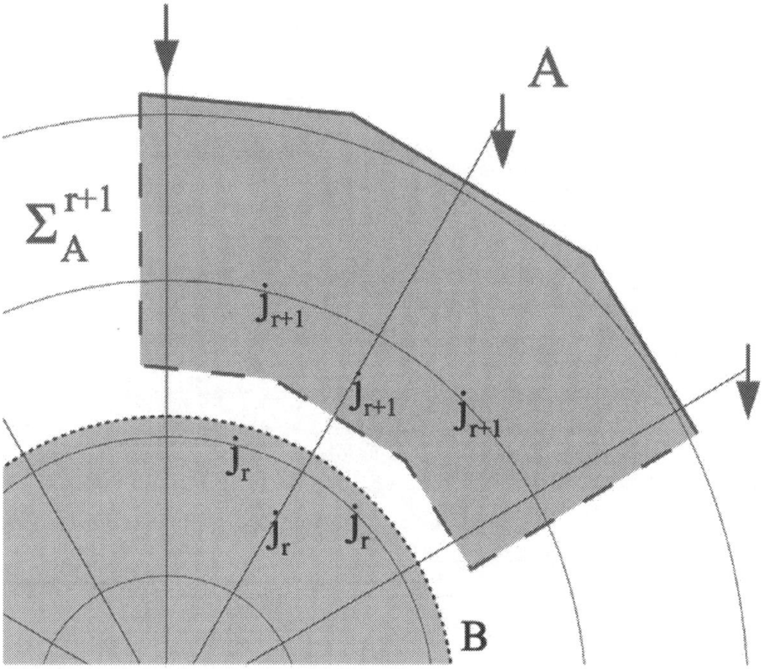

Figure 8.8 Black hole-like behaviour for the entanglement entropy between boundary subregions, when bulk entropy grows and the boundary region is extended.

entanglement scaling. This strategy has been applied for quantum gravity states in [29], and outlined in its general formal aspects in [33].

Here we just outline the main steps in the analysis.

Given the full algebra, one can identify algebraic subsystems, i.e. two subsets of observables separately 'testing' the two subsets of quantum dofs one is interested in mapping, to be considered as 'input' and 'output' (it could be bulk and boundary or two portions of the boundary). These are not subalgebras, in general. One can then define a map between input-output operator spaces via the Choi-Jamiolkowski isomorphism.

The isometry condition on such map, our proxy for holography or complete transmission of information, can be shown to follow from a certain set of necessary conditions, corresponding to the requirement that the operator map defines for a quantum channel.

These conditions can be identified using the same random tensor network techniques, in the regime of large spins (which minimizes key quantum fluctuations), and translated again also in conditions on the entropy of the underlying quantum gravity state.

One obtains again a Ryu-Takanayagi-like formula for the Renyi entropy of the quantum state depending on a set of minimal surfaces, one for each spin sector, thus one for each superposed quantum geometry, and with a definition of 'area spectrum' for minimal surfaces that can be related to but differ from the one used in canonical loop quantum gravity or simplicial quantum geometry.

In general, the necessary conditions for isometry amount to have negligible correlations between boundary data and intertwiner bulk data, and on specific peaking properties of the quantum state around a subset of spin sectors. These same conditions may actually become also sufficient ones upon additional constraints on the quantum states.

To summarize, for both boundary-bulk and boundary-boundary maps, holographic behaviour appears to require the bulk Hilbert space to be comparatively small with respect to the boundary one (as measured by the dimensionality of their respective subspaces), and the total boundary area (scaling with the size of spin spaces assigned to its links) to be approximately constant across different subsectors of spin assignments.

Again, the importance of these results is not so much in the details of their conclusions, but in the very fact that (quantum) geometric properties and quantum information theoretic properties are deeply intertwined and can be studied also in a non-spatiotemporal, purely combinatorial and algebraic context, hopefully shedding light on the emergence of holographic (and gravitational) behaviour at macroscopic scales as well as on its fundamental origin.

8.3 Quantum gravity processes as quantum causal histories (or not)

We now turn to the quantum dynamics of the quantum gravity structures we considered in Section 8.2, and that we characterized in quantum information-theoretic terms. Our main point is that a similar quantum information-theoretic characterization can be provided also for the quantum gravity processes they are subject to, and that information theoretic tools can be applied to the analysis of their (quantum) causal properties. As in Section 8.2, we consider a subset of quantum gravity formalisms, sharing many of their constitutive structures, and focus on their shared elements rather than their differences.

8.3.1 Quantum causal processes of atoms of space: Quantum spacetime as a quantum circuit

A general scheme for quantum processes respecting minimal causality conditions, and to which we can try to fit or adapt fundamental dynamical processes for our 'atoms pf quantum space', is represented by the formalism of quantum causal histories. The version we refer to here is the one in [34, 35], with its initial development in a quantum gravity context to be found in [36, 37].[2]

Possible dynamical processes are given by a set of 'events' together with an order relation between pairs of them; these are also the constitutive elements of a directed graph. In 4D quantum gravity models based on (quantum) simplicial geometry, fundamental events may be taken to correspond to 4-simplices, while order relations between pairs of them correspond to their shared 3-simplices. The directed graph would then correspond to the dual 1-skeleton of the oriented simplicial 4-complex. Note that this realization implies a restriction to 5-valent directed graphs (Fig. 8.9). For Lorentzian models, the order relations can be given a causal interpretation. An important special case is represented by partially ordered sets (posets), which are directed graphs that are also irreflexive, i.e. do not contain closed causal loops. Posets are also called, in the quantum gravity literature, *causal sets* and are the basic entities in the causal set approach to quantum gravity [39].

Clearly, this structure can be decomposed into elementary 'evolution steps', corresponding to the possible orientation assignments of the 5-valent nodes: 5 links outgoing, 1 link incoming/4 links outgoing, 2 links incoming/3 links outgoing and their inverses.

The quantum process corresponding to each directed graph is obtained by an assignment of Hilbert spaces to the links (and tensor products of Hilbert spaces for unordered sets of links) and elementary 'evolution' operators to the nodes; in addition one can also include 'gluing operators' to the links,

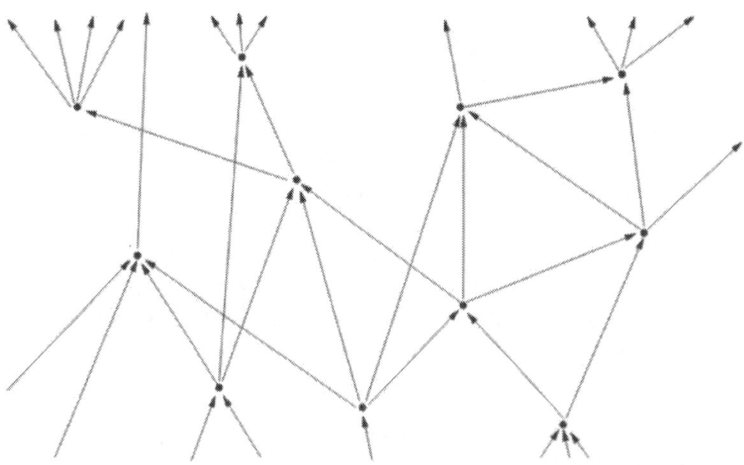

Figure 8.9 An example of a 5-valent directed graph.

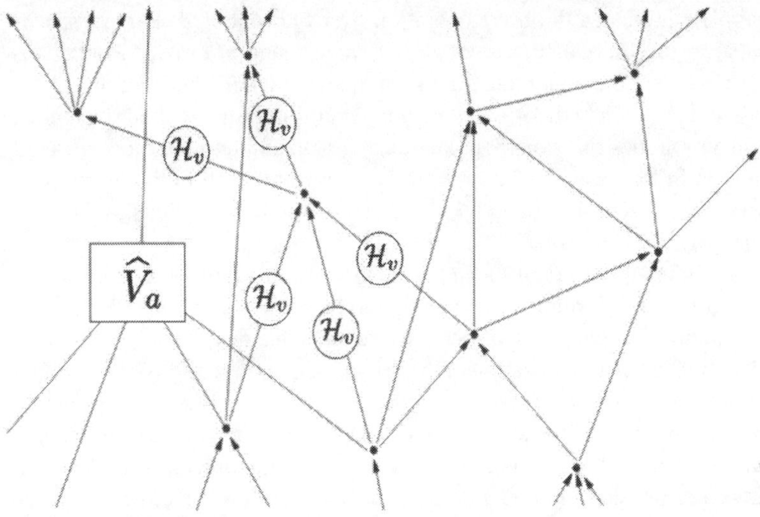

Figure 8.10 An example of an elementary quantum process.

enforcing a prescription for the 'transmission of information' from one node/event to another (Fig. 8.10). [3]

A specification of a quantum dynamics and thus a specific quantum gravity model corresponds to the assignment of a quantum (probability) amplitude to each quantum process (together with any restriction on the class of allowed processes, be it on the combinatorial structure or the associated operators and Hilbert spaces). Such amplitude is defined by the chosen node operator and gluing operator, in turn characterized by the corresponding kernels

$$\mathcal{V}_n : \bigotimes_{e\in\partial n} \mathcal{H}_v^e \longrightarrow \mathbb{C}$$
$$\mathcal{P}_e : \mathcal{H}_v^e \otimes \mathcal{H}_v^{e*} \longrightarrow \mathbb{C}$$

where we have left implicit the dualization of the Hilbert spaces required to reflect the different orientations of the links associated to nodes. Notice that the specification of a gluing operator requires a 'doubling' of the Hilbert space associated to the links of the directed graph, distinguishing one copy of it associated to one of the connected nodes from the (dualized) one associated to the other connected node.[4] Given these building blocks, the quantum amplitude associated to the given process Γ can be defined as:

$$\mathcal{A}(\Gamma) = \underset{e\in\Gamma}{Tr}\left(\prod_{e\in\Gamma}\mathcal{P}_e\prod_{n\in\Gamma}\mathcal{V}_n\right) \tag{8.7}$$

where Tr indicates the trace operation over the (doubled) Hilbert spaces associated to the links of the process. Different quantum gravity formalisms (notably canonical loop quantum gravity, spin foam models, lattice gravity path integrals and group field theories) share this general structure of their dynamical quantum amplitudes in a covariant language. For example, for a more detailed presentation in the case of spin foam models (adapted also to the group field theory language) see [44], in addition to the existing introductions to all these quantum gravity formalisms.

In fact, in quantum gravity formalisms that use this kind of discrete structures, the complete quantum dynamics should be completed with the definition of a continuum limit/approximation, since reliance on any given cellular complex represents at best a truncation of the full set of quantum degrees of freedom of the fundamental theory. Also in accordance with the superposition principle and the interpretation of the above discrete structures as possible quantum processes, this limit is encoded within a sum over all complexes/processes within the allowed class. The full quantum dynamics is therefore encoded in a partition function (or transition amplitude, if the complexes have boundaries) of the general form:

$$Z = \sum_{\Gamma} w(\Gamma)\mathcal{A}(\Gamma) \tag{8.8}$$

where again the specification of the additional weights $w(\Gamma)$ concurs to the definition of the specific model being considered.

Two points are worth emphasizing about this construction. First, these quantum processes, like the quantum states discussed in the previous section, can be reformulated in the language of tensor networks, and this has been applied to the study of renormalization of quantum gravity dynamics from a lattice gauge theory perspective [45]. Second, at this level, thus before any extraction of coarse-grained or otherwise effective continuum dynamics, the connection of these dynamical quantum amplitudes to that of gravitational theories can only be with lattice gravity, i.e. some discretized version of gravitational physics. This connection depends strongly on the specific quantum gravity formalism being considered. In spin foam models, thus also in group field theories, the connection can be shown (modulo remaining technical open issues) in the form of an exact equivalence with lattice gravity path integrals for first order gravity theories[5] [44], or in a semi-classical, asymptotic regime with second order gravity [46] (in terms of area variables [47]).

We are interested in identifying the requirements on such quantum processes that make them *bona fide* quantum causal histories, that is, that make each quantum process encore a properly causal unitary evolution. We do not discuss here to what extent this is something we should expect or require, from fundamental quantum gravity processes, besides a few short remarks.

The required properties have been identified in [34, 35]. Consider the evolution operator $E_{\alpha\beta} : \mathcal{H}_\alpha \rightarrow \mathcal{H}_\beta$, obtained by composing all the elementary operators V connecting two complete a-causal subsets α and β of links, that is subsets of links that are all causally unrelated within each subset and one the complete causal future (past) of the other, where the Hilbert spaces $\mathcal{H}_{\alpha,\beta}$ are the tensor product of the Hilbert spaces associated to the constituting links. For the quantum process to define a quantum causal history, the operators $E_{\alpha\beta}$ should be: (a) reflexive: $E_{\alpha\alpha} = I_\alpha$, (b) antisymmetric: $E_{\alpha\beta}E_{\beta\alpha} = I_\alpha \Leftrightarrow E_{\alpha\beta} = E_{\beta\alpha} = I_\alpha$; (c) transitive: $E_{\alpha\beta}E_{\beta\gamma} = E_{\alpha\gamma}$; (d) unitary: $\sum_\beta E_{\alpha\beta}E^\dagger_{\alpha\beta} = \sum_\beta E_{\alpha\beta}\bar{E}_{\beta\alpha} = I_\alpha$.

One can then check or impose that the relevant evolution operators of quantum gravity models of interest satisfy these requirements. However, one could also question their necessity. Indeed, we have stated that the full quantum gravity dynamics should rather be given by a sum over all such quantum processes. Therefore, one could argue that the operators satisfying proper causality conditions are those obtained by performing this sum, that is

$$\mathcal{E}_{\alpha\beta} = \sum_c \lambda_c E^c_{\alpha\beta} : \mathcal{H}_\alpha \longrightarrow \mathcal{H}_\beta \tag{8.9}$$

between the same complete a-causal subsets. In other words, even accepting that the fundamental quantum gravity dynamics should be expressed in the language of quantum causal histories, one could argue that it is the full transition amplitudes that should satisfy causality constraints, and not the possible individual quantum processes.

It is easy to see [51] that these two perspectives are not equivalent, and in fact, they are not even consistent with one another. More precisely, while imposing 'micro-reflexivity'(i.e. reflexivity of individual processes) implies full reflexivity (i.e. reflexivity of the evolution obtained upon summing over microprocesses) and the same is true for antisymmetry, the other two requirements are much more problematic. First, one could make micro-transitivity compatible with transitivity of the full evolution, by defining the latter more generally as $\sum_\beta \mathcal{E}_{\alpha\beta}\mathcal{E}_{\beta\gamma} = \mathcal{E}_{\alpha\gamma}$, which is just the standard composition of quantum (transition) probability amplitudes; but micro-transitivity itself appears to be too strong, a requirement from the discrete quantum gravity perspective; indeed, taking into account also the dual lattice formulation of quantum processes, micro-transitivity is equivalent to a condition of partial triangulation invariance of quantum dynamics, i.e. a requirement that the quantum amplitudes associated to individual lattices are partially independent of the chosen lattice. To the extent in which gravity is not a topological field theory with no local propagating degrees of freedom, this is a suspicious condition since it may imply a partial triviality of quantum dynamics. Second, one can actually verify that unitarity of the full evolution implies that the micro-evolution *must not be unitary*. So one has to make a choice.

It remains a valid goal to have the full quantum gravity dynamics, obtained via a sum over elementary quantum processes, defining a ('coarse-grained') quantum causal history. This would be interesting from the perspective we are exploring in this contribution because it would mean that quantum information and quantum computation can be found at the very heart of quantum gravity also at the dynamical level. The same interest would remain if instead one insists on the elementary processes be themselves quantum causal histories.

Indeed, it is a general result [52] that a quantum causal history admits a unitary evolution between its acausal surfaces if and only if it can be represented as a quantum computational network, i.e. a quantum circuit.

The idea of spacetime as a quantum circuit would then find a concrete realization, if quantum gravity evolution can be formulated in terms of quantum causal histories, at the elementary or coarse-grained level, in the above language or in the more refined one of observable algebras and CP maps [35]. See also [53, 54] and the cited literature on process matrix formalism and indefinite causal structures for related directions.

Let us now look in more detail at a couple of features of quantum gravity processes that have to be realized in order for them to define quantum causal histories, and thus quantum circuits, and see which obstacles one has to face to do so.

8.3.2 *Causal hiccups and causal indifference in QG processes*

As discussed, a proper representation in terms of quantum circuits requires: (a) the absence or "irrelevance"of causal loops; (b) suitable conditions ensuring unitarity of evolution operators.

Concerning closed causal loops, there are three possible strategies that can be followed in constructing quantum gravity models: (a) define a quantum dynamics (amplitudes) that eliminates causal loops altogether; (b) define a quantum dynamics (amplitudes) that suppresses causal loops, by assigning them subdominant contributions, in the relevant regimes; for example, one could consider admitting causal loops in the theory, provided they do not spoil expected semi-classical or continuum physics; (c) define a quantum dynamics (amplitudes) that only allows 'harmless'. Let us stress that the directed graphs underlying all current spin foam models and lattice gravity path integrals (or group field theory perturbative amplitudes) contain closed loops of order relations, i.e. causal loops. Thus the issue is of concrete relevance for such quantum gravity formalisms. While the first two options may be technically very challenging but are conceptually straightforward, when exactly a causal loop is physically harmless requires a more careful analysis. The issue has been studied for example in [52], in the quantum gravity context, following earlier work by David Deutsch [55] in a general quantum mechanical context. The upshot is that causal loops are either entirely disruptive or entirely harmless to paraphrase. More precisely, they either prevent the standard formulation of quantum mechanics to be applicable or, when they do not, they lead to no observable changes, since they simply end up contributing an extra subspace to the ordinary causality-respecting system. Moreover, by applying suitable (Deutsch) criteria, the causally well-behaved region of the process decouples entirely from the causal loop if the quantum dynamics remains linear, and if it does not stay linear, then the causal loop does not carry independent degrees of freedom. For more details we refer again to [52].

Let us now discuss unitarity of quantum evolution and, before that, the very dependence of the quantum gravity transition amplitudes from the order of their arguments, thus a notion of past/future relating the quantum states it depends on, which is in many ways a prerequisite for it. The issue is: given two ('initial' and 'final') quantum states, which kind of quantum amplitude do we define, via the gravitational path integral? Let us first recall a few facts about quantum gravity path integrals, which can be verified at the formal level in great generality [56, 57], and have to be then realized concretely in more rigorous manner by quantum gravity approaches, including the ones based on discrete structures that we have focused on here. Different quantum gravity amplitudes can be defined starting from the same proper-time truncation of the full path integral (in canonical form), obtained via appropriate gauge-fixing of the general expression:

$$K[h^2_{ij}, h^1_{ij}; N(\tau_2 - \tau_1)] = \int \mathcal{D}h_{ij}(x,\tau)\mathcal{D}\pi^{ij}(x,\tau)e^{iS(h_{ij},\pi^{ij});N} \qquad (8.10)$$

where the amplitude depends on the fixed metric data on the two (past/future) boundaries. This expression could be required to correspond to (the matrix elements of) a unitary evolution operator in proper time. However, this is not the physical (transition) amplitude for quantum gravity, since the lack of integration over proper time (lapse) means that we have not yet imposed any conditions encoded in the Hamiltonian constraint of the (canonical) theory, thus no full quantum gravitational dynamics (the Einstein's equations are indeed encoded in the Hamiltonian constraint), besides enforcing at best some semi-classical restriction (due to the appearance of the gravitational action in the expression). One can then define a 'causal' transition amplitude (the analogue for quantum gravity of what would be, for a relativistic particle, the Feynman propagator, by integrating the lapse (proper time) over the full positive range:

$$\mathcal{K}[h^2_{ij}, h^1_{ij}] = \int_{N(x)=0}^{N(x)=+\infty} \mathcal{D}[N(x)(\tau_2 - \tau_1)]\, K[h^2_{ij}, h^1_{ij}; N(\tau_2 - \tau_1)]\,. \quad (8.11)$$

This is the canonical counterpart of the straightforward Lagrangian gravitational path integral $\mathcal{K}[h^2_{ij}, h^1_{ij}] = \int \mathcal{D}g_{\mu\nu} e^{iS(g_{\mu\nu})}$. It is invariant under Lagrangian (covariant) diffeomorphisms and indeed switches to its own complex conjugate under switch of spacetime orientation; in other words, it does register an ordering between its two arguments, the boundary quantum states. It does not, however, give a solution to the canonical Hamiltonian constraint, i.e. it is not invariant under canonical symmetries (the canonical Dirac algebra, counterpart of covariant diffeomorphisms, which are a subset of the canonical ones). A solution of the canonical Hamiltonian constraint is instead obtained by integrating the lapse (proper time) over the full (positive and negative) real values:

$$C[h^2_{ij}, h^1_{ij}] = \int_{N(x)=-\infty}^{N(x)=+\infty} \mathcal{D}[N(x)(\tau_2 - \tau_1)] \, K[h^2_{ij}, h^1_{ij}; N(\tau_2 - \tau_1)] \, . \quad (8.12)$$

This indeed defines (formally) a physical scalar product between canonical quantum gravity states, solving all the constraints of the theory (or, equivalently, the matrix elements of the projector operator onto such solutions). Its Lagrangian counterpart would look like $C[h^2_{ij}, h^1_{ij}] = \int \mathcal{D}g_{\mu\nu} \left[e^{iS(g_{\mu\nu})} + e^{-iS(g_{\mu\nu})} \right] = \int \mathcal{D}g_{\mu\nu} \, \cos(S(g_{\mu\nu}))$. This quantity does *not* register the spacetime orientation and it is symmetric under its switch, not encoding any ordering among it arguments, the 'initial' and 'final'quantum states.

Spin foam models (equivalently, the perturbative transition amplitudes of group field theories) aim to be discretized and thus mathematically better defined realization of the gravitational path integral. Which of the above quantities do they actually realize?

All the most studied spin foam models are discrete counterparts of the path integral for gravity formulated as a constrained topological BF theory. It turns out that, like their continuum counterpart (and the path integral for topological BF theory itself) they are invariant under switch of spacetime (lattice) orientation, more precisely the inversion of the orientation of their constitutive simplicial structures; this invariance is in fact realized locally at the level of each node or 4-simplex contribution $\mathcal{V}$ to the total amplitude $\mathcal{A}$ as well as at the level of lower-dimensional simplices (e.g. triangles) in the simplicial complex. Recall that the 1-skeleton of this simplicial complex corresponds to the directed graph whose order relations have a tentative causal interpretation (in Lorentzian models). The orientation independence of the spin foam amplitudes thus implies that none of the most studied spin foam models defines a proper quantum causal history (a quantum circuit) and a unitary quantum gravity dynamics. This would remain true even if one was able to remove causal loops from the underlying directed graph.

It is possible to construct 'properly causal' spin foam models, but a suitable restriction of their amplitudes so that they register faithfully the orientation of the underlying complex. This has been done first for the Barrett-Crane model in [51], and more recently by similar procedures for the EPRL model in [58, 59]. These restricted models are therefore candidates for the realization of the 'causal propagator' for quantum gravity and for a formulation in terms of quantum causal histories and quantum circuits. However, these causality-inspired constructions are all rather *ad hoc* and we still lack a systematic construction procedure of spin foam models from first principle (rather than by restricting by hand a-causal models) taking into account causality restrictions, as well as a more complete analysis of the properties of the present 'ad-hoc' ones.

8.4 Conclusions

We have argued that both semi-classical considerations and quantum gravity formalisms suggest, in different ways, that spacetime and gravity may be emergent, collective, not fundamental notions, and that the universe may be a (peculiar, background independent) quantum many-body system of pre-geometric quantum entities, some yet to be unraveled 'atoms of space'. In particular, there are intriguing indications that topology and geometry may emerge from the entanglement among such fundamental quantum entities. More generally, an intriguing possibility is that quantum spacetime physics may be formulated, in its most fundamental level, entirely in the language of quantum information.

A variety of quantum gravity formalisms share the same combinatorial and algebraic quantum structures as quantum states: quantized simplicial structures and spin networks. We have outlined the ways in which such quantum states can be described in quantum information-theoretic terms. More precisely, we have summarized how these quantum states: (a) can be seen as generalized tensor networks and realize a precise discrete entanglement/geometry (and topology) correspondence; (b) can be framed as information channels (or quantum circuits); (c) can be used to define bulk/boundary and boundary-to-boundary maps, for which one can then identify conditions for holographic behaviour. This could indicate an avenue towards understanding the microscopic origin of holographic behaviour in quantum gravity.

We have then discussed how, in the same quantum gravity formalisms, dynamical quantum processes can be recast as quantum causal histories, provided some key properties are implemented in their amplitudes, and then again as quantum circuits; we have also pointed out some of the challenges faced to implement the required properties. Again, the main point is that quantum information tools and language may be the appropriate ones to formulate the quantum dynamics of the microscopic constituents of the universe, when geometry and fields fail.

Before we go on to comment on some more conceptual aspects of these conclusions, we point out that (tensorial) group field theory framework, on top of providing a completion of lattice gravity path integrals and spin foam models (thus sharing the same quantum amplitudes) and a second quantized framework for spin network states (thus a convenient Fock space structure for their Hilbert space) [60, 61], provides also a number of almost standard field theoretic tools to study them and in particular to extract effective continuum gravitational physics from them [62]. This means that, even in this more abstract, non-spatiotemporal, pre-geometric context, one can apply quantum field theoretic techniques to the quantum information structures we presented above, to analyse their formal properties and to unravel their physical meaning.

Some philosophical considerations: is the universe a quantum computer? To conclude let us offer some thoughts on different ways in which we could interpret the relevance of quantum information language and tools for encoding the fundamental microstructure of spacetime and the universe.

A straightforward attitude is to give to this fact an ontological basis: the universe is a quantum computer. In this case, the fact that quantum information is the appropriate language to understanding is no surprise: it is just the language representing how it fundamentally operates, and that constitutes its basic laws. From the epistemological point of view, this attitude follows from and it is grounded on a straightforward scientific realism: the world is out there and entirely independent, in its properties, of our epistemic activities, which achieve (at best) a faithful (albeit partial) representation of the way the world is. It is also tied to a realist and ontologically committed view on laws of nature: they are what governs the physical world, i.e. the rules by which it functions and evolves.

Challenges against all the above are numerous and the philosophical debate about each of the above points is old, intricate and interesting. Here we want to make two brief comments about this 'the universe is a quantum computer' view, implicitly based on the attitude we just summarized.

A first immediate one is that all the realist views on laws face challenges at different levels from quantum gravity, and in particular from the very possibility of space, time and geometry being emergent notions. The quantum information structures we discussed in this contributions can be seen as encoding these challenges, but at the same time offering tentative ways to meet (if not solve) them. This is discussed in [63].

A second one is that one can adopt (and try to develop further) a less ontologically committed view on the object of our scientific theories, i.e. the physical world, including laws of nature, and thus a weaker version of (scientific) realism. One can adopt a more epistemic view on physical laws based on a more substantial role given to epistemic agents, taken to be irreducible and not negligible (outside convenient idealizations). This epistemic, agent-based view would tie well with a more participatory form of realism, in which what is real is, roughly speaking, only the result and content of the interaction between the world and the epistemic agents, none of which is independently real outside of such interaction. This weaker form of realism would be the only kind of ontological commitment allowed by the epistemic premises. This view may have implications for (and be tested with) the interpretation of quantum mechanics, as well as the construction and interpretation of theories of quantum spacetime and geometry.

More generally, it would lead to a view in which the universe is (to a large extent) what we think it is (or what we model it as), in the sense that it is our epistemic constructions that *make reality*, rather than simply represent it. Obviously this is just a vague statement, to be made more precise and articulated, but it is maybe interesting to see how a similar view changes how we may interpret the role of (quantum) information theoretic structures in fundamental quantum gravity. In a fundamental quantum gravity context, we argued, we have no spacetime notions to rely on; we have to think the world (and model it) without spacetime. We are then left with combinatorics and algebra as mathematical language, and with information processing, rather than definite, ontologically grounded objects (whose ontological characterization would normally *assume* space and time), as the only 'dynamical' content. This reflects the more basic, more irreducible structures in our thinking, which at the same time correspond (given the above view of what it is to be 'real') to the more basic structures 'in the world'. (Quantum) Computers are abstract models of (quantum) information processing, and of our own thinking. To the extent in which the universe is (largely) what we think it is (in the sense outlined above), and we think like (quantum) computers, it is not so surprising, perhaps, that the quantum (non-spatiotemporal) universe is naturally modelled as a quantum computer.

These rather vague considerations are offered, here, only as potentially useful indications of philosophical avenues to explore and develop further, on the basis of the scientific developments in quantum gravity, that we have summarized in this contribution.

Acknowledgements

We acknowledge support through the Grant PR28/23 ATR2023-145735 (funded by MCIN /AEI /10.13039/501100011033). We also thank the organizers and the participants of the "QG and computation" workshop in Sydney for a very interesting event and for many stimulating discussions, as well as the editors of this volume for their patience.

Notes

1. Interestingly, this more traditional perspective is also, historically and in part of the present community, the one adopted to interpret some of the same quantum gravity formalisms that we suggest can be fruitfully understood from an emergent spacetime perspective.
2. Of course, the abstract characterization of quantum dynamics and quantum causality, as well as its generalization to a context in which geometry and thus causal relations are themselves dynamical and subject to superposition, is a hot topic in quantum foundations, with many recent developments. See [40–43] for a small sample. Obviously, we are not going to review any of that.
3. In fact, a more complete and consistent definition of a quantum process in this language, of a quantum causal history in particular, is given in terms of an assignment of algebras of operators and completely positive maps [35]. We give here a simplified earlier construction, which is sufficiently indicative of the general points we want to make.
4. Notice that this doubling is necessary if one wants to allow for a composition of processes or, from the point of view of their dual cellular complexes, the composition of different cellular complexes along shared boundaries.
5. Usually, with gravity formulated as a constrained topological theory [48]. But see [49, 50] for models constructed following alternative strategies.

References

[1] Bianchi, E., and Livine, E. R. (2023). Loop quantum gravity and quantum information. *In* Encyclopedia of Mathematical Physics (2nd ed.).

[2] Colafranceschi, E. (2022). Emergent spacetime properties from entanglement. PhD Thesis, University of Nottingham.

[3] Butterfield, J., and Isham, C. J. (1999). Space-time and the philosophical challenge of quantum gravity. *In* Callender, C. and Huggett, N. (Eds.), *Physics Meets Philosophy at the Planck Scale: Contemporary Theories in Quantum Gravity*, pp. 33–89. Cambridge University Press.

[4] Seiberg, N. (2006). Emergent spacetime. *In* Proceedings of the 23rd Solvay Conference in Physics: The Quantum Structure of Space and Time, pp. 163–178.

[5] Padmanabhan, T. (2015). Emergent gravity paradigm: Recent progress. *Mod. Phys. Lett. A*, 30(03), 1540007.

[6] Oriti, D. (2018). Levels of spacetime emergence in quantum gravity. https://arxiv.org/abs/1807.04875arXiv:1807.04875.

[7] Carlip, S. (2014). Challenges for emergent gravity. *Stud. Hist. Phil. Sci. B*, 46, 200–208.

[8] Van Raamsdonk, M. (2017). Lectures on gravity and entanglement. *In* Theoretical Advanced Study Institute in Elementary Particle Physics: New Frontiers in Fields and Strings, pp. 297–351.

[9] Swingle, B. (2018). Spacetime from entanglement. *Ann. Rev. Condensed Matter Phys.*, 9, 345–358.

[10] Rangamani, M., and Takayanagi, T. (2017). *Holographic Entanglement Entropy*. Lecture Notes in Physics, Vol. 931. Springer.

[11] Pereira, R. (2010). Spinfoams from simplicial geometry. PhD. Thesis, Université de la Méditerranée Aix-Marseille II.

[12] Perez, A. (2013). The spin foam approach to quantum gravity. *Living Rev. Rel.*, 16, 3.

[13] Baytaş, B., Bianchi, E., and Yokomizo, N. (2018). Gluing polyhedra with entanglement in loop quantum gravity. *Phys. Rev. D*, 98(2), 026001.

[14] Colafranceschi, E., and Oriti, D. (2021). Quantum gravity states, entanglement graphs and second-quantized tensor networks. *JHEP*, 07, 052.

[15] Orús, R. (2014). A practical introduction to tensor networks: Matrix product states and projected entangled pair states. *Annals Phys.*, 349, 117–158.

[16] Verstraete, F., Murg, V., and Cirac, J. I. (2008). Matrix product states, projected entangled pair states, and variational renormalization group methods for quantum spin systems. *Adv. Phys.*, 57, 143–224.

[17] Cirac, J. I., Pérez-García, D., Schuch, N., and Verstraete, F. (2021). Matrix product states and projected entangled pair states: Concepts, symmetries, theorems. *Rev. Mod. Phys.*, 93(4), 045003.

[18] Singh, S., McMahon, N. A., and Brennen, G. K. (2018). Holographic spin networks from tensor network states. *Phys. Rev. D*, 97(2), 026013.

[19] Tagliacozzo, L., Celi, A., and Lewenstein, M. (2014). Tensor networks for lattice gauge theories with continuous groups. *Phys. Rev. X*, 4(4), 041024.

[20] Evenbly, G., and Vidal, G. (2015). Tensor network renormalization. *Phys. Rev. Lett.*, 115(18), 180405.

[21] Chen, Q., and Livine, E. R. (2021). Loop quantum gravity's boundary maps. *Class. Quant. Grav.*, 38(15), 155019.

[22] Czelusta, G., and Mielczarek, J. (2021). Quantum simulations of a qubit of space. *Phys. Rev. D*, 103(4), 046001.

[23] Singh, S., Pfeifer, R. N. C., and Vidal, G. (2010). Tensor network decompositions in the presence of a global symmetry. *Phys. Rev. A*, 82(5), 050301.

[24] Han, M., and Huang, S. (2017). Discrete gravity on random tensor network and holographic Rényi entropy. *JHEP*, 11, 148.

[25] Chirco, G., Oriti, D., and Zhang, M. (2018). Group field theory and tensor networks: Towards a Ryu–Takayanagi formula in full quantum gravity. *Class. Quant. Grav.*, 35, 115011.

[26] Chirco, G., Goessmann, A., Oriti, D., and Zhang, M. (2020). Group field theory and holographic tensor networks: Dynamical corrections to the Ryu–Takayanagi formula. *Class. Quant. Grav.*, 37, 095011.

[27] Colafranceschi, E., Chirco, G., and Oriti, D. (2022). Holographic maps from quantum gravity states as tensor networks. *Phys. Rev. D*, 105, 066005.

[28] Chirco, G., Colafranceschi, E., and Oriti, D. (2022). Bulk area law for boundary entanglement in spin network states: Entropy corrections and horizon-like regions from volume correlations. *Phys. Rev. D*, 105, 046018.

[29] Colafranceschi, E., Langenscheidt, S., and Oriti, D. (2024). Holographic properties of superposed quantum geometries. *Phys. Rev. D*, 110, 046024.

[30] Akers, C., and Wei, A. Y. (2024). Background independent tensor networks. *SciPost Phys.*, 17, 090.

[31] Hayden, P., Nezami, S., Qi, X.-L., Thomas, N., Walter, M., and Yang, Z. (2016). Holographic duality from random tensor networks. *JHEP*, 11, 009.

[32] Qi, X.-L., Yang, Z., and You, Y.-Z. (2017). Holographic coherent states from random tensor networks. *JHEP*, 08, 060.

[33] Langenscheidt, S., Colafranceschi, E., and Oriti, D. (2024). Channel-State duality with centers. https://arxiv.org/abs/2404.16004 arXiv:2404.16004.

[34] Markopoulou, F. (2000). Quantum causal histories. *Class. Quant. Grav.*, 17, 2059–2072.

[35] Hawkins, E., Markopoulou, F., and Sahlmann, H. (2003). Evolution in quantum causal histories. *Class. Quant. Grav.*, 20, 3839–3854.

[36] Markopoulou, F., and Smolin, L. (1997). Causal evolution of spin networks. *Nucl. Phys. B*, 508, 409.

[37] Markopoulou, F., and Smolin, L. (1998). Quantum geometry with intrinsic local causality. *Phys. Rev. D*, 58, 084032.

[38] Dowker, F., and Surya, S. (2024). The causal set approach to the problem of quantum gravity. In *Handbook of Quantum Gravity*. https://doi.org/10.1007/978-981-19-3079-9_70-1DOI:10.1007/978-981-19-3079-9_70-1.

[39] Oreshkov, O., Costa, F., and Brukner, Č. (2012). Quantum correlations with no causal order. *Nature Commun.*, 3, 1092.

[40] Goswami, K., Giarmatzi, C., Kewming, M., Costa, F., Branciard, C., Romero, J., et al. (2018). Indefinite causal order in a quantum switch. *Phys. Rev. Lett.*, 121, 090503.

[41] Baumann, V., Krumm, M., Guérin, P. A., and Brukner, Č. (2022). Noncausal Page-Wootters circuits. *Phys. Rev. Res.*, 4, 013180.

[42] Mrini, L., and Hardy, L. (2024). Indefinite causal structure and causal inequalities with time-symmetry. Preprint.

[43] Hardy, L. (2005). Probability theories with dynamic causal structure: A new framework for quantum gravity. Preprint.

[44] Finocchiaro, M., and Oriti, D. (2020). Spin foam models and the Duflo map. *Class. Quant. Grav.*, 37, 015010.

[45] Dittrich, B., Mizera, S., and Steinhaus, S. (2016). Decorated tensor network renormalization for lattice gauge theories and spin foam models. *New J. Phys.*, 18, 053009.

[46] Asante, S. K., Dittrich, B., and Haggard, H. M. (2020). Effective spin foam models for four-dimensional quantum gravity. *Phys. Rev. Lett.*, 125, 231301.

[47] Dittrich, B., and Padua-Argüelles, J. (2024). Twisted geometries are areametric geometries. *Phys. Rev. D*, 109, 026002.

[48] Bianchi, E., and Hellmann, F. (2013). The construction of spin foam vertex amplitudes. *SIGMA*, 9, 008.

[49] Baratin, A., Flori, C., and Thiemann, T. (2012). The Holst spin foam model via cubulations. *New J. Phys.*, 14, 103054.

[50] Dekhil, R., Laudonio, M., and Oriti, D. (2025). A state sum for four-dimensional Lorentzian quantum geometry in terms of edge vectors. https://arxiv.org/abs/2501.10115arXiv:2501.10115.

[51] Livine, E. R., and Oriti, D. (2003). Implementing causality in the spin foam quantum geometry. *Nucl. Phys. B*, 663, 231.

[52] Livine, E. R., and Terno, D. R. (2007). Quantum causal histories in the light of quantum information. *Phys. Rev. D*, 75, 084001.

[53] Arrighi, P., and Martiel, S. (2017). Quantum causal graph dynamics. *Phys. Rev. D*, 96, 024026.

[54] Arrighi, P., Durbec, A., and Wilson, M. (2024). Quantum networks theory. *Quantum*, 8, 1508.

[55] Deutsch, D. (1991). Quantum mechanics near closed timelike lines. *Phys. Rev. D*, 44, 3197.

[56] Teitelboim, C. (1982). Quantum mechanics of the gravitational field. *Phys. Rev. D*, 25, 3159.

[57] Halliwell, J. J., and Hartle, J. B. (1991). Wave functions constructed from an invariant sum over histories satisfy constraints. *Phys. Rev. D*, 43, 1170.

[58] Engle, J., and Zipfel, A. (2016). Lorentzian proper vertex amplitude: Classical analysis and quantum derivation. *Phys. Rev. D*, 94, 064024.

[59] Bianchi, E., and Martin-Dussaud, P. (2024). Causal structure in spin foams. *Universe*, 10(4), 181.

[60] Oriti, D. (2011). The microscopic dynamics of quantum space as a group field theory. In C. Callender and N. Huggett (Eds.), *Foundations of Space and Time: Reflections on Quantum Gravity*, pp. 257–320.

[61] Oriti, D. (2014). Group field theory and loop quantum gravity. https://arxiv.org/abs/1408.7112arXiv:1408.7112.

[62] Oriti, D. (2021). Tensorial group field theory condensate cosmology as an example of spacetime emergence in quantum gravity. https://arxiv.org/abs/2112.02585arXiv:2112.02585.

[63] Lam, V., and Oriti, D. (2024). The quantum gravity seeds for laws of nature. *Eur. J. Phil. Sci.*, 14, 63.

Part 4

Conceptual Issues for Computational Foundations

9 Do Simples Exist?: The Light Shed by Computation on an Ancient Question

Gemma De les Coves

9.1 Overture

Should a theory of quantum gravity be fundamentally discrete or continuous? That is, should it posit discrete building blocks of spacetime, or matter, or should it be ultimately continuous? Given that a theory of quantum gravity ought to describe our physical reality in what we may consider it to be its most fundamental level, the question seems tantamount to:

9.1.1 Do simples exist?

A simple is something that cannot be divided. I am referring to simples of spacetime, as fundamental particles such as electrons or quarks that cannot be divided. So this is asking whether there is a bottom level of reality, it enquires for the nature of the fabric of reality, to invoke David Deutsch's book [1]. As you can imagine, this question has been considered for the breath of our philosophical history, and many views have been proposed—often amusing, other times moving.

I would like to approach this question in accordance with the theme of the present collection of essays. Namely, how does computation illuminate the fundamental discreteness or continuity of the physical world? Does computation tell us anything about the nature of the smallest?

In contrast with general relativity, and more generally the description of spacetime in theoretical physics, in the theory of (classical) computation all actresses are finite and discrete, with the potential infinity reserved for the length

DOI: 10.4324/9781003546139-13

of the tape. As a matter of fact, I regard the theory of computation as a principled way to reach the potential infinite from the finite, as I will try to argue. But what is the physical load of the theory of computation?

We tend not to regard computation as a physical law, and thus rank its lessons lower in importance than those provided by, say, general relativity or quantum field theory. At the same time, we believe that the theory of computation captures some deep facts about our world, notably via the Church–Turing thesis. Now,

9.1.2 Should we endow computation with ontological value? (⋆)

If our search for a theory of quantum gravity is guided by the ontological commitments to the mathematics of spacetime, this will result in fundamental continuity, whereas if it is guided by the principles of information and its processing (i.e. computation), this will result in fundamental discreteness.

My short answer to the question whether simples exist is 'I don't know'. I would love to know. Instead, I will approach (⋆) through various considerations. A guiding thread will be that introducing a finiteness in the infinite is akin to the Pythagorean idea of imposing a *peras* in the *apeiron*, that is, a bound in the unbounded.

To that end, I shall assume that there exists a physical reality, and shall refer to it as the World. I will adopt a scientific realist attitude towards fundamental physics, where I presuppose that our fundamental physical theories are engaged in the project of coming to understand the world as it is independent of our experience of it. This entails that mathematical monists, who posit that only (certain) mathematical structures exist (such as [2]), will not be on board. In a sense, the question underlying this essay concerns the role that a mathematical abstraction (namely that of computation) ought to play in our development of a theory of quantum gravity. When facing the relation between the abstract and the World, I will not take an intuitionist's viewpoint, by which mathematical statements have meaning.

I will start sharing the view that computation is a principled way towards potential infinity (§9.2), as well as the two main lessons that originate from it: unreachability (the failure to reach infinity) and universality (the ability to talk about oneself). Then I will turn to the relation between computation and the World (§9.3). I will close with some reflections on human finitude (§9.4).

This is a good place for a disclaimer. I do not understand quantum physics. I believe it is fair to say that there is a massive disagreement with regard to that understanding, ranging from agential realism [3], Everettian quantum mechanics, to various forms of reality of the wavefunction [4]. I am not an expert on gravity or quantum gravity. Neither am I a philosopher. I will approach this question with the attitude that I believe emanates from scientific endeavour: that of humbleness and ambition.

9.2 Computation: A principled way towards potential infinity

Discreteness is etched in fire at the heart of information theory. Computation theory, which is concerned with the processing of information, is therefore also fundamentally discrete. The potential infinity is reserved for the fact that sequences of finite alphabets, such as bit sequences, are arbitrarily long, or equivalently that the tape of a Turing machine is unbounded. The consequences of this discreteness stretch across any intellectual endeavour that aims to spell out the information degrees of freedom and their processing, be it to elucidate the computation carried out by the evolution of a system or to quantify resources. To the best of my knowledge, there is no comparable (hyper)computation theory in the continuum [5].

Surprisingly, quantum information theory invokes the actual infinity via the complex numbers, and consequently so does quantum computation (see page 193 for a definition of the actual infinity). This is despite the fact that it purports to describe a reality where certain observables (or: variables) are quantized. And, in quantum information theory, it is quintessential that each orthogonal basis of the Hilbert space has potential reality. This is what distinguishes a high dimensional classical information variable from a quantum bit.

9.2.1 Reaching the infinite from the finite

Kleene said that ‘an algorithm is a finite answer to an infinite number of questions’. I see computation as a principled way towards potential infinity. I am using the Aristotelian distinction between potential infinity and actual infinity [6] (see also [7]). In modern terms, the potential infinity stands for the unbounded, or for an endless process that can continue indefinitely but is never completed. It is something that is never realized as a whole—for example, the sequence of natural numbers. The actual infinity, by contrast, is a fully realized, completed totality that exists all at once—for example, the set of all natural numbers.[1] We will occasionally also distinguish between the infinite by division and by addition. The infinite by division is embodied in $2^{\aleph_0}$, the cardinal that measures the size of $\mathbb{R}$ and any of its segments. The infinite by addition is embodied in $\aleph_0$, the cardinal that measures the size of $\mathbb{N}$.

The mathematical structure of computation does not stand isolated from others, but quite the contrary. There is, first of all, a well-known equivalence between computation and formal systems, by which the unfolding of an algorithm on an input string is seen as the proof that a given (well-formed) formula in an axiomatic system is a theorem or a negation thereof (see e.g. [2] or [8, App. B]). The equivalence of computation with formal languages is even more intimate. Every algorithm accepts a formal language. Such a language can be generated by a so-called generative grammar (see e.g. [9]), which can be classified in the Chomsky hierarchy of grammars. That grammar can be seen as an active description of the language (because the grammar generates the strings), whereas the transition rules of the head of the Turing machine can be seen as a passive description thereof (because the machine accepts the strings). In either case, in my eyes both of them are a generalization of induction: a finite set of rules that can be applied an unbounded number of times. They are a finite handle towards the potential infinite, a principled path towards the unbounded. Consider again axiomatization. Its appeal is precisely that it purports to trap an infinite wealth of information or wisdom in a finite, manageable stock of basic self-evident principles [6]. Such a stock (whether self-evident or not) is the finite 'mould' that is to be applied a finite but arbitrary number of times.

As an illustration, consider the set of bit strings of any finite length, $\{0, 1\}^*$, where * denotes the Kleene star. Consider a Boolean function $f : \{0, 1\}^* \to \{0, 1\}$ and the corresponding subset of elements mapped to 1, $S = \{x \in \{0, 1\}^* \mid f(x) = 1\}$. The members of S are specified in one of two ways: by providing some condition that they, and they alone, satisfy; or by enumerating them, The first is equivalent to giving an algorithm for them, that is, providing a Turing machine that accepts every bit string x such that $f(x) = 1$, and that rejects every x such that $f(x) = 0$. If there is such a finite decision procedure for determining whether a given natural number belongs to a given set of natural numbers, then the set S is 'tidy' enough for there to be a number-predicate that defines it. Namely, it is recursive. In the vast majority of cases—as a matter of fact, in almost every case—this is not so. We must enumerate every member of S (or invoke a Turing machine with an oracle). The set is too wild to admit a finite description. We shall return to this failure in page 197.

Some have seen physical interactions as a grammar of some aspects of the World [10, 11]. In that view, local interactions are akin to the transition rules of an automaton, or the generative grammar of a formal language. The question that concerns us in this chapter is whether such a view can be upheld at a more fundamental level.

9.2.2 Introducing a **peras** *in the* **apeiron**

I find some of the early struggles with the infinite inspiring for our goals. I believe they are similar, in a modern guise, to our argument. In the following, I will rely on Moore [6].

The Greek word '*peras*' is usually translated as 'limit' or 'bound'. '*To apeiron*' refers to that which has no *peras*, the unlimited or unbounded. Anaximander initiated the enquiry: what is the 'principle' of all things, the basic stuff of which all things are made? He invoked *to apeiron* as a basic cosmological principle. In radical constrast, the Pythagoreans believed that determinacy was the effect of a general imposition of the *peras* on the *apeiron*. It was as if *to apeiron* determined what the possibilities were, and the *peras* was imposed on it to determine which of them was to be realized. Conflict between opposites could be resolved by the *peras* holding them in a harmonious balance; everything that ultimately made sense was because a *peras* was imposed on *apeiron*. The world was a system of structures built within a void, that structure it, order it and give it definite shape. And the world continues to 'breathe in' and, at the same time, to subjugate, the surrounding *apeiron*.[2]

In my eyes, identifying the finite *within* the infinite is akin to planting the seed of the *peras* in the void of *apeiron*. Presumably, this is what notions of computation ought to do on the rebellious infinite posited by our description of spacetime.

This could be formalized as follows. Given a metric space (X, d), where X is a collection of points and d a distance on them, fix a positive $\epsilon > 0$. Consider the set of points $Y \subset X$ such that $d(y_i, y_j) > \epsilon$ for all $y_i, y_j \in Y$. Then Y is equivalent (by some notion of equivalence) to a discrete set. Such Y could serve as a code, that is, hold a notion of information. Note that ϵ could be defined over a continuous or a discrete set. The notion of equivalence will depend on the purposes of formalization.

I would now like to reflect on the introduction of margins in various settings. I will start with physics and continue with natural language.

9.2.3 *Introducing margins in physics*

In physics, a margin can be understood as the boundary of a finite region. Introducing boundary conditions entails the emergence of discrete energy levels, as only integer multiples of half wavelengths are allowed, giving rise to standing waves. In condensed matter physics, the introduction of a lattice results in periodicity. In either case, they can be seen as introducing a *peras* in the *apeiron*.

In quantum theory, the situation is less clear. The wavefunction is a continuous object (with a controversial metaphysical status), whose energy levels are discrete (and ultimately dependent on the finiteness of $\hbar$). Some of the struggles in early quantum mechanics have to do precisely with a tension between the discrete and the continuous, as microscopic phenomena had both wave and particle aspects, although these aspects are never revealed simultaneously and depend on the experimental context [3].

9.2.4 *Natural languages*

In my eyes, another principled partition of the continuum is provided by natural languages, such as English or Catalan. Each language sets a scheme of description and classification of what may be considered to be a continuum of facts and affairs. Such a continuum is discretized into concepts (lexical items), which are combined recursively by means of grammar (or syntax). They provide a handle of the World. The recursive nature of grammar, acting on the discretized items, allows to describe not only the World but also fictional worlds, fantasies, imaginations, counterfactuals. Natural language partitions the World and has the means to recombine the blocks. The partition of the World is akin to imposing a *peras* on the *apeiron* of the World. The recursive recombination opens the door to potential infinity of situations. Together, they enable the capability of describing (or even considering) that which has not been perceived.

Others have expressed similar thoughts. Quine argued that it is only because there are infinitely many things that we need to operate with the fundamental notion of a thing at all, for this notion is used principally in making generalizations [12]. If there were only finitely many things, we could specify one by one what each was like.

For Leibniz, thought is the prime example of the unification in one perspective of the multiplicity of the universe, the perception of many things at the same time [13]. Naming something is, in some sense, a unification of many perceptions. Leibniz considered symbolic thinking both a strength and a weakness of human cognition. A strength because it immensely extends the putative grip of thought on reality and thereby its ability to operate. But a weakness because this grip is slippery. If find this a beautiful portrait of the situation.

As a matter of fact, Leibniz aimed to create an artificial human language, his sought-for *Characteristica universalis*, that would rely on a clear and unambiguous correspondence between signs and concepts. Such a system would allow to reduce thinking to calculating. It would allow to extend the lessons of computation to the cognitive realm. We shall return to this fascinating perspective, even if only obliquely, in the Coda.

Let us now return to our main thread and examine the scope of the potential infinite. More precisely, the scope of natural numbers. How expressive are they?

9.2.5 The long reach of natural numbers

As we saw above, the natural numbers embody the potential infinite. The tape of a Turing machine operates over natural numbers; the rules of a generative grammar can be applied a natural number of times. Now, how significant are the natural numbers when compared to other kinds of mathematical sets? How representative are they of types of infinity? A priori, they seem to be a tiny initial segment of the endless sequence of ordinals. In this sense, they constitute almost a negligible drop in the Set-hierarchical ocean [6]. Yet, a foundational result in set theory, the Löwenheim–Skolem theorem, brings to the foreground the unsuspected long reach of the natural numbers. This theorem entails that, however many true sentences from a language you are presented with, you can never rule out the hypothesis that the things being described were natural numbers. It transpires that no amount of set theory can force us to recognize that its subject matter comprises any more than that, let alone that it is the full panoply of sets with all their complex interrelations. It is as if, when we consider the hierarchy of Sets, we are forced to conclude that all but a very tiny portion of it (namely, the natural numbers) is irrelevant. It would be all the same if nothing other than this infinitesimal portion. In this sense, the natural numbers are very expressive.

What does the Löwenheim–Skolem theorem imply for the power of Turing machines? Unfortunately not much, as far as I understand. Suppose that M is a model (perhaps big, not countable) and N is a model whose universe are the natural numbers, and it satisfies the same statements as M. Then there is a Turing machine which determines a computable set of N. The Löwenheim–Skolem theorem provides an isomorphism between N and a subset of M, but the process of obtaining N from M is not effective, and hence not computable, as it needs the axiom of choice. This implies that the subset of M determined by the Turing machine (via the isomorphism) needs not be computable.

In summary, the natural numbers have a longer reach than it may seem at first sight, but this reach may be slippery.

9.2.6 *Failing at reaching the infinite*

Reaching the infinite from the finite must necessarily fail in its totality, or else the finite and the infinite would be equivalent. The failure crystallizes in the form of uncomputable problems, of which the halting problem is the most renowned. Computability brings the distinction between the finite and the infinite to the fore: it singles out those parts of the infinite that can be reached from the finite. In the words of page 194, most infinite sets are too wild to admit a finite description. This failure has been called the referent—expression tension in [8]. We called it unreachability [14].

There are deeply satisfactory manners to understand this failure. To the best of my knowledge, they are all embraced by the little known Lawvere's fixed point theorem (see [15, 16] or [14, p. 42]). Instances of this theorem include the uncomputability of the halting problem, Gödel's incompleteness theorems, Cantor's theorem, Russell's paradox in set theory and more. In my words, the theorem proves that certain mathematical structures are not 'closed' but can be transcended.[3] Examples of these structures include the set problems computable by a universal Turing machine, the set of properties for which there exists a set that collects all elements with that property, or the set of provable statements within a formal system. The transcendence is proven by means of a contradiction created via self-reference and negation. Equivalently, the contrapositive of the theorem proves that certain structures must have a fixed point. This can be interpreted as the fact that they are not too expressive; they cannot reach too far. The contraspositive of the theorem does not rely on a contradiction, but directly generates the fixed point.

As an example of the reach and depth of the previous claims, take Gödel's first incompleteness theorem (see e.g. [17] or [6]). Let us refer to a finite, consistent collection of axioms as an axiomatic base. Gödel established that given any axiomatic base A for set theory, there is some true set-theoretical statement s such that s cannot be proved using A. In other words, any axiomatic base has certain limitations. It follows that set-theoretical truth is different from provability using this or that particular axiomatic base. Gödel showed that no axiomatic system would ever be strong enough to enable us to prove every truth about sets—unless it was inconsistent, in which case it would enable us to prove anything whatsoever. In the case of Gödel's theorem, truth transcends provability via a sentence that says 'I am unprovable'. Such a paradox results in a contradiction that we usually do not want to admit, which forces us to reject a desired assumption. As per Gödel's theorem, this is the assumption that the formal system was consistent and complete. According to Cantor's theorem, there exists a surjection from the set to the power set. As per the halting problem, there exists a Turing machine that solves it. According to Russell's paradox, for any property there exists a set of elements with that property.

The profound consequences of Gödel's theorem in logic have spilled into the human mind. Some have argued that it follows from Gödel's theorem that the mind's capabilities are beyond those of a computer (e.g. [17] or [18, 19]). Others have provided more nuanced accounts (e.g. [20, 21]). I now believe that some of these issues have a bearing on intimate matters, as I will argue in the Coda.

The other side of the coin of unreachability is a beautiful structure called universality, to which we now turn. This will shed light on the role of finiteness from other angles.

9.2.7 Jumping to universality

Uncomputability proofs rely on the contradiction of self-reference and negation, or more broadly, self-reference composed with a function without a fixed point. The dual perspective provided by the contrapositive of Lawvere's theorem, i.e. the fixed point statement, gives rise to self-reference. I believe that self-reference is behind the results of universality. Take for example the existence of universal Turing machines. That is a notable statement: there is a fixed machine—an object with fixed transition rules, a fixed algorithm—that can run any possible algorithm. An object (a Turing machine) can *talk about*, or interpret, other objects like itself (other Turing machines). Usually, objects can interpret simpler objects, where 'simpler' is defined with respect to an often unspoken hierarchical structure. In the case of Turing machine, they classify bit strings. Universality ensues when a machine is expressive enough to read the description of another machine in the form of a bit string and mimic it. Simpler types of automata than Turing machines cannot interpret the code of other objects like themselves. This includes linear bounded automata, pushdown automata and finite state automata. Trying to understand the essence of such expressions of universality brought us to the investigations in [14], where we also wanted to understand the relation to universal spin models [22] and other similar structures.

We thus come to the tension between universality and unreachability: Turing machines can talk about other Turing machines (the plus side) but this allows to prove the many functions they cannot compute (the flip side). The core of the problem is the following. If 'universal' means 'all-encompassing', can something universal encompass a negation of itself? No, it can't, if it has to do so consistently. Hence this universality is frustrated, only partial, but nonetheless important. Enriched systems (such as Turing machines with an oracle for the halting problem) will possess a new kind of universality which will be frustrated by a new attack of self-reference and negation. This can be superseded by a super-super-Turing machine which is susceptible to a similar attack, and so on *ad infinitum* (ironically). This gives rise to the arithmetic hierarchy. Universality and unreachability keep chasing each other. For formal systems, the analogue of a universal Turing machine would be a weakly representative predicate (see e.g. [8, App. B]).

Now, Deutsch uses the term 'universality' in a rather broad way (unfortunately not formalized), and claims that the jump to universality can only happen in discrete systems because only these are capable of error correction [23]. I have been trying to understand this idea for years. He may be saying that the possibility of universality requires a code, which is something discrete. Without this code any possibility or number would be a valid element, disallowing the possibility of error correction. In other words, one needs to introduce a *peras* in the *apeiron*.

It may help to consider Hegel's perspective. For Hegel, to be finite is for it to be a mere aspect of the whole, something limited and up against an 'other' [6]. A finite thing's other both defines and negates it; it determines both what the thing is and what it is not. The thing is finite precisely because it can be delineated and set apart in this way. This may be the property we are after: the elements of the code must be discrete so that there is a finite distance between them and there is something 'other' in between that makes error correction possible.

Finally, similar ideas have recently been put forward in biology [8], where the inventions of so-called codes of life are understood as transitions from the continuous to the discrete. As a matter of fact, their guiding question is similar to ours: do computational principles drive evolutionary transitions? They argue that they do, and interpret major evolutionary transitions (origin of life, formation of eukaryotic cells, emergence of multicellularity, etc.) through the lens of computation theory. More precisely, biological organisms are seen as hierarchical dynamical systems that generate regularities in their phase-spaces through interactions with their environment. These emergent regularities can be interpreted as (higher-level) information patterns which may influence the (lower-level) organisms via downward causation. These loops of causation are known as *tangled hierarchies* or *strange loops* [18]. Such loops can nurture self-modelling capabilities which improve the efficiency of organisms' replication. Once such an encoding is adopted, the tangled hierarchies generate tensions (computational inconsistencies) between what is encodable within the current setup and what is possible, that is, realizable in the current environment. Within the discussion of page 197, the former corresponds to the reach of the finite from the infinite, whereas the latter corresponds to the infinite. An evolutionary transition resolves these tensions by expanding the problem-space, at the cost of generating new tensions in the expanded space, in a continual process. Ultimately, this gives rise to the biological arrow of time.

9.3 Computation and the World

After having shared what I feel are the most valuable abstract lessons of computation, let me now turn to the relation between computation and the World. We must first face some metaphysical issues.

9.3.1 The interfaces of physics with metaphysics and abstractions

Whose job is it—the physicist's or metaphysician's—to tell us whether the world is fundamentally discrete or continuous? For much of our history, the distinction between the two was blurred, merging in the beautiful notion of natural philosophy. I assume that the metaphysical and physical inquiry are intimately related, and in fact that metaphysical inquiry should begin in science. Such a stance is called naturalism or physicalism [24], and starts with the recognition that it is within science (or physics) and not in some prior philosophy, that reality is to be identified and described. It is worth noting the difficulty to explain common sense facts from physicalism, such as that we seem to have thoughts, free will or that time passes. I will ignore these important issues for the present essay.[4]

Another concern with science guiding metaphysics (as in naturalism) is that some kinds of representations used in science are not intended to have ontological import. This is made explicit by the *indispensability argument*, associated to Putnam and Quine [4]. This argument about realism for mathematical entities goes as follows:

1 We ought to have ontological commitments to all that is indispensable to our best scientific theories.

2 Mathematical entities are indispensable to our best scientific theories.

Therefore,

3 We ought to have ontological commitment to mathematical entities.

I personally do not subscribe to 1. On the other hand, I would count computation as one of our best theories, although I am not sure if it qualifies as a scientific theory. Depending on one's position in the above matters, one may weigh differently one's answers to the question at the heart of this chapter namely what ontological value should we attribute to the fundamental discreteness of computation?

There is yet another interface we should consider—that between abstractions and the World. Should mathematics guide physics, or vice versa? Since computation is equivalent to formal systems (page 193), asking about the ontological value of computation is parallel to enquiring about the relation between certain mathematical structures and the World. The mathematical structure of computation and information is discrete, whereas that of Riemannian manifolds is continuous. So we are asking whether the discrete structures featuring in computation ought to be endowed with ontological value. Recall that mathematical monists would claim that only (certain) mathematical structures exist, whereas the view perhaps closer to our experience contends that mathematical abstractions do not exist in the World in the sense that atoms do.

9.3.2 The Church–Turing thesis

If we are to ponder about the physical import of computation, the most relevant link between the World and computation, it seems to be, is given by the Church–Turing thesis. It posits that any physical process is captured by the running of a Turing machine. It entails that the abstract, theoretical model of a Turing machine (or any other equivalent model, such recursively enumerable functions, λ calculus or Post systems) is capable of modelling every possible physical process. In other words, the running of a Turing machine unveils the common nature of any dynamical physical process.[5] This thus seems to uncover something very deep (and useful) about the unfolding in time of any physical situation. On the other hand, I am not sure what computation has to say about the kinematics, which I would regard as the set of possibilities of physical situations.

Note that the Church–Turing thesis would be rendered false if we could 'witness' any form of infinity in the World, i.e. somehow trade a finite thing for an infinite one. This would allow us, for example, to solve the halting problem, as we could transform the infinite waiting time into a finite amount. A footprint of infinity could be witnessed from the implementation of the non-local game proposed by the MIP* = RE result [25], which would allow to distinguish whether a quantum state is of finite or infinite dimension (see also [7]).

9.3.3 *The physical relevance of computational complexity*

While I find the Church–Turing thesis very insightful,[6] I am unsure about the physical relevance of computational complexity theory, which is broadly concerned with quantifying the resources required by a Turing machine to solve a problem. Consider the separation between easy and hard problems, as given by P and NP, or between NP-solvable-in-practice and EXP, or between Computable (Recursive) and Semicomputable (Recursive Enumerable). How relevant are these classifications to the World? The classes rely on worst case complexity and asymptotic statements. Closely knit is the notion of a reduction, by which computational problems are classified in these classes. While mathematically beautiful and well-behaved (for example, reductions are transitive), the image of a reduction is usually tiny. Take for example the recent result MIP* = RE [25]. The reduction from an RE−complete problem, such as halting, to a problem in MIP* is so contrived that it makes you wonder what we are really learning about the World. Another example are recent works (see e.g. [26–28]) showing that problems in physics or quantum information are undecidable—similar concerns apply.

In my eyes, the problem is that many foundational choices in the theory of computational complexity are motivated by their mathematical soundness and beauty. The most important of them, it seems to me, is granting the potential infinity, which is integral to the theory. The whole theory of computation becomes trivial if the number of instances is finite, i.e. if the length of the tape is bounded. Such a language can be recognized by a dummy Finite State Automaton, where 'dummy' denotes that the answer to each instance is hardwired in the transition rules of the machine. This highlights the absurdity of only distinguishing between finite and potential infinite, as opposed to finite-but-small and finite-but-big.[7] In summary, we should be weary of drawing physical conclusions from the lessons of computational complexity.

Where I believe the distinction between the finite and the potential infinite is quintessential (and not reducible to a finite-but-small versus finite-but-big distinction) is in the philosophical problem of induction. Stating that something is the case for a finite number of cases is qualitatively different than stating it for any number of cases (the unbounded). The latter does not follow from the former. In this light, the problem of induction can be seen as the problem of jumping from the finite to the potential infinite. Hume argued that it cannot be solved.

Before addressing the heart of the matter (the existence of simples for spacetime), let me briefly zoom out and examine aspects in which the finite and infinite are similar or dissimilar.

9.3.4 *Similarities between the finite and the infinite*

Consider first the mereology of the finite and the infinite, that is, the relation between parts and whole. One hallmark of the infinite is that it can be put in one-to-one correspondence with a proper part of itself, whereas the finite cannot. In this aspect, the infinite is dissimilar from the finite. Yet, they are similar in another mereological aspect. In other words, neither a finite nor infinite set can be put in one-to-one correspondence with its power set. There is no bijection from a set S to its power set $\wp(S)$, regardless of whether S be finite or infinite. This had been long known for finite sets (because $|\wp(S)| = 2^{|S|}$), and Cantor showed that this is also true for infinite sets, leading to an (infinite) gradation of cardinalities of the infinite. If we identify $\wp(S)$ with the set of attributes of S (see [7]), we could say that no finite set can encompass its own attributes, and no infinite set can do so either. This can be proven with the liar paradox, or Lawvere's theorem, or the diagonalization argument (all equivalent; see page 197). The powerful attack of self-reference and negation applies to both the finite and the infinite.

Another similarity between the finite and the infinite is that they both allow for local triviality. I define locality triviality as the fact that change be trivial in a small enough vicinity. In other words, it entails that there is a neutral centre of oscillation. In my eyes, local triviality is a funding principle in mathematics and theoretical physics. For the former, it is often embodied in the form of an identity operation, which plays a crucial role in algebraic structure (a vector space, a monoid, a group, a manifold...) as well as in analysis (the notion of continuity, the limit of infinitesimal change...). For the latter, the paradigm would be the harmonic oscillator, or any description of a physical phenomenon relying on the aforementioned mathematical structures. My point here is that local triviality can be discrete or continuous—a group can be finite or infinite, such as a reflection or a rotation, or a vector space can have a finite or infinite number of elements. Local triviality, thus, can be instantiated both in finite and infinite structures and will not help us separate them apart.

Probably the most important example of locality triviality is that of persistence and change over time. A compelling account of the passing of time faces the challenge that (1) things change, but (2) not much. They change but somewhat persist. The second point implies that there is a sense of continuity; if it were not the case, temporal evolution would give rise to a rugged landscape of disconnected spatial slices, which may not allow to develop any knowledge at all. Historically, accounts have swung from one extreme to another [very roughly, everything changes (Heraclitus), to nothing does and change is an appearance (Parmenides)]. The challenge is that change appears to be both a unity and a multiplicity. I imagine that both a discrete and a continuous account of time would feature local triviality. And local triviality can be expressed both in the finite and the infinite, as argued earlier.

9.3.5 Do simples exist?

It is high time we face the central question 'Do simples of spacetime exist?' which we are in fact replacing by 'Does computation hint at the existence of such simples?'

If a simple is something that cannot be divided, all I can do is remark the various notions of divisibility proposed by Holden [29]. He defines an extended entity as

- *physically divisible* if and only if its spatially distinct parts can be broken apart by natural processes and separated from one another;
- *metaphysically divisible* if and only if it is logically possible that its spatially distinct parts could exist separately from one another;
- *formally divisible* if and only if it has parts that can be distinguished by their spatial properties, regardless of whether those parts can be separated from one another;
- *intellectually divisible* if and only if a mind could represent it in thought as containing diverse parts and regardless of whether these parts are genuinely spatially distinct.

It transpires that we are discussing if spacetime is composed of physically indivisible units. Or if, on the contrary, spacetime is physically divisible *ad infinitum.*

An insightful distinction is that between the actual parts and the potential parts doctrine [29]. In the former, all parts into which a body can be divided are already present in the body prior to division. They are fully fledged concrete existents. In the latter, division creates these parts. It thus requires an ever-increasable but always actually finite number of parts.

If we were to posit the infinite physical divisibility of spacetime, I presume it would be set in the potential parts doctrine. The actual parts doctrine together with infinite divisibility leads to contradictions ensued by the paradoxes of the infinite. Some can be found in page 205. These paradoxes, however, disappear for the potential parts doctrine.

However, the potential parts doctrine relies on the potential infinite, and each potential infinite presupposes an actual infinite, if rigorously applied mathematically. This is formulated in the *Domain Principle* by Priest [12] (following Hallett):

> For every potential infinity there is a corresponding actual infinity.

The key idea is that the domain cannot itself be something variable, since otherwise each fixed support for the study would collapse. The domain is a definite, actually infinite series of values; such totalization is conceptually unavoidable. These considerations spill beyond mathematics into conceptual struggles with the infinite, as explained by Priest, but I am unsure how they impact on Holden's arguments.

Be it as it may, I believe that no form of infinity can be supported in the physical world, as I tried to argue in [7]. It seems to me that a thing that exists in the World must have a beginning and an end. One must be able to attribute it an energy. If it has no margins, any energy attribution would be inconsistent. This brings to the fore the importance of margins. I thus seem to align with the Pythagoreans, vouching for the imposition of a bound in the unbounded, a *peras* in the *apeiron*.

Finally, let me note that the discreteness or continuity of time would make a difference in the following argument. Metaphysicians ponder over the ontology of time. That is, whether only the present exists, or also the past and the future [24, 30–32]. One criticism to the view that only the present exists (called presentism) is that the present seems to be so thin that it can barely exist. If we imagine the present like an immense knife separating the past from the future, the past and the future would squeeze the blade of the knife to a point where it would be so thin that it would hardly be anything on its own. The ontology of the present would be vanishingly small. This criticism, however, seems to rely on the assumption that time is accurately described by a real parameter, and thus that it is fundamentally continuous. If, on the other hand, time were fundamentally discrete, the ontology of the present would not be vanishingly small. That is, if simples of time existed, presentism would be spared of this criticism. Note that presentism has other problems, notably its apparent incompatibility with relativity.

As a final remark, I would like to recognize the success of infinities.

9.3.6 The embezzling success of infinities

Both geometry and analysis lean heavily on the infinite. Calculus involves generalizations about finite quantities, but the whole enterprise succeeds only because there are infinitely many of them. As a matter fact, study of what is finite is sometimes only possible in an infinite framework (cf. the Domain Principle in page 204). As a consequence, both the infinitesimally small and the infinitely large have led to embezzling success on the physical theories that rely on their beautiful mathematical properties.

But we must warn against the perils of infinities. The infinite is not a determined or finished whole, but it is riddled with paradoxes. There are paradoxes of the infinitely small, the infinitely big, or the one and the many (see e.g. [6]

or [33]). I enjoy the light cast by Priest on the infinite: he sees the infinite as the paradox of reaching beyond the limit of iteration, and analyzes the situation in parallel to trying to reach beyond the limits of thought, the limits of cognition, and the limits of conception [12]. Consider, for example, the relations between a line and the points in it. Mathematics encourages the idea that continuity can be built up out of points, whereas the most that can be built up out of points is a series of infinitely repeated discontinuities. Continuity is something basic, and points are just a mathematical fiction wrought from it [6]. Bergson held that mathematics involved a fundamental falsification of continuity because of its commitment to the notion of a point. Gödel said that summing up all the points we still do not get the line; rather the points form some kind of scaffold on the line.

My impression is that physics has a mixed relation with infinity. Sometimes it enjoys its mathematical beauty and other times it suffers from its 'mishaviour'. For example, physics often relies on idealizations, which serve as a fulcrum from which to reach the physically relevant case [7]. Think of the role played by the thermodynamic limit, zero temperature, or pure states, in relation to finite size systems, finite temperature or mixed states, respectively. Indeed, according to Ney, an idealization is a false assumption introduced in a theory in order to make it simpler to use [24]. The point is that the ideal case often involves an infinity, whereas the physically relevant case does not. So, in these scenarios, resorting to the infinite is helpful, mainly because it admits a simpler mathematical description. But other times, theories defined directly at infinity involve some divergencies. In order to tame these divergences, one usually discretizes an infinite object (such as space or time or momentum) and carefully takes the limit to infinity (often called the continuum limit). Lattice gauge theories follow this strategy, as well as some approaches to quantum gravity such as (causal) dynamical triangulation. The goal is to rid the first infinity of some of its 'weeds'.

9.4 Coda: Human finitude

I would like to close with a remark on our very awkward existence. One cannot help but wonder how our mind engages in such deep discussions from within our perishable bodies. How can we conceive of infinity from within our finite beings? We seem to transcend our limitations by means of reason. There is much to say about that; I find the final part of [6] very moving, on which I will now lean heavily. I have also reflected upon the wound of infinity within the human condition in [34].

Some have argued that the concept of infinite is an *a priori* that belongs to a kind of native mental lens through which we view things, and have further posited the infinitude of reason (Kant). I don't feel compelled by this view.

Hilbert, instead, said that there is no such thing as the infinite, but we can proceed as if it existed. He added that nowhere is the infinite realized, it is neither present in nature nor admissible as a foundation in our rational thinking, but its role is merely that of an idea. This is more in line with the unspoken position of this essay. Existentialists (such as Jaspers) have argued that one's consciousness of one's own finitude, contrasted with the infinitude that lay beyond one's horizon, relates to the absurdity of life. I may agree with the analysis but not the conclusion.

I believe that the tension between our finitude and the vastness of the world is parallel to what has been at the centre of our discussion, that is, the tension between the controlled approach to the potential infinity provided by computation versus the untamed infinites of the reals in our descriptions of spacetime. After all, the very definition of a Turing machine was inspired by the external behaviour of a human computer: it is an abstraction thereof. A Turing machine mimics a human computer with a finite memory and a finite set of rules therein, equipped with a notebook (of unbounded length; that's questionable) with a written input that she can overwrite according to the rules. In other words, the way Turing machines approach potential infinity is similar to the way humans do it.[8] Both may be paths towards imposing a *peras* in the *apeiron*. For cognition to be possible, the world must be 'tamed' by margins that allow us to classify it and thus comprehend it. This puts an existential gloss on the ideas of this chapter.

There is an aspect in which each of us is infinite. While we are aware that we only live for a finite time, our experience of life is total, absolute. (I am assuming that we cannot invoke any sort of existence after or before this one—a very sober assumption, in my eyes.) In other words, I cannot experience life without being alive. Therefore, as far as my experience is concerned, my life has no boundaries. In this respect, my life resembles the infinite. To quote Wittgenstein, 'Death is not an event in life: we do not live to experience death.' Yet, at the same time, I am aware of my own finitude. All I can do is try to incarnate this contradiction with some dignity.

Notes

1. Some totalities are immeasurably big, too big to be regarded as genuine sets at all; the totality of all ordinals is an example. The collective infinitude of all sets is potential, not actual, because there is no such thing as the set of all sets, or the largest infinity. Cantor described them as inconsistent totalities; he said they are manies too big to be regarded as ones [6].
2. The regular cycles of the planets, the recurring patterns in nature, the finely proportioned structures in the physical world—for the Pythagoreans, these all betokened rhyme and reason, that which is comprehensible and good, that which has a *peras*. The *apeiron*, by contrast, was something abhorrent. They believed that because it had no end in the sense of limit (*peras*), it equally had no end in the sense of purpose or destiny (*telos*). Integral to this

picture were the natural numbers. Nowadays, destiny has no place in scientific theories, and moral considerations are disentangled from physical ones.

3. I use the word 'transcendence' because it must be possible to cast the nucleus of Lawvere's theorem in terms of Priest's Schema T, which is a tension between existence, closure and transcendence [12].
4. As you may have noticed, this text continuously stumbles upon issues bordering the central one that I do not know how to resolve. As a consequence, the writing tends to be rather apophatic, that is, point out that which is not.
5. More than in any other case, this connection has been prominent in the direction *from* abstraction *to* the World, leading to the construction of computers and ultimately the revolution of our information age.
6. I am saying nothing about the strong Church–Turing thesis, by which Turing machines can simulate any physical process efficiently.
7. More broadly, in mathematics many distinctions are trivial for finite objects. For example, all norms are equivalent in finite dimensional vector spaces. Or the question whether commuting Hilbert spaces are equivalent to tensor product Hilbert spaces (recently resolved in the negative [25]) is trivial for finite dimensional ones.
8. This analogy has been made by others too, e.g. by Pavlovic recently [35].

References

[1] Deutsche, D. (1997) *The Fabric of Reality*. Penguin Books.

[2] Tegmark, M. (2008). The mathematical universe. *Foundations of Physics*, 38, 101–150.

[3] Barad, K. (2007). *Meeting the Universe Halfway*. Duke University Press.

[4] Ney, A., & Albert, D. Z. (Eds.). (2013). *The Wavefunction: Essays on the Metaphysics of Quantum Mechanics*. Oxford University Press.

[5] Longley, J. R., & Normann, D. (2015). *Higher-Order Computability*. Springer.

[6] Moore, A. W. (2019). *The Infinite* (3rd ed.). Routledge.

[7] De les Coves, G. (2024b). Quantum theory: Ideals, infinities and pluralities. In M. Cuffaro & S. Hartmann (Eds.), *Open Systems: Physics, Metaphysics, and Methodology*. Oxford University Press.

[8] Prokopenko, M., Davies, P. C. W., Harré, M., Heisler, M., Kuncic, Z., Lewis, G. F., Livson, O., Lizier, J. T., & Rosas, F. E. (2024). Biological arrow of time: Emergence of tangled information hierarchies and self-modelling dynamics.

[9] Kozen, D. C. (1997). *Automata and Computability*. Springer.

[10] Reinhart, T., & De las Cuevas, G. (2022). The Grammar of the Ising Model: A New Complexity Hierarchy. *arXiv preprint arXiv:2208.08301*. `https://arxiv.org/abs/2208.08301`

[11] Stengele, S., Drexel, D., & De las Cuevas, G. (2023). Classical spin Hamiltonians are context-sensitive languages. *Proceedings of the Royal Society A*, 479, 20220553.

[12] Priest, G. (2001). *Beyond the Limits of Thought*. Clarendon Press.

[13] Antognazza, M. R. (2016). *Leibniz: A Very Short Introduction*. Oxford University Press.

[14] Gonda, T., Reinhart, T., Stengele, S., & De les Coves, G. (2024). A framework for universality in physics, computer science and beyond. *Compositionality (to appear)*. arXiv:2307.06851. `https://arxiv.org/abs/2307.06851`

[15] Lawvere, F. W. (1969). Diagonal arguments and cartesian closed categories. *Lecture Notes in Mathematics*, 92, 134–145.

[16] Yanofsky, N. S. (2003). A universal approach to self-referential paradoxes, incompleteness and fixed points. *Bulletin of Symbolic Logic*, 9, 362–386.

[17] Nagel, E., & Newman, J. R. (1958). *Gödel's Proof*. Routledge.

[18] Hofstadter, D. (1979). *Gödel, Escher, Bach*. Basic Books.

[19] Penrose, R. (1989). *The Emperor's New Mind*. Oxford University Press.

[20] Raatikainen, P. (2005). On the philosophical relevance of Gödel's incompleteness theorems. *Revue Internationale de Philosophie*, 4, 513.

[21] Fraser, P. T., Solé, R., & De las Cuevas, G. (2021). Why can the brain (and not the computer) make sense of the liar paradox? *Frontiers in Ecology and Evolution*, 9, 802300.

[22] Reinhart, T., Engel, B., & De les Coves, G. (2024). The Structure of Emulations in Classical Spin Models: Modularity and Universality. *arXiv preprint arXiv:2407.13428*. `https://arxiv.org/abs/2407.13428`

[23] Deutsch, D. (2011). *The Beginning of Infinity*. Allen Lane. `https://www.thebeginningofinfinity.com`

[24] Ney, A. (2014). *Metaphysics: An Introduction*. Routledge.

[25] Ji, Z., Natarajan, A., Vidick, T., Wright, J., & Yuen, H. (2020). MIP* = RE. *arXiv preprint arXiv:2001.04383*. `https://arxiv.org/abs/2001.04383`

[26] Cubitt, T. S., Pérez-García, D., & Wolf, M. M. (2015). Undecidability of the spectral gap. *Nature*, 528, 207.

[27] De las Cuevas, G., Cubitt, T. S., Cirac, J. I., Wolf, M. M., & Pérez-García, D. (2016). Fundamental limitations in the purifications of tensor networks. *Journal of Mathematical Physics*, 57, 071902.

[28] Klingler, A., van der Eyden, M., Stengele, S., Reinhart, T., & De las Cuevas, G. (2023). Many bounded versions of undecidable problems are NP-hard. *SciPost Physics*, 14, 173. `https://scipost.org/SciPostPhys.14.6.173`

[29] Holden, T. (2004). *The Architecture of Matter*. Oxford University Press.

[30] Loux, M. J., & Crisp, T. M. (2017). *Metaphysics: A Contemporary Introduction* (4th ed.). Routledge.

[31] Carroll, J. W., & Markosian, N. (2010). *An Introduction to Metaphysics*. Cambridge Introductions to Philosophy.

[32] G. De les Coves. (2025) Do the past and the future exist? Time, space and determinism. In R. Sole and C. Lalueza-Fox, editors, *The Invention of Time* (To appear, 2025).

[33] Rayo, A. (2019). *On the Brink of Paradox: Highlights from the Intersection Between Mathematics and Philosophy*. MIT Press.

[34] De les Coves, G. (2024a). Being human: The wound of infinity. An intimate encounter of science and culture. In J. Ramoneda & R. Garrigasait (Eds.), *In an Open Field: Contemporary Cultural Practices*. Institut Ramon Llull & Arcàdia.

[35] Pang, J. C., Aquino, K. M., Oldehinkel, M., Robinson, P. A., Fulcher, B. D., Breakspear, M., & Fornito, A. (2023). Geometric constraints on human brain function. *Nature*, 618, 566–574.

10 On the Nature of Time

Stephen Wolfram

10.1 The computational view of time

Time is a central feature of human experience. But what actually is it? In traditional scientific accounts it's often represented as some kind of coordinate much like space (though a coordinate that for some reason is always systematically increasing for us). But while this may be a useful mathematical description, it's not telling us anything about what time in a sense "intrinsically is".

We get closer as soon as we start thinking in computational terms. Because then it's natural for us to think of successive states of the world as being computed one from the last by the progressive application of some computational rule. And this suggests that we can identify the progress of time with the "progressive doing of computation by the universe".

But does this just mean that we are replacing a "time coordinate" with a "computational step count"? No, it doesn't because of the phenomenon of computational irreducibility [1]. With the traditional mathematical idea of a time coordinate one typically imagines that this coordinate can be "set to any value", and that then one can immediately calculate the state of the system at that time. But computational irreducibility implies that it's not that easy. Because it says that there's often essentially no better way to find what a system will do than by explicitly tracing through each step in its evolution.

In the pictures on the left there's computational reducibility, and one can readily see what state will be after any number of steps t. But in the pictures on the right there's (presumably) computational irreducibility, so that the only way to tell what will happen after t steps is effectively to run all those steps (Fig. 10.1).

And what this implies is that there's a certain robustness to time when viewed in these computational terms. There's no way to "jump ahead" in time;

DOI: 10.4324/9781003546139-14

Figure 10.1 Computational reducibility versus irreducibility.

the only way to find out what will happen in the future is to go through the irreducible computational steps to get there.

There are simple idealized systems (say with purely periodic behavior) where there's computational reducibility, and where there isn't any robust notion of the progress of time. But the point is that – as the Principle of Computational Equivalence implies – our universe is inevitably full of computational irreducibility which in effect defines a robust notion of the progress of time [2].

10.2 The role of the observer

That time is a reflection of the progress of computation in the universe is an important starting point. But it's not the end of the story. For example, here's an immediate issue. If we have a computational rule that determines each successive state of a system, it's at least in principle possible to know the whole future of the system. So given this why, do we have the experience of the future only "unfolding as it happens"?

It's fundamentally because of the way we are as observers [3]. If the underlying system is computationally irreducible, then to work out its future behavior, an irreducible amount of computational work is required. But it's a core feature of observers like us that we are computationally bounded. So we can't do all that irreducible computational work to "know the whole future" – and instead we're effectively stuck just doing computation alongside the system itself, never able to substantially "jump ahead", and only able to see the future "progressively unfold".

In essence, therefore, we experience time because of the interplay between our computational boundedness as observers, and the computational irreducibility of underlying processes in the universe. If we were not computationally bounded, we could "perceive the whole of the future in one gulp" and we wouldn't need a notion of time at all. And if there wasn't underlying computational irreducibility, there wouldn't be the kind of "progressive revealing of the future" that we associate with our experience of time.

A notable feature of our everyday perception of time is that it seems to "flow only in one direction" – so that, for example, it's generally much easier to remember the past than to predict the future. And this is closely related to the second law of thermodynamics, which (as I've argued at length elsewhere [4]) is once again a result of the interplay between underlying computational irreducibility and our computational boundedness. Yes, the microscopic laws of physics may be reversible (and indeed if our system is simple – and computationally reducible – enough of this reversibility may "shine through"). But the point is that computational irreducibility is in a sense a much stronger force.

Imagine that we prepare a state to have orderly structure. If its evolution is computationally irreducible, then this structure will effectively be "encrypted" to the point where a computationally bounded observer can't recognize the structure. Given underlying reversibility, the structure is in some sense inevitably "still there" – but it can't be "accessed" by a computationally bounded observer. And as a result, such an observer will perceive a definite flow from orderliness, in what is prepared, to disorderliness, in what is observed. (In principle one might think it should be possible to set up a state that will "behave anti-thermodynamically" – but the point is that to do so would require predicting a computationally irreducible process, which a computationally bounded observer can't do.)

One of the longstanding confusions about the nature of time has to do with its "mathematical similarity" to space. And indeed ever since the early days of relativity theory, it has been seemed convenient to talk about "spacetime" in which notions of space and time are bundled together.

But in our Physics Project [5] that's not at all how things fundamentally work. At the lowest level the state of the universe is represented by a hypergraph [6] which captures what can be thought of as the "spatial relations" between discrete "atoms of space". Time then corresponds to the progressive rewriting of this hypergraph [7].

And in a sense the "atoms of time" are the elementary "rewriting events" that occur. If the "output" from one event is needed to provide "input" to another, then we can think of the first event as preceding the second event in time – and the events as being "timelike separated". And in general we can construct a causal graph that shows the dependencies between different events [8].

So how does this relate to time – and spacetime? As we'll discuss below, our everyday experience of time is that it follows a single thread. And so we tend to want to "parse" the causal graph of elementary events into a series of slices that we can view as corresponding to "successive times". As in standard relativity theory [9], there typically isn't a unique way to assign a sequence of such "simultaneity surfaces", with the result that there are different "reference frames" in which the identifications of space and time are different.

The complete causal graph bundles together what we usually think of as space with what we usually think of as time. But ultimately the progress of time is always associated with some choice of successive events that "computationally build on each other". And, yes, it's more complicated because of the possibilities of different choices. But the basic idea of the progress of time as "the doing of computation" is very much the same. (In a sense time represents "computational progress" in the universe, while space represents the "layout of its data structure".)

Very much as in the derivation of the second law (or of fluid mechanics from molecular dynamics), the derivation of Einstein's equations for the large-scale behavior of spacetime from the underlying causal graph of hypergraph rewriting depends on the fact that we are computationally bounded observers [10]. But even though we're computationally bounded, we still have to "have something going on inside", or we wouldn't record – or sense – any "progress in time".

It seems to be the essence of observers like us – as captured in my recent Observer Theory [3] – that we equivalence many different states of the world to derive our internal perception of "what's going on outside". And at some rough level we might imagine that we're sensing time passing by the rate at which we add to those internal perceptions. If we're not adding to the perceptions, then in effect time will stop for us – as happens if we're asleep, anesthetized or dead.

It's worth mentioning that in some extreme situations, it's not the internal structure of the observer that makes perceived time stop; instead it's the underlying structure of the universe itself. As we've mentioned, the "progress of the universe" is associated with successive rewriting of the underlying hypergraph. But when there's been "too much activity in the hypergraph" (which physically corresponds roughly to too much energy-momentum), one can end up with a situation in which "there are no more rewrites that can be done" – so that in effect some part of the universe can no longer progress, and "time stops" there [11]. It's analogous to what happens at a spacelike singularity (normally associated with a black hole) in traditional general relativity. But now it has a very direct computational interpretation: one's reached a "fixed point" at which there's no more computation to do. And so there's no progress to make in time.

10.3 Multiple threads of time

Our strong human experience is that time progresses as a single thread. But now our Physics Project suggests that at an underlying level, time is actually in effect multithreaded, or, in other words, that there are many different "paths of history" that the universe follows [12]. And it is only because of the way we as observers sample things that we experience time as a single thread.

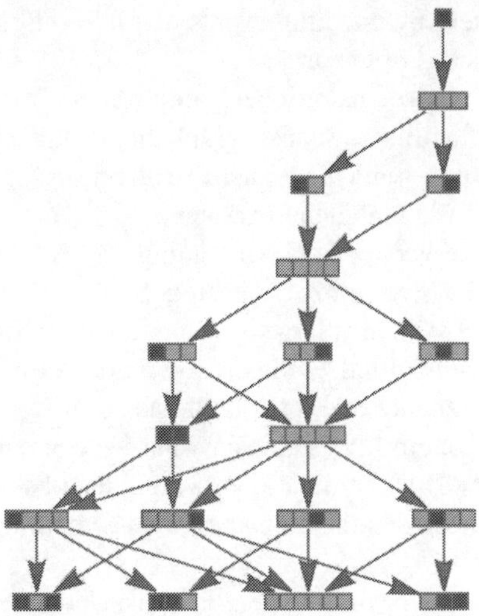

Figure 10.2 A multiway graph.

At the level of a particular underlying hypergraph the point is that there may be many different updating events that can occur, and each sequence of such updating event defines a different "path of history". We can summarize all these paths of history in a multiway graph in which we merge identical states that arise (Fig. 10.2):

But given this underlying structure, why is it that we as observers believe that time progresses as a single thread? It all has to do with the notion of branchial space [13], and our presence within branchial space. The presence of many paths of history is what leads to quantum mechanics; the fact that we as observers ultimately perceive just one path is associated with the traditionally-quite-mysterious phenomenon of "measurement" in quantum mechanics [14].

When we talked about causal graphs above, we said that we could "parse" them as a series of "spacelike" slices corresponding to instantaneous "states of space" – represented by spatial hypergraphs. And by analogy we can similarly imagine breaking multiway graphs into "instantaneous slices". But now these slices don't represent states of ordinary space; instead they represent states of what we call branchial space.

Ordinary space is "knitted together" by updating events that have causal effects on other events that can be thought of as "located at different places in space". (Or, said differently, space is knitted together by the overlaps of the elementary light cones of different events.) Now we can think of branchial space as being "knitted together" by updating events that have effects on events that end up on different branches of history.

(In general there is a close analogy between ordinary space and branchial space, and we can define a multiway causal graph that includes both "spacelike" and "branchlike" directions – with the branchlike direction supporting, not light cones, but what we can call entanglement cones.)

So how do we as observers parse what's going on? A key point is that we are inevitably part of the system we're observing. So the branching (and merging) that's going on in the system at large is also going on in us. That means we have to ask how a "branching mind" will perceive a branching universe. Underneath, there are lots of branches and lots of "threads of history". And there's a lot of computational irreducibility (and even what we can call multicomputational irreducibility [15]). But computationally bounded observers like us have to equivalence most of those details to wind up with something that "fits in our finite minds".

We can make an analogy to what happens in a gas. Underneath, there are lots of molecules bouncing around (and behaving in computationally irreducible ways). But observers like us are big compared to molecules, and (being computationally bounded) we don't get to perceive their individual behavior, but only their aggregate behavior – from which we extract a thin set of computationally reducible "fluid-dynamics-level" features.

And it's basically the same story with the underlying structure of space. Underneath, there's an elaborately changing network of discrete atoms of space. But as large, computationally bounded observers we can only sample aggregate features in which many details have been equivalenced, and in which space tends to seem continuous and describable in basically computationally reducible ways.

So what about branchial space? Well, it's basically the same story. Our minds are "big", in the sense that they span many individual branches of history. And they're computationally bounded so they can't perceive the details of all those branches, but only certain aggregated features. And in a first approximation what then emerges is in effect a single aggregated thread of history.

With sufficiently careful measurements we can sometimes see "quantum effects" in which multiple threads of history are in evidence. But at a direct human level we always seem to aggregate things to the point where what we perceive is just a single thread of history – or in effect a single thread of progression in time.

It's not immediately obvious that any of these "aggregations" will work. It could be that important effects we perceive in gases would depend on phenomena at the level of individual molecules. Or that to understand the large-scale structure of space, we'd continually be having to think about detailed features of atoms of space. Or, similarly, that we'd never be able to maintain a "consistent view of history", and that instead we'd always be having to trace lots of individual threads of history.

But the key point is that for us to stay as computationally bounded observers we have to pick out only features that are computationally reducible – or in effect boundedly simple to describe.

Closely related to our computational boundedness is the important assumption we make that we as observers have a certain persistence [16]. At every moment in time, we are made from different atoms of space and different branches in the multiway graph. Yet we believe we are still "the same us". And the crucial physical fact (that has to be derived in our model) is that in ordinary circumstances there's no inconsistency in doing this.

So the result is that even though there are many "threads of time" at the lowest level – representing many different "quantum branches" – observers like us can (usually) successfully still view there as being a single consistent perceived thread of time.

But there's another issue here. It's one thing to say that a single observer (say a single human mind or a single measuring device) can perceive history to follow a single, consistent thread. But what about different human minds or different measuring devices? Why should they perceive any kind of consistent "objective reality"?

Essentially the answer, I think, is that they're all sufficiently nearby in branchial space. If we think about physical space, observers in different parts of the universe will clearly "see different things happening". The "laws of physics" may be the same – but what star (if any) is nearby will be different. Yet (at least for the foreseeable future) for all of us humans, it's always the same star that's nearby.

And so it is, presumably, in branchial space. There's some small patch in which we humans – with our shared origins – exist. And it's presumably because that patch is small relative to all of branchial space that all of us perceive a consistent thread of history and a common objective reality.

There are many subtleties to this, many of which aren't yet fully worked out. In physical space, we know that effects can in principle spread at the speed of light. And in branchial space the analog is that effects can spread at the maximum entanglement speed (whose value we don't know, though it's related by Planck unit conversions to the elementary length and elementary time [17]). But for maintaining our shared "objective" view of the universe, it's crucial that we're not all going off in different directions at the speed of light. And of course the reason that doesn't happen is that we don't have zero mass. And indeed presumably nonzero mass is a critical part of being observers like us.

In our Physics Project it's roughly the density of events in the hypergraph that determines the density of energy (and mass) in physical space (with their associated gravitational effects). And similarly it's roughly the density of events in the multiway graph (or in branchial graph slices) that determines the density of action – the relativistically invariant analog of energy – in branchial space (with its associated effects on quantum phase). And though it's not yet completely clear how this works, it seems likely that once again when there's mass, effects don't just "go off at the maximum entanglement speed in all directions", but instead stay nearby.

There are definitely connections between "staying at the same place", believing one is persistent, and being computationally bounded. But these are what seem necessary for us to have our typical view of time as a single thread. In principle we can imagine observers very different from us – say with minds (like the inside of an idealized quantum computer) capable of experiencing many different threads of history. But the Principle of Computational Equivalence suggests that there's a high bar for such observers. They need to be able to deal with not only computational irreducibility but also multicomputational irreducibility, in which one includes both the process of computing new states and the process of equivalencing states.

And so for observers that are "anything like us" we can expect that once again time will tend to be as we normally experience it, following a single thread, consistent between observers.

(It's worth mentioning that all of this only works for observers like us "in situations like ours". For example, at the "entanglement horizon" for a black hole [18] – where branchially-oriented edges in the multiway causal graph get "trapped" – time as we know it in some sense "disintegrates" because an observer won't be able to "knit together" the different branches of history to "form a consistent classical thought" about what happens.)

10.4 Time in the ruliad

In what we've discussed so far we can think of the progress of time as being associated with the repeated application of rules that progressively "rewrite the state of the universe". In the previous section we saw that these rules can be applied in many different ways, leading to many different underlying threads of history.

But so far we've imagined that the rules that get applied are always the same – leaving us with the mystery of "Why those rules, and not others?" But this is where the ruliad comes in [19]. Because the ruliad involves no such seemingly arbitrary choices: it's what you get by following all possible computational rules.

One can imagine many bases for the ruliad. One can make it from all possible hypergraph rewritings or all possible (multiway) Turing machines. But in the end it's a single, unique thing: the entangled limit of all possible computational processes. There's a sense in which "everything can happen somewhere" in the ruliad. But what gives the ruliad structure is that there's a definite (essentially geometrical) way in which all those different things that can happen are arranged and connected.

So what is our perception of the ruliad? Inevitably we're part of the ruliad – so we're observing it "from the inside". But the crucial point is that what we perceive about it depends on what we are like as observers. And my big surprise in the past few years has been that assuming even just a little about what we're like as observers immediately implies that what we perceive of the ruliad follows the core laws of physics we know. In other words, by assuming what we're like as observers, we can in effect derive our laws of physics.

The key to all this is the interplay between the computational irreducibility of underlying behavior in the ruliad, and our computational boundedness as observers (together with our related assumption of our persistence). And it's this interplay that gives us the second law in statistical mechanics, the Einstein equations for the structure of spacetime and (we think) the path integral in quantum mechanics. In effect what's happening is that our computational boundedness as observers makes us equivalence things to the point where we are sampling only computationally reducible slices of the ruliad, whose characteristics can be described using recognizable laws of physics.

So where does time fit into all of this? A central feature of the ruliad is that it's unique – and everything about it is "abstractly necessary". Much as given the definition of numbers, addition and equality, it's inevitable that one gets $1 + 1 = 2$. Similarly given the definition of computation, it's inevitable that one gets the ruliad. Or, in other words, there's no question about whether the ruliad exists; it's just an abstract construct that inevitably follows from abstract definitions.

And so at some level this means that the ruliad inevitably just "exists as a complete thing". And so if one could "view it from outside", one could think of it as just a single timeless object, with no notion of time.

But the crucial point is that we don't get to "view it from the outside". We're embedded within it. And, what's more, we must view it through the "lens" of our computational boundedness. And this is why we inevitably end up with a notion of time.

We observe the ruliad from some point within it. If we were not computationally bounded, then we could immediately compute what the whole ruliad is like. But in actuality we can only discover the ruliad "one computationally bounded step at a time" – in effect progressively applying bounded computations to "move through rulial space".

So even though in some abstract sense "the whole ruliad is already there", we only get to explore it step by step. And that's what gives us our notion of time, through which we "progress".

Inevitably, there are many different paths that we could follow through the ruliad. And indeed every mind (and every observer like us) – with its distinct inner experience – presumably follows a different path. But as we described for branchial space, the reason we have a shared notion of “objective reality” is presumably that we are all very close together in rulial space; we form in a sense a tight “rulial flock”.

It’s worth pointing out that not every sampling of the ruliad that may be accessible to us conveniently corresponds to exploration of progressive slices of time. Yes, that kind of “progression in time” is characteristic of our physical experience, and our typical way of describing it. But what about our experience, say, of mathematics?

The first point to make is that just as the ruliad contains all possible physics, it also contains all possible mathematics [20]. If we construct the ruliad, say from hypergraphs, the nodes are now not “atoms of space”, but instead abstract elements (that in general we call emes [21]) that form pieces of mathematical expressions and mathematical theorems. We can think of these abstract elements as being laid out now not in physical space, but in some abstract metamathematical space.

In our physical experience, we tend to remain localized in physical space, branchial space, etc. But in “doing mathematics”, it’s more as if we’re progressively expanding in metamathematical space, carving out some domain of “theorems we assume are true”. And while we could identify some kind of “path of expansion” to let us define some analog of time, it’s not a necessary feature of the way we explore the ruliad.

Different places in the ruliad in a sense correspond to describing things using different rules. And by analogy to the concept of motion in physical space, we can effectively “move” from one place to another in the ruliad by translating the computations done by one set of rules to computations done by another. (And, yes, it’s nontrivial to even have the possibility of “pure motion” [22].) But if we indeed remain localized in the ruliad (and can maintain what we can think of as our “coherent identity”), then it’s natural to think of there being a “path of motion” along which we progress “with time”. But when we’re just “expanding our horizons” to encompass more paradigms and to bring more of rulial space into what’s covered by our minds (so that in effect we’re “expanding in rulial space”), it’s not really the same story. We’re not thinking of ourselves as “doing computation in order to move”. Instead, we’re just identifying equivalences and using them to expand our definition of ourselves, which is something that we can at least approximate (much like in “quantum measurement” in traditional physics) as happening “outside of time”. Ultimately, though, everything that happens must be the result of computations that occur. It’s just that we don’t usually “package” these into what we can describe as a definite thread of time.

10.5 So what in the end is time?

From the paradigm (and Physics Project ideas) that we've discussed here, the question "What is time?" is at some level simple: time is what progresses when one applies computational rules. But what's critical is that time can in effect be defined abstractly, independent of the details of those rules, or the "substrate" to which they're applied. And what makes this possible is the Principle of Computational Equivalence, and the ubiquitous phenomenon of computational irreducibility that it implies.

To begin with, the fact that time can robustly be thought of as "progressing", in effect in a linear chain, is a consequence of computational irreducibility – because computational irreducibility is what tells us that computationally bounded observers like us can't in general ever "jump ahead"; we just have to follow a linear chain of steps.

But there's something else as well. The Principle of Computational Equivalence implies that there's in a sense just one (ubiquitous) kind of computational irreducibility. So when we look at different systems following different irreducible computational rules, there's inevitably a certain universality to what they do. In effect they're all "accumulating computational effects" in the same way. Or in essence progressing through time in the same way.

There's a close analogy here with heat. It could be that there'd be detailed molecular motion that even on a large scale worked noticeably differently in different materials [23]. But the fact is that we end up being able to characterize any such motion just by saying that it represents a certain amount of heat, without getting into more details. And that's very much the same kind of thing as being able to say that such-and-such an amount of time has passed, without having to get into the details of how some clock or other system that reflects the passage of time actually works.

And in fact there's more than a "conceptual analogy" here because the phenomenon of heat is again a consequence of computational irreducibility [4]. And the fact that there's a uniform, "abstract" characterization of it is a consequence of the universality of computational irreducibility.

It's worth emphasizing again, though, that just as with heat, a robust concept of time depends on us being computationally bounded observers. If we were not, then we'd able to break the second law by doing detailed computations of molecular processes, and we wouldn't just describe things in terms of randomness and heat. And similarly, we'd be able to break the linear flow of time, either jumping ahead or following different threads of time.

But as computationally bounded observers of computationally irreducible processes, it's basically inevitable that – at least to a good approximation – we'll view time as something that forms a single one-dimensional thread.

In traditional mathematics-based science, there's often a feeling that the goal should be to "predict the future" – or in effect to "outrun time". But computational irreducibility tells us that in general we can't do this, and that the only way to find out what will happen is just to run the same computation as the system itself, essentially step by step. But while this might seem like a letdown for the power of science, we can also see it as what gives meaning and significance to time. If we could always jump ahead, then at some level nothing would ever fundamentally be achieved by the passage of time (or, say, by the living of our lives) [24]; we'd always be able to just say what will happen, without "living through" how we got there. But computational irreducibility gives time and the process of it passing, a kind of hard, tangible character.

So what does all this imply for the various classic issues (and apparent paradoxes) that arise in the way time is usually discussed?

Let's start with the question of reversibility. The traditional laws of physics basically apply both forward and backward in time. And the ruliad is inevitably symmetrical between "forward" and "backward" rules. So why is it then that in our typical experience time always seems to "run in the same direction"?

This is closely related to the second law, and once again it's a consequence of our computational boundedness interacting with underlying computational irreducibility. In a sense what defines the direction of time for us is that we (typically) find it much easier to remember the past than to predict the future. Of course, we don't remember every detail of the past. We only remember what amounts to certain "filtered" features that "fit in our finite minds". And when it comes to predicting the future, we're limited by our inability to "outrun" computational irreducibility.

Let's recall how the second law works. It basically says that if we set up some state that's "ordered" or "simple", then this will tend to "degrade" to one that's "disordered" or "random". (We can think of the evolution of the system as effectively "encrypting" the specification of our starting state to the point where we – as computationally bounded observers – can no longer recognize its ordered origins.) But because our underlying laws are reversible, this degradation (or "encryption") must happen when we go both forward and backward in time (Fig. 10.3):

But the point is that our "experiential" definition of the direction of time (in which the "past" is what we remember, and the "future" is what we find hard to predict) is inevitably aligned with the "thermodynamic" direction of time we observe in the world at large. And the reason is that in both cases we're defining the past to be something that's computationally bounded (while the future can be computationally irreducible). In the experiential case the past is computationally bounded because that's what we can remember. In the thermodynamic case it's computationally bounded because those are the states we can prepare.

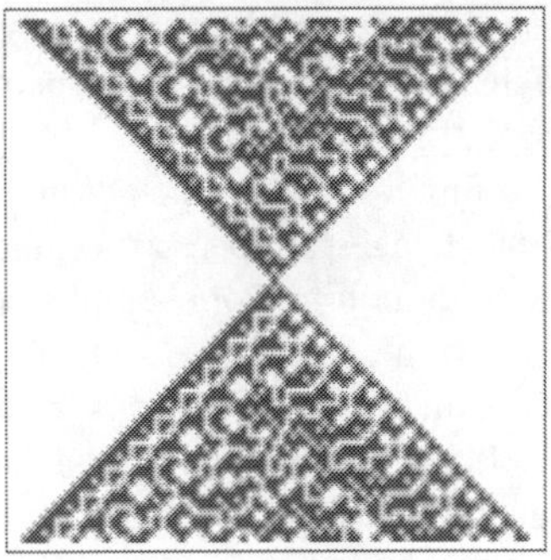

Figure 10.3 Illustration of "encryption" when going forward and backward in time.

In other words, the "arrows of time" are aligned because in both cases we are in effect "requiring the past to be simpler".

So what about time travel? It's a concept that seems natural – and perhaps even inevitable – if one imagines that "time is just like space". But it becomes a lot less natural when we think of time in the way we're doing here: as a process of applying computational rules.

Indeed, at the lowest level, these rules are by definition just sequentially applied, producing one state after another – and in effect "progressing in one direction through time". But things get more complicated if we consider not just the raw, lowest-level rules, but what we might actually observe of their effects. For example, what if the rules lead to a state that's identical to one they've produced before (as happens, for example, in a system with periodic behavior)? If we equivalence the state now and the state before (so we represent both as a single state). Then we can end up with a loop in our causal graph (a "closed timelike curve") [25]. And, yes, in terms of the raw sequence of applying rules, these states can be considered different. But the point is that if they are identical in every feature, then any observer will inevitably consider them the same.

But will such equivalent states ever actually occur? As soon as there's computational irreducibility, it's basically inevitable that the states will never perfectly match up. And indeed for the states to contain an observer like us (with "memory", etc.), it's basically impossible that they can match up.

But can one imagine an observer (or a "timecraft") that would lead to states that match up? Perhaps somehow it could carefully pick particular sequences of atoms of space (or elementary events) that would lead it to states that have "happened before". And indeed in a computationally simple system this might be possible. But as soon as there's computational irreducibility, this simply isn't something one can expect any computationally bounded observer to be able to do. And, yes, this is directly analogous to why one can't have a "Maxwell's demon" observer that "breaks the Second Law" [26]. Or why one can't have something that carefully navigates the lowest-level structure of space to effectively travel faster than light [27].

But even if there can't be time travel in which "time for an observer goes backwards", there can still be changes in "perceived time", say as a result of relativistic effects associated with motion. For example, one classic relativistic effect is time dilation, in which "time goes slower" when objects go faster. And, yes, given certain assumptions, there's a straightforward mathematical derivation of this effect. But in our effort to understand the nature of time, we're led to ask what its physical mechanism might be. And it turns out that in our Physics Project it has a surprisingly direct – and almost "mechanical" – explanation.

One starts from the fact that in our Physics Project space and everything in it is represented by a hypergraph which is continually getting rewritten. And the evolution of any object through time is then defined by these rewritings. But if the object moves, then in effect it has to be "re-created at a different place in space" – and this process takes up a certain number of rewritings, leaving fewer for the intrinsic evolution of the object itself, and thus causing time to effectively "run slower" for it. (And, yes, while this is a qualitative description, one can make it quite formal and precise, and recover the usual formulas for relativistic time dilation.)

Something similar happens with gravitational fields. In our Physics Project, energy-momentum (and thus gravity) is effectively associated with greater activity in the underlying hypergraph. And the presence of this greater activity leads to more rewritings, causing "time to run faster" for any object in that region of space (corresponding to the traditional "gravitational redshift").

More extreme versions of this occur in the context of black holes. (Indeed, one can roughly think of spacelike singularities as places where "time ran so fast that it ended".) And in general – as we discussed above – there are many "relativistic effects" in which notions of space and time get mixed in various ways.

But even at a much more mundane level, there's a certain crucial relationship between space and time for observers like us. The key point is that observers like us tend to "parse" the world into a sequence of "states of space" at successive "moments in time". But the fact that we do this depends on some quite specific features of us, and in particular our effective physical scale in space as compared to time.

In our everyday life we're typically looking at scenes involving objects that are perhaps tens of meters away from us. And given the speed of light that means photons from these objects get to us in less than a microsecond. But it takes our brains milliseconds to register what we've seen. And this disparity of timescales is what leads us to view the world as consisting of a sequence of states of space at successive moments in time.

If our brains "ran" a million times faster (i.e. at the speed of digital electronics), we'd perceive photons arriving from different parts of a scene at different times, and we'd presumably no longer view the world in terms of overall states of space existing at successive times.

The same kind of thing would happen if we kept the speed of our brains the same, but dealt with scenes of a much larger scale (as we already do in dealing with spacecraft, astronomy, etc.).

But while this affects what it is that we think time is "acting on", it doesn't ultimately affect the nature of time itself. Time remains that computational process by which successive states of the world are produced. Computational irreducibility gives time a certain rigid character, at least for computationally bounded observers like us. And the Principle of Computational Equivalence allows there to be a robust notion of time independent of the "substrate" that's involved: whether us as observers, the everyday physical world, or, for that matter, the whole universe.

References

[1] Wolfram, Stephen. (2002). Computational Irreducibility. In *A New Kind of Science*, chap. 12, "The Principle of Computational Equivalence." Champaign, IL: Wolfram Media. `https://www.wolframscience.com/nks/chap-12--the-principle-of-computational-equivalence#sect-12-6--computational-irreducibility`.

[2] Wolfram, Stephen. (2002). The Principle of Computational Equivalence. In *A New Kind of Science*, chap. 12. Champaign, IL: Wolfram Media. `https://www.wolframscience.com/nks/chap-12--the-principle-of-computational-equivalence/`.

[3] Wolfram, Stephen. (2023). Observer Theory. *Stephen Wolfram Writings*, December. `https://writings.stephenwolfram.com/2023/12/observer-theory/`.

[4] Wolfram, Stephen. (2023). Computational Foundations for the Second Law of Thermodynamics. *Stephen Wolfram Writings*, February. `https://writings.stephenwolfram.com/2023/02/computational-foundations-for-the-second-law-of-thermodynamics/`.

[5] Wolfram Physics Project. (n.d.). *Wolfram Physics Project*. `https://www.wolframphysics.org/`.

[6] Wolfram, Stephen. (2020). Finally We May Have a Path to the Fundamental Theory of Physics… and It's Beautiful: What Is Space? *Stephen Wolfram Writings*, April 14. `https://writings.stephenwolfram.com/2020/04/`.

[7] Wolfram, Stephen. (2020). Finally We May Have a Path to the Fundamental Theory of Physics... and It's Beautiful: Time. *Stephen Wolfram Writings*, April 14. https://writings.stephenwolfram.com/2020/04/.

[8] Wolfram, Stephen. (2020). Finally We May Have a Path to the Fundamental Theory of Physics... and It's Beautiful: The Graph of Causal Relationships. *Stephen Wolfram Writings*, April 14. https://writings.stephenwolfram.com/2020/04/.

[9] Wolfram, Stephen. (2020). Finally We May Have a Path to the Fundamental Theory of Physics... and It's Beautiful: Deriving Special Relativity. *Stephen Wolfram Writings*, April 14. https://writings.stephenwolfram.com/2020/04/.

[10] Wolfram, Stephen. (2020). Finally We May Have a Path to the Fundamental Theory of Physics... and It's Beautiful: General Relativity and Gravity. *Stephen Wolfram Writings*, April 14. https://writings.stephenwolfram.com/2020/04/.

[11] Wolfram, Stephen. (2020). Cosmology, Expansion, and Singularities. In *A Class of Models with the Potential to Represent Fundamental Physics*. Wolfram Physics Project. https://www.wolframphysics.org/technical-introduction/potential-relation-to-physics/cosmology-expansion-and-singularities.

[12] Wolfram, Stephen. (2020). Finally We May Have a Path to the Fundamental Theory of Physics... and It's Beautiful: The Inevitability of Quantum Mechanics. *Stephen Wolfram Writings*, April 14. https://writings.stephenwolfram.com/2020/04/.

[13] Wolfram, Stephen. (2020). The Concept of Branchial Graphs. In *A Class of Models with the Potential to Represent Fundamental Physics*. Wolfram Physics Project. https://www.wolframphysics.org/technical-introduction/the-updating-process-for-string-substitution-systems/the-concept-of-branchial-graphs/.

[14] Wolfram, Stephen. (2023). Observer Theory: The Case of Quantum Mechanics. *Stephen Wolfram Writings*, December. https://writings.stephenwolfram.com/2023/12/observer-theory#the-case-of-quantum-mechanics.

[15] Wolfram, Stephen. (2022). Games and Puzzles as Multicomputational Systems: Humanizing Multicomputational Processes. *Stephen Wolfram Writings*, June. https://writings.stephenwolfram.com/2022/06/games-and-puzzles-as-multicomputational-systems#humanizing-multicomputational-processes.

[16] Wolfram, Stephen. (2021). What Is Consciousness? Some New Perspectives from Our Physics Project. *Stephen Wolfram Writings*, March. https://writings.stephenwolfram.com/2021/03/what-is-consciousness-some-new-perspectives-from-our-physics-project/.

[17] Wolfram, Stephen. (2020). Units and Scales. In *A Class of Models with the Potential to Represent Fundamental Physics*. Wolfram Physics Project. https://www.wolframphysics.org/technical-introduction/potential-relation-to-physics/units-and-scales/.

[18] Wolfram, Stephen. (2020). Event Horizons and Singularities in Spacetime and Quantum Mechanics. In *A Class of Models with the Potential to Represent Fundamental Physics*. Wolfram Physics Project. https://www.wolframphysics.org/technical-introduction/potential-relation-to-physics/event-horizons-and-singularities-in-spacetime-and-quantum-mechanics/.

[19] Wolfram, Stephen. (2021). The Concept of the Ruliad. *Stephen Wolfram Writings*, November. https://writings.stephenwolfram.com/2021/11/the-concept-of-the-ruliad/.

[20] Wolfram, Stephen. (2022). The Physicalization of Metamathematics and Its Implications for the Foundations of Mathematics. *Stephen Wolfram Writings*, March. https://writings.stephenwolfram.com/2022/03/.

[21] Wolfram, Stephen. (2022). The Physicalization of Metamathematics and Its Implications for the Foundations of Mathematics: Metamathemes. *Stephen Wolfram Writings*, March. https://writings.stephenwolfram.com/2022/03/.

[22] Wolfram, Stephen. (2022). On the Concept of Motion. *Stephen Wolfram Writings*, March. https://writings.stephenwolfram.com/2022/03/on-the-concept-of-motion/.

[23] Wolfram, Stephen. (2023). How Did We Get Here? The Tangled History of the Second Law of Thermodynamics: What Is Heat? *Stephen Wolfram Writings*, January. https://writings.stephenwolfram.com/2023/01/.

[24] Wolfram, Stephen. (2002). The Phenomenon of Free Will. In *A New Kind of Science*, chap. 12, section 7. Champaign, IL: Wolfram Media. https://www.wolframscience.com/nks/chap-12--the-principle-of-computational-equivalence#sect-12-7--the-phenomenon-of-free-will.

[25] Wolfram, Stephen. (2020). Finally We May Have a Path to the Fundamental Theory of Physics… and It's Beautiful: Black Holes,

Singularities, etc. *Stephen Wolfram Writings*, April 14. https://writings.stephenwolfram.com/2020/04/.

[26] Wolfram, Stephen. (2023). Computational Foundations for the Second Law of Thermodynamics: Maxwell's Demon and the Character of Observers. *Stephen Wolfram Writings*, February. https://writings.stephenwolfram.com/2023/02/computational-foundations-for-the-second-law-of-thermodynamics/#maxwells-demon-and-the-character-of-observers.

[27] Wolfram, Stephen. (2020). Faster than Light in Our Model of Physics: Some Preliminary Thoughts. *Stephen Wolfram Writings*, October. https://writings.stephenwolfram.com/2020/10/faster-than-light-in-our-model-of-physics-some-preliminary-thoughts/.

11 The Prototime Interpretation of Quantum Mechanics

Susan Schneider and Mark Bailey

11.1 Introduction

At the heart of contemporary physics is a contradiction between the study of the very big and the very small – between the supermassive structures (e.g., black holes) in Einstein's theory of general relativity and the subatomic arena of quantum mechanics. Work in the field of quantum gravity (QG) tries to resolve this contradiction, and increasingly, it is claiming something astonishing: the fundamental ingredients of reality are not spatiotemporal. Instead, spacetime emerges from something more fundamental, something defined in terms of a mathematical structure that dispenses with any spatiotemporal metric [1–4]. Just as the transparency of water is not found in a single H_2O molecule, at the finest level of resolution, spacetime drops out of the picture.

Herein, we sketch and explore a position in which spacetime emerges from a quasi-temporal reality called "prototime". According to this position, there is time in the sense of spacetime, as well as a different, more fundamental, "prototemporal" dimension or parameter from which spacetime emerges. (Some may instead wish to think of this dimension simply as a "parameter", instead of a dimension because it regards dimensions as spatiotemporal entities.) Our chapter is tentative and exploratory. The argument form is inference to the best explanation. We claim that the Prototime Interpretation (PI) is worthy of further consideration as a superior explanation for perplexing quantum phenomena such as delayed choice, superposition, the wave-particle duality and nonlocality[1] [5, 6]. In Section 11.2, we introduce PI. Section 11.3 identifies its advantages. Section 11.4 discusses several implications of the view, such as its deterministic nature[2].

DOI: 10.4324/9781003546139-15

11.2 The Prototime Interpretation (PI)

Our point of departure is the simple fact that a quantum system that is entangled in a "pure state" has zero Von Neumann entropy when the system is considered as a whole, where Von Neumann entropy is a common measure of the entropy of quantum systems[3]. (A quantum system, *S*, is in a "pure state" when it is in a precise, well-defined state, being described by a single wave function that contains all the information about *S*.) We further appeal to the following uncontroversial point:

1 **Entanglement connectivity:** Fundamental particles can be entangled, even across vast spatial distances. When two particles, *a* and *b*, are entangled, their properties become correlated such that the state of one particle is instantaneously linked to the state of the other.

This is the "spooky action at a distance", that Einstein referred to, and, bizarre as it is, it has been demonstrated in numerous experiments. Entanglement connectivity is a detectable phenomenon within our universe. It is neither spatiotemporally nor causally isolated from the 4D world. It is not happening in some unrelated, inaccessible, parallel universe but from a part of our universe that we do not yet understand.

Now consider a very controversial claim. For the purpose of argument we suppose, *controversially*, that entanglement connectivity is a causal phenomenon.

1 **Assumption:** Entanglement connectivity is causal. An entangled state, *a*, either directly causes a change in a particle *b*, or, the states of *a* and *b* are jointly caused by, or mediated by, one or more other state(s) at the prototemporal level.

As Hume observed, empirical investigation of any phenomenon does not actually detect a cause; it merely detects a correlation because causation is not something that can be seen directly in the world; it is only inferred [7, 8]. But normally, a causal relation is an obvious avenue to consider given the presence of a reliable correlation. For saying that there is merely a correlation, rather than a causal relation, calls for explanation as well. And indeed, the idea that entanglement connectivity is a mere correlation is bizarre. However, while it is bizarre to merely assert a correlation, there is an important consideration in its favor, for, of course, the presence of a causal relation between entangled states at vast distances would contradict relativity theory, involving superluminal signaling [9]. For this reason, the above assumption (2) is very controversial, to say the least.

Notice, however, (2) can be true if it does not lead to spooky action at a distance. More specifically, we propose the following:

1 **5D-ism:** The universe has at least one added dimension (or parameter) – one in which entanglement connectivity happens. This is not an extended spatial dimension but a parameter of prototemporal connectivity.

According to 5D-ism, the classical, everyday reality we experience exists on the 4D "surface" of a larger 5D universe. The universe has at least five dimensions: three spatial, one temporal, and at least one added parameter or degree of freedom that is nonspatiotemporal, underlying entanglement connectivity. If assumption (2) is correct, this supposed causal connection is not a phenomenon that makes sense merely on the assumption of 4D spacetime, the initial conditions and the relativistic laws. Indeed, it is quite puzzling from a relativistic framework, as noted. We propose that it may require at least one added parameter, or degree of freedom, that is neither spatial nor temporal – at least where "temporal" is used in the "Einsteinian spacetime" sense. (Herein, for uniformity we use "temporal" and "time" in the sense of Einsteinian spacetime.) Of course, the standard view is that quantum entanglement involves instantaneous correlations, but due to the No Signaling Principle, it doesn't allow for faster than light transmission of information. It is not possible to use quantum entanglement to send messages superluminally. However, this does not preclude causation in prototime, as entities in prototime are not ones in which the constraints of spacetime apply. Although the standard picture merely asserts that there is a correlation between entangled states, there is nothing to rule out the possibility that at the level of prototime, there is a causal relation between entangled particles, or the particles' states are jointly caused by, or mediated by, other state(s) in prototime. However, macroscopic observers cannot use quantum entanglement to send messages faster than the speed of light, as per the No Signaling Principle.

How does time emerge from a more fundamental prototemporal reality? Physicists and philosophers have long puzzled over the problem of time's arrow, the puzzle of why time moves forward, not backward, given that physical laws seem symmetrical. A popular response to the problem of time's arrow involves appealing to the phenomenon of entropy [10–12]. In thermodynamics, entropy is the measure of the disorder in a system. According to the second law of thermodynamics, the total entropy of an isolated system will inevitably increase over time. This means that systems will naturally evolve from ordered states to more disordered states.

This common approach to time's arrow is particularly suggestive in light of the phenomenon of entanglement. For entangled systems in a pure state – systems with zero von Neumann entropy and that have not decohered and interacted with the environment – may not really be in spacetime at all. This is because measurement (with decoherence) introduces entropy, and time's arrow, into the system. An entangled quantum system that is in a pure state would not be one in which time's arrow applies. It is only through the process of decoherence that the particles become integrated with spacetime itself. Another way to put the point is that during measurement (with decoherence) the environment "measures" the system and this disturbs it, causing the system to lose its superposition. Doing this introduces thermodynamic entropy into the system and the system transitions to a classical state. An entangled system in a pure state is not in spacetime, but the act of measurement (with decoherence) introduces classical entropy and time's arrow into the system.

While this point is speculative, there is a body of work lending insight into how time's arrow emerges that is compatible with our approach [13]. Quantum Darwinism (QD) is a well-respected hypothesis that explains the emergence of classical reality from quantum possibilities. In brief, QD is dependent on the interaction of (quantum) superpositions that ultimately converge to some stable (classical) state. Some states are more stable than others; these more stable states are known as "pointer states". For example, a measurement might be a pointer state, which causes the measured particle to decohere to a stable, measured state. All quantum objects interact in this same manner, becoming entangled with each other as they interact, ultimately converging to stable, classical states through the process of decoherence [14]. Because the number of decohered states that is available to any quantum object greatly exceeds the number of available "pure" unentangled quantum states, in practice, classical objects don't interact and suddenly enter into a quantum state. In this manner, QD ultimately gives rise to classical temporal ordering. In sum, from quantum decoherence, entropy and time's arrow ultimately emerge from an aspatial, protemporal arena.

So, according to this view, the phenomenon of quantum entanglement plays a crucial role in our understanding of time's arrow. While classical views link the progression of time to the dispersal of energy and increasing entropy, the modern understanding of QD sees quantum entanglement as the driving force [15–17]. As particles become more and more entangled, systems move toward equilibrium, which gives the appearance of time moving in a specific direction. This quantum perspective not only offers a more fundamental explanation for the arrow of time, but also helps bridge the gap between classical and quantum thermodynamics.

Recent experiments have found some support for QD. For example, two teams, one at Sapienza University of Rome and another at the University of Science and Technology of China, employed photons to simulate quantum systems and their environments. They noted that even a single photon can serve to act as an environment, introducing decoherence and selection, and that the information about the quantum system saturates quickly as more and more of the environment is considered. Further, another experiment (led by Fedor Jelezko at Ulm University in Germany) used a nitrogen atom in a diamond's crystal lattice as the quantum system. The atom's unpaired electron can interact with surrounding carbon atoms. The findings confirmed that the state of the nitrogen atom is "recorded" in its surroundings multiple times, which is consistent with QD's predictions. While these experiments align with QD, they don't conclusively prove QD is the only explanation for the emergence of classicality. However, these tests are still significant steps in understanding the bridge between the quantum and classical worlds [17].

In sum, ours could be a universe with two time-like dimensions, one that involves time in the familiar sense of spacetime and in which time has a direction or arrow, and a different prototemporal dimension that lacks a direction or arrow. This fifth dimension or parameter is a non-spatial arena, yet prototime involves causation between events. Because time possesses a definite direction upon decoherence, positing a timeless or prototemporal arena in this context does not introduce time paradoxes. This position is novel, and unusual, but we believe it is worthy of consideration, as it seriously takes the possibility that entanglement relations confer an added dimension to reality, one that causally determines events in the 4D world.

11.3 Advantages

So, for what reason do we have to take this view seriously? It is important to note that the existing explanations of quantum phenomena are difficult to adjudicate. Unfortunately, key claims still remain untestable and/or rely on controversial philosophical assumptions, such as with the many-world interpretation's appeal to branching parallel universes. PI faces these same hurdles, and it awaits more formalism. To its credit, it draws from leading trends in physics, such as spacetime emergence and QD. We further believe PI is worthy of further consideration as a superior explanation for the following well-confirmed yet bizarre phenomena that we now outline. Further, where other theories offer explanations that are equally satisfying, this view may be more parsimonious than leading contenders, such as string-theoretic views of quantum phenomena that require commitments to entities like branes and several extra spatial dimensions.

We propose that PI offers the following advantages:

1 **PI provides a richer understanding of superposition:** Quantum superposition is a puzzling phenomenon in which a particle doesn't exist

in a single state but exists as a superposed combination of all possibilities, until measurement or observation, at which point the particle has a determinate state. The prototemporal dimension introduces a fundamental timeless level in which the particle is effectively "everywhere, all at once". That is, according to PI, the particle does not need to be in a determinate state because there is no singular moment in time, at the prototemporal level, in which it must occupy a determinate state. For time's arrow is not in play. Instead, the particle is in a superposition of all states until a measurement is performed and the particle interacts with the familiar, time-bound universe. This interaction situates the particle in time, forcing it to adopt a definite state.

2 **PI provides a unique perspective on the No Signaling Principle:** The standard view says that quantum entanglement involves instantaneous correlations only; due to the No Signaling Principle, it does not allow for faster than light transmission of information. Quantum entanglement cannot send messages to macroscopic observers superluminally. We uphold the No Signaling Principle. But notice that "speed of light" is a spatiotemporal notion, requiring both a distance and time metric, both of which are not present at the level of prototime. The phenomenon of time's arrow arises only when systems interact with the environment. Although the standard view must assert that there is merely a correlation between entangled states, to avoid violating the no signaling principle, there is actually nothing to rule out the possibility that at the level of prototime, there is a causal relationship between entangled states, (perhaps mediated by something else at that level), not just correlations. However, macroscopic observers cannot use quantum entanglement to send messages faster than the speed of light.

3 **PI rejects "spooky action at a distance":** Related to (2) above is the concern that entanglement seemingly involves instantaneous correlations across vast distances, which seems like superluminal communication or what Einstein famously called "spooky action at a distance" [9]. Because PI proposes a non-spatiotemporal arena in which entanglement communication might occur, the instantaneous correlations do not violate the luminal speed limit. Further, no distance metric exists in prototime, as it is aspatial, and so there is no "distance" over which spooky action could occur.

4 **The wave-particle duality**: Particles are known for exhibiting both wave-like and particle-like features, depending on how they are measured. This phenomenon is an expected feature of PI because particles exist fundamentally in a cloud of potential states at the prototemporal

level. Only through interacting with the 4D world do they exhibit particle-like (spatial and temporal) behaviors, a duality that is a manifestation of a particle's existence in two different time like structures.

5 **The double-slit experiment**: The double-slit experiment involves particles passing through two slits, generating an interference pattern on a screen. When one attempts to measure which slit the particle traveled through, bizarrely, the pattern of interference goes away, as if the particle somehow decided to behave in a particle-like manner, and not a wavelike manner, based on the fact that it was measured. PI says that this bizarre behavior is actually expected because all possible paths exist until the point of measurement and decoherence. At that point, the particle goes into a determinate state in spacetime.

6 **Delayed-choice phenomena:** The double-slit experiment discussed above can be modified to become a delayed-choice experiment in which the choice of whether to measure the path of the particle is made *after* the particle passes through the slits and yet *before* the particle hits the screen. Astonishingly, the outcome on the screen seems to depend on the choice that is made after the particle passed through the slits, seeming like the particle "decides" how it should behave based on an event that has not yet occurred [18, 19]. PI says that the "choice" made upon measurement is actually an outcome of being in the prototemporal state until a measurement is made.

Now let us turn to an important objection to our claim that PI offers an explanatory advantage with respect to the above phenomena. One can object that these same advantages could be provided by the more straightforward, familiar position that takes spacetime to emerge from an entirely timeless, aspatial reality, rather than from prototime, an idea which, the objector will point out, is unclear. What is the notion of "quasi-time", after all?

We will call this more common position the "Timeless Reality" view. Even setting aside the issue of parsimony, which arises for string theoretic versions of the view, we believe the timeless reality position is flawed. (Explaining the flaw will also help us flesh out the notion of prototime a bit more.) The problem with the timeless reality view is that it is difficult to see how a fundamental timeless level can yield the universe we experience. All around us is the phenomenon of change – we introspect changes in our conscious states, and both our inner experience and scientific work on consciousness provide details on how objects and properties in the world change and evolve. In contrast to this, the literature on timeless reality often appeals to highly mathematical views of reality, and this can lead to a sort of mathematical Platonism gone mad, where the entire universe is seen as an abstract entity, like an equation [20]. Schneider has elsewhere expressed concerns with this approach because it does not explain how there is a concrete, empirical world in which change occurs [6]. Here, the natural question to ask of this sort of view is: what are mathematical entities? The field of philosophy of mathematics studies this question, and there are longstanding controversies about the nature of mathematical properties. If one is a Platonist, it is not clear how abstract mathematical entities can causally interact with the physical world, for a purely aspatial and atemporal reality lacks any kind of concreteness, seemingly casting its lot with a metaphysics disconnected from the concrete, causal world. If one has in mind some form of nominalism about mathematical entities, however, then one needs to explain how they are defining their nominalism; it cannot be in terms of spacetime or macroscopic phenomena like human classificatory systems, on pain of circularity. Entities like spacetime, minds and classificatory systems are all presumably ultimately determined by the base level, not the other way around.

Because prototime is not time in the familiar sense of spacetime, in which time has an arrow, it is unsurprising that prototime is hard for humans to grasp. But there are resources in the field of contemporary analytic metaphysics that can help. To begin with, a metaphysical picture of base reality needs some fundamental elements that go beyond abstracta. There must be something in one's fundamental ontology that makes sense of causation and change. Notice that the fundamental level that the PI posits is not one without causal relations. Again, entanglement connectivity is real, and this phenomenon cannot be explained by information transfer *within* spacetime itself. If Assumption (2) is correct, there is entanglement causation that exists in a different, additional, dimension or parameter that is not just ordinary spacetime. A purely atemporal picture would not seem robust enough to accommodate this underlying causal phenomena, as far as we can tell.

Some philosophers, such as Barry Loewer and David Lewis, contend that fundamental physical reality consists in a spatiotemporal mosaic of properties that are essentially non-dispositional, ("categorical" or sometimes "categorical" properties). Laws of nature and causal relations are merely patterns that supervene on this more fundamental mosaic [21]. While these views were not developed in the context of debates about spacetime emergence, this same neo-Humean "categorialist" view of property natures remains influential. According to this neo-Humean ontology, causation and change supervene upon an underlying acausal, non-dispositional reality. As important and influential as this line of thinking is, however, this kind of ontology, especially when paired with highly mathematical views of fundamental reality, would not provide the needed explanation of how change could exist [6, 22].

In contrast to this neo-Humean position, it has been observed that empirical properties seem to be dispositional: properties in nature have some causative effect on something else. We commonly talk about, and identify, properties in terms of what they do – by how those properties impact us, other objects and our measurement instruments. For example, the notion of electron charge is meaningless without some force or field acting on that charge. If the charge did not interact with anything else, its existence would be, at a bare minimum, permanently epistemically unavailable to us. Further, we could postulate an infinite number of properties that have no causal powers – properties that don't actually do anything at all. However, this would be unparsimonious. Therefore, it seems reasonable to assume that empirical properties have at least partly causal natures [23].

Similar discussions have appeared in the philosophy of science literature. Ontic structural realism (OSR), postulated by James Ladyman and and Don Ross, is a view that treats the notion of structure as being primitive, where information transfer through structured interactions mediate causation. In this view, reality is fundamentally nothing but patterns all the way down [24]. This raises a similar issue: that the laws merely articulate structures, and at the fundamental level, they are highly mathematical. But what do the laws relate? That is, what underlying entities are we describing with our highly mathematical physical theories? A common objection to OSR is the mistaken assumption that it views the world as purely mathematical, which would be incongruent with physicalism and any distinction between the concrete and the abstract. The response is that the relational structures are, in fact, real (properties or something else). While they can be mapped to isomorphic mathematical abstracta, that doesn't negate the existence of a physical structure to which they map. P. M. Ainsworth puts forth an interpretation of OSR where properties and relations are ontologically primitive, but objects are not [25]. In the following section we raise a similar approach, one that appeals to bundle theory.

Thus far, we have discussed several explanatory advantages to PI. Because PI does not invoke extra spatial dimensions, and because spacetime emerges from a base reality consisting in prototemporal dispositional properties, we believe the position is parsimonious. Further, in contrast to the Timeless Reality view, which may explain the above quantum phenomena (e.g., superposition) and in some versions, may stand to be equally parsimonious (invoking the same number of spatial dimensions as PI), PI is better able to explain change, claiming that reality consists in causally interacting prototemporal properties. These are dispositional properties that are defined as being capable of giving rise to spatiotemporal phenomenon. This helps flesh out prototime, illustrating why it is a sort of "quasi-time" that has dynamic features. It is not time in the sense of spacetime, yet it is nevertheless a causal arena, having events that are instantiations of dispositional properties indexed to a prototemporal (and nonspatiotemporal) metric. Now let us explore the metaphysical framework in more detail.

11.4 Determinism, digital physics and the simulation hypothesis

PI is deterministic. Recall our assumption for the purpose of the following argument:

Entanglement connectivity is causal: An entangled state, a, either directly causes a change in a particle b, or, the states of a and b are jointly caused by, or mediated by, one or more other state(s) at the prototemporal level.

Hidden variable theories claim that the probabilistic nature of quantum mechanics stems from a hidden variable that we have yet to uncover, and that quantum systems are actually deterministic [26]. Consider the behavior of any two entangled states; such states are commonly observed to follow a pattern that seems deterministic. For example, measuring one instantaneously impacts the other. Further, the value of one particle is non-randomly correlated with, and indeed, on our view, in some sort of causal relation with, the state of its entangled particle. These facts, when combined with the above assumption and the view that spacetime emerges from entanglement, suggest that PI is deterministic. Quantum events are the output of the deterministic function conforming to the probabilistic predictions of standard quantum mechanics.

A natural question is whether one can derive the standard probabilities of quantum mechanics from the underlying prototemporal structure – a deterministic function from states of an entangles system, S, to states in spacetime. Since we cannot access future states in the 4D manifold, it is impossible to access the complete details of the deterministic structure of the universe. To make matters worse, the universe consists in a complex web of entangled states bearing connectivity R to each other, so the "entanglement object" is one singular, enormously complex, entanglement object that underlies all the spacetime. (We shall call this "The Megaobject".) Yet from the vantage point of a hypothetical omniscient being having upon the Megaobject, a larger pattern may be evident, upon considering the past, present and future states of entangled particles. From our vantage point, however, massive intractability looms.

Yet many body experiments have cleverly isolated more complex quantum systems, even entangling a tardigrade [27]. But the question is: what future states are relevant? An obvious candidate is measurement. Here, delayed choice cases may be instructive, for if our theory is correct, the future choice, together with the past and present states of the system, provides the hidden variable that maps to outcomes in our spacetime. The choice made in the future is in the elements of the prototemporal structure and it does match the outcome we observe. This explains why the particle seems to "know" the future measurement setting.

There are other exciting implications of determinism as well. It is possible that spacetime and its occupants are epiphenomenal aspects of the prototemporal level; just as philosophers have entertained that consciousness is itself epiphenomenal, being determined by, and supervenient on, more basic physical properties but itself causally inert, so too, the locus of causal action may be at the prototemporal level, and the 4D world, including our own consciousness, are merely epiphenomenal features of it. This is a major departure from our current worldview, and much of physics, which takes spacetime as the primary arena for causal action.

Now let us turn to a related matter. Thus far, our discussion envisions a universe in which all the spacetime emerges from the quantum decoherence of entanglement objects at the more basic level. Given this, it is natural to ask: is reality itself effectively a quantum computer? Further, might we be in some sort of simulation? While we cannot delve into this matter in detail, we believe this matter calls attention to the need for a richer metaphysical understanding of quantum phenomena.

Digital physics, the intriguing concept suggesting that the universe is, or at least operates like, a computer program, is of interest to many in light of the simulation hypothesis, artificial life, the import of information theory and more. The core proposition of digital physics, that all phenomena can ultimately be described by information processing or computational rules, together with Nick Bostrom's simulation argument, raises important questions about whether the aforementioned "base level" is that of a computer simulation and whether we might even be faced with an epistemic situation in which we cannot determine, as subjects residing in spacetime, whether a certain approach to digital physics is right, as opposed to a simulation hypothesis – a sort of underdetermination of theory by all the available evidence.

Indeed, we might appeal to computer simulations to explore the space of theories, to try to resolve the issue, where actual experiments are unavailable. However, perhaps there is no possible function that we could derive that maps the base computation to certain emergent subroutines. Being in a simulation, we may be limited in our ability to build a computer capable of universal computation to the same fidelity as the computational universe in which it exists. It would be like trying to build a simulated computer that would be more powerful than the computer running the actual simulation. Interestingly, a machine cannot compute itself in more than real time, according to Stephen Wolfram's principle of computational irreducibility [28]. Otherwise, infinite computational speed would be possible. Furthermore, some processes are not inherently mappable to their outcomes using deterministic functions. For instance, the emergence of markets in an economy is dependent on the local interactions of the market participants; however, this relationship can't be compressed to a deterministic mathematical relationship – it requires stochastic simulation or direct observation to derive any insight. This is what we refer to as **algorithmic incompressibility**. It is possible that this is simply an epistemic issue, due to our ignorance of the math and physics required to fully describe this type of system; or it could be of metaphysical origin, representing a fundamental limit to our ability to deterministically compute certain phenomena. If computational irreducibility holds and algorithmic incompressibility has an epistemic limit, the computational speed limit of the 4D universe might be one that is set by a base reality computing at a finite speed – perhaps suggestive of a simulated reality[4].

It is worth noting that a process ontology is compatible with a simulation hypothesis because the program can be implemented by properties having their causal powers essentially. On our view, the causal powers of the properties are determined by the role the properties play with respect to the other properties they are entangled with, where the state of one property is instantaneously connected to the state of another, regardless of the "distance" between them in the prototemporal realm. This is a mechanism for property interaction in the absence of a normal time dimension. Some in contemporary metaphysics may prefer to claim that base reality "realizes" macroscopic events, rather than causing them directly. However, if the 4D world is in a computer simulation generated by the base reality, then it may be more appropriate to claim that the base reality causes events in the 4D world (what we might call this "upward causation"). This upward causation from the base to the spatiotemporal would be a form of genuine emergence, one without downward causation, perhaps, but one in which the base level causes, rather than realizes higher-level events.

11.5 Conclusion

PI draws from the idea that spacetime emerges from entanglement, which is causally connected through a nonspatiotemporal parameter, called "prototime". We have claimed that time's arrow emerges from entropy arising during quantum decoherence. We have urged that the prototime view deserves consideration as a framework that may address a range of perplexing phenomena in quantum mechanics, such as superposition, delayed choice and spooky action at a distance. It offers a deterministic perspective that suggests the probabilistic nature of quantum mechanics is due to our limited epistemic access to the prototemporal arena.

11.6 Acknowledgments

We are grateful to Misha Klopukh, Michael Ostroff, Edwin Turner, Dean Rickles, George Musser and Michael Huemer for feedback on this paper.

Notes

1. For an earlier framing of the Prototime Interpretation and implications for consciousness and the self see Schneider, "Emergent Spacetime, the Megastructure Problem, and the Metaphysics of the Self," Philosophy East and West 74 (2024): 314–32. She has previously stressed that the very same entities that fundamental physics investigates, these entities that spacetime emerges from, may very well be the very same ingredients that give rise to consciousness (Schneider, Susan. "Idealism, or Something Near Enough." In Idealism: New Essays in Metaphysics, edited by Tyron Goldschmidt and Kenneth L. Pearce, Oxford University Press, 2017. pp. 234-256).
2. In a related target paper for a forthcoming Journal of Consciousness Studies Special Issue, we employ the prototime interpretation to develop a new version of panpsychism, which we call "Superpsychism." (Schneider and Bailey, forthcoming.) According to Superpsychism, the fundamental physical level has a more advanced form of consciousness than spacetime occupants, in the sense that it exhibits maximal coherence, zero entropy and holistic integration of conscious states. The position differs from Cosmopsychism, for whereas cosmospychists like Goff (Goff, Philip. (2017). Consciousness and Fundamental Reality. Oxford University Press) and Nagasawa and Wager (Nagasawa, Yujin, and Khai Wager. (2015). Panpsychism and Priority Cosmopsychism. In T. Alter and Y. Nagasawa (Eds.), Consciousness in the Physical World: Perspectives on Russellian Monism (pp. 113–134). Oxford University Press.) locate the fundamental unit of consciousness in the very biggest element, we claim the greatest form of consciousness inheres in the holistically entangled structure, a structure that is not even spatiotemporal and which underlies spacetime itself.
3. Individual subsystems of an entangled system have non-zero Von Neumann entropy but a system as a whole in a pure state (whether entangled or not) has zero Von Neumann entropy, reflecting a state with maximum knowledge/no uncertainty.
4. Herein, we have been referring to a "base" level for the purpose of discussion, but it is important to bear in mind that for all we know, there is yet a more basic level, and indeed, it is conceptually possible that it is turtles all the way down.

References

[1] George Musser, Spooky Action at a Distance: The Phenomenon That Reimagines Space and Time–and What It Means for Black Holes, the Big Bang, and Theories of Everything, First edition (New York: Scientific American/Farrar, Straus and Giroux, 2015).

[2] Nathan Seiberg, "Emergent Spacetime," in The Quantum Structure of Space and Time, 2007, 163–213, https://doi.org/10.1142/9789812706768_0005.

[3] Brian Swingle, "Spacetime from Entanglement," Annual Review of Condensed Matter Physics 9, no. 1 (March 10, 2018): 345–58, https://doi.org/10.1146/annurev-conmatphys-033117-054219.

[4] Christian Wüthrich, Baptiste Le Bihan, and Nick Huggett, eds., Philosophy Beyond Spacetime: Implications from Quantum Gravity, 1st ed. (Oxford University Press, 2021), https://doi.org/10.1093/oso/9780198844143.001.0001.

[5] Susan Schneider, “Emergent Spacetime, the Megastructure Problem, and the Metaphysics of the Self,” Philosophy East and West 74 (2024): 314–32, https://api.semanticscholar.org/CorpusID:269395209.

[6] Susan Schneider, “Idealism, or Something Near Enough,” in Idealism: New Essays in Metaphysics, ed. Tyron Goldschmidt and Kenneth L. Pearce (Oxford University Press, 2017), . pp. 234-256. https://doi.org/10.1093/oso/9780198746973.003.0017.

[7] Hume, David. ”An enquiry concerning human understanding.” In *Seven masterpieces of philosophy*, pp. 183-276. Routledge, 2016.

[8] David Hume, “A Treatise of Human Nature”, Edited by David Fate Norton, and Mary J. Norton (Oxford: Oxford University press, 2000).

[9] A. Einstein, B. Podolsky, and N. Rosen, “Can Quantum-Mechanical Description of Physical Reality Be Considered Complete?,” Physical Review 47, no. 10 (May 15, 1935): 777–80, https://doi.org/10.1103/PhysRev.47.777.

[10] Giulia Rubino, Gonzalo Manzano, and Časlav Brukner, “Quantum Superposition of Thermodynamic Evolutions with Opposing Time’s Arrows,” Communications Physics 4, no. 1 (November 26, 2021): 251, https://doi.org/10.1038/s42005-021-00759-1.

[11] Seth Lloyd, “Pure State Quantum Statistical Mechanics and Black Holes” (arXiv, July 1, 2013), http://arxiv.org/abs/1307.0378.

[12] Wojciech Hubert Zurek, “Quantum Darwinism,” Nature Physics 5, no. 3 (March 2009): 181–88, https://doi.org/10.1038/nphys1202.

[13] Carroll, Sean M. (2010). *From Eternity to Here: The Quest for the Ultimate Theory of Time*. Dutton.

[14] Noah Linden et al., “Quantum Mechanical Evolution towards Thermal Equilibrium,” Physical Review E 79, no. 6 (June 4, 2009): 061103, https://doi.org/10.1103/PhysRevE.79.061103.

[15] W. H. Zurek, “Environment-Induced Superselection Rules,” Physical Review D 26, no. 8 (October 15, 1982): 1862–80, https://doi.org/10.1103/PhysRevD.26.1862.

[16] Wojciech Hubert Zurek, “Quantum Darwinism,” Nature Physics 5, no. 3 (March 2009): 181–88, https://doi.org/10.1038/nphys1202.

[17] Ming-Cheng Chen et al., “Emergence of Classical Objectivity of Quantum Darwinism in a Photonic Quantum Simulator” (arXiv, August 5, 2019), http://arxiv.org/abs/1808.07388.

[18] Wheeler, John A. (1978). The ‘Past’ and the ‘Delayed-Choice’ Double-Slit Experiment. In A.R. Marlow (Ed.), Mathematical Foundations of Quantum Theory (pp. 9–48). Academic Press.

[19] John Von Neumann, Robert T. Beyer, and Nicholas A. Wheeler, Mathematical Foundations of Quantum Mechanics, New edition (Princeton: Princeton University Press, 2018).

[20] Max Tegmark, Our Mathematical Universe: My Quest for the Ultimate Nature of Reality, First edition (New York: Alfred A. Knopf, 2014).

[21] Barry Loewer and University of Arkansas Press, "Humean Supervenience:," Philosophical Topics 24, no. 1 (1996): 101–27, https://doi.org/10.5840/philtopics199624112.

[22] Schneider, Susan. "Does the mathematical nature of physics undermine physicalism?." Journal of Consciousness Studies 24, no. 9-10 (2017): 7–39.

[23] John Heil, The Universe as We Find It (Oxford: Clarendon Press, 2012).

[24] James Ladyman et al., Every Thing Must Go: Metaphysics Naturalized (Oxford; New York: Oxford University Press, 2007).

[25] Peter Mark Ainsworth, "What Is Ontic Structural Realism," Studies in History and Philosophy of Modern Physics 41 (2010): 50–57, https://api.semanticscholar.org/CorpusID:119776505.

[26] David Bohm and Basil James Hiley, The Undivided Universe: An Ontological Interpretation of Quantum Theory (London New York: Routledge, 1993).

[27] K. S. Lee et al., "Entanglement between Superconducting Qubits and a Tardigrade," New Journal of Physics 24, no. 12 (December 1, 2022): 123024, https://doi.org/10.1088/1367-2630/aca81f.

[28] Stephen Wolfram, A New Kind of Science (Champaign (Ill.): Wolfram, 2002).

12 Ruliology: Linking Computation, Observers, and Physical Law

Dean Rickles, Hatem Elshatlawy, and Xerxes D. Arsiwalla

12.1 Introduction: Foundations of physical theory

> If we had invented the digital computer before inventing graph paper, we might have a very different theory of the universe today.
>
> Jacques Vallee, *Dimensions*[1,2]

Often when thinking about the modeling of reality in physical theories, we employ an abstract space that is supposed to represent all possible states of a system, a modal arena, one point of which will correspond to the present state of a 'real-world' system. This provides the kinematical structure of a theory when forces are ignored (yielding a larger space of possibilities than is physically allowed) and the dynamics when forces are included (yielding the so-called nomologically posssible states). In general, what is not in the space is not a possibility; and what is not a possibility is not in the space. This can be a *universal* state space as in the geometrodynamics of John Wheeler, where it is known as "superspace" [40]. Here, "points" of the space are 3-dimensional geometric configurations (i.e. Riemannian geometries on a 3-manifold) of the universe and histories are then represented as trajectories (paths through the space, generating spatiotemporal worlds), corresponding to possible universes—the space of 3-geometries is understood as the space of 3-metrics "quotiented" by the diffeomorphism group (the invariance group of general relativity), identifying

DOI: 10.4324/9781003546139-16

those metrics differing by elements of that group.[3] Furthermore, in quantum geometrodynamics, we envisage a wave function over this configuration space which assigns amplitudes for the various types of states of the universe.[4] While presented as a rather fundamental description of physics, even the superspace point of view clearly stands several rungs up on the ontological ladder, presupposing several layers of deeper structure.[5]

The "Physics Project" recently initiated by Stephen Wolfram [59, 60, 62, 63, 67] aims to describe how all other levels of structure are built from the ground up, that is, from ontological ground zero. The basic structure is not a set of elements as such, but a *totality* that can then be decomposed to generate possible universes, including our own. This structure is called the "Ruliad" or "Rulial space" and is usually expressed informally as the result of carrying out a process (or, rather, many such) to infinity, yielding "the entangled limit of all possible computations": it is what is generated by carrying out all possible rules in every possible way [64, 65]. It is computationally exhaustive. Like the universal state space of geometrodynamics or moduli spaces, paths of the Ruliad correspond to possible histories of universe (though unlike the former two, the Ruliad is a purely syntactic structure, defined independently of any *a priori* geometric notions).

Ruliology, a term coined by Stephen Wolfram, studies the intricate structure of rule space and investigates how different rules, including apparently very simple ones, can lead to diverse and complex behaviors. It represents the study of the Ruliad, a profound and encompassing framework within the Physics Project that serves as the theoretical foundation for understanding the myriad of computational universes. Instead of treating reality as a mere collection of isolated entities, Ruliology embraces the idea of a vast, interconnected web of all conceivable computations, executed through every possible rule. This intricate and boundless space is not just a speculative novelty; it provides a comprehensive map from which individual universes, including ours, can be derived. Such a perspective challenges traditional views on the nature of reality and paves the way for a more unifying, computational understanding of the cosmos. Given its ambitious scope, the potential depth of insight into the nature of rules, and its profound implications, Ruliology demands rigorous exploration and merits earnest attention in the broader scientific discourse.

It is perhaps worth remarking up front on the similarities to David Deutsch's notion of a "constructor" here [21] because in that case, as in Wolfram's approach, one is demonstrating existence through a constructive (computationally conceived) procedure—both also find some insufficiency in the orthodox Turing machine model of a universal computer as a model of reality. What is possible can be constructed *physically* from some rule (or "task" in Deutsch's terminology), and what cannot be constructed is impossible (i.e. there is no such constructor up to the task). Note, also, that as constructive theories, they have an end-goal in mind (namely, that which is to be constructed), and so both contain teleological elements.[6] The key idea of constructor theory is, then, simply that the focus of fundamental theory should be which transformations of some medium or substrate into another such state can be caused to occur, as well, by implication, as those which cannot be so caused. Given substrate independence, the focus becomes the *transformations* themselves as the ontological core of the theory. The precise nature of this, essentially, modal structure consisting of counterfactuals has yet to be adequately nailed down, since while the transformations themselves are always grounded in some physical substrate, the counterfactuals, as non-actual by definition, are clearly not (though see [39] for a discussion of some of the options and problems).

Other related constructivist approaches which lend similar precedence to processes over substrates include "Assembly Theory", "Process Theories", and "Intuitionistic Physics". The first of these [20], focused on the detection of life, is based on *rules of assembly*, which take into account the number of independent parts and their connections, such that as the number increases, the need for memory increases, which enables the reconstruction of the whole from locally stored rules. Process theories (the second approach) are founded in the framework of category theory and seek to formalize physical operations in diagrammatic terms in which the diagrams (representing morphisms between objects) are expressed as objects and transformations within an appropriate monoidal category [1, 18, 19, 31–33]. Intuitionistic Physics (the third approach) seeks a formalization of physical observations and measurements based on intuitionistic logic, rather than classical logic, to escape, for one, the fact that a physics based on real numbers will face the problem that we will never be able to grasp them, requiring as they do infinite Shannon information to specify their non-repeating decimal expansions [22, 29] (note that intuitionistic mathematics involves a temporal, step-wise process, rather than an eternal, Platonic structure, and this will be important for our later claims about the essential limitations of the Ruliad *qua* fundamental theory).

The substrate-independent approach described above, in constructor theory, is more or less what Einstein once called the "principle theory" method (see, e.g. [24]). Rather than dealing with what things are made of, in terms of composition, the method looks at the higher-level principles that any and all things must obey, regardless of their physical constitution.[7] This transcends particular physical theories and provides a theory of theories: a meta-theory. Wolfram's approach shares this feature of being a theory of theories—one might call it a theory of *all* theories. Among other things, such a meta-theory bears the burden of having to explain how physical notions of space, time, matter, laws, and observers arise, which we identify as the core elements of physical theories. Wheeler recognized this challenge in the 1970s and coined the term "pregeometry" in part to address these very issues.[8] Note, also, that in delving to this deeper level, one can evade the usual problems with treating quantum mechanics as a universal theory, in which one can model not just microscopic systems but also the very agents using quantum mechanics [27]. In the Wolfram case at least, quantum mechanics is a feature of the structure itself (the so-called multiway description), a consequence, rather than the structure being explicitly constructed to capture quantum features from the outset—the same can be said of the other theories that fall out of the structure without being inserted *ad hoc*.

In this chapter, we attempt to explain the Wolfram model, focusing more on conceptual issues than formal details, and aim to bring out the role of observers as it appears in the model, showing how it is essential for making sense of standard physics as well as mathematics. The relationship between the Ruliad and the observer is also explicated. We use the basic idea of second-order cybernetics to elucidate a deeper understanding of how the Wolfram model includes the observer, explaining the very generation of the world we see as a kind of observer-selection effect, much like a reference frame in relativistic theories. This allows us, moreover, to provide an account of the nature of "physical laws" as sampling-invariance. However, the Wolfram model also lets us see in a new light a fundamental limitation of trying to gain fundamental knowledge of the world from the standpoint of the observer doing modeling. Let us begin with an account of the basic elements of the Wolfram model.

12.2 The Wolfram model as an abstract rewriting system and the concept of the ruliad

When you come to a fork in the road, take it!

Yogi Berra

This section provides a literature overview of the basic abstract rewriting constructions that constitute the Wolfram model, with a particular emphasis on their non-deterministic aspects, as captured via *multiway systems*.[9] We discuss the concept of the Ruliad, representing the entangled limit of everything that is computationally possible, which emerges as a key theoretical construct associated to the Wolfram model [64, 65]. We begin with a preliminary description of the Wolfram model in terms of diagrammatic rewriting rules acting on hypergraphs. The Wolfram model represents a discrete framework that posits structures such as continuous spacetime geometries which may potentially emerge from large-scale limits of the underlying discrete structures [7, 23, 31, 59, 60]. Furthermore, the evolution of these structures is dictated by various forms of rewriting rules, such as those based on graphs, hypergraphs, or strings.[10] To illustrate this, a Wolfram model hypergraph can be represented abstractly as a finite collection of ordered or unordered relations (hyperedges) between labeled nodes, as defined below[11] and shown in Fig. 12.1.

Definition 12.1: A wolfram model hypergraph $H = (V, E)$ is characterized by a finite set of hyperedges E that belong to the non-empty subset of the power

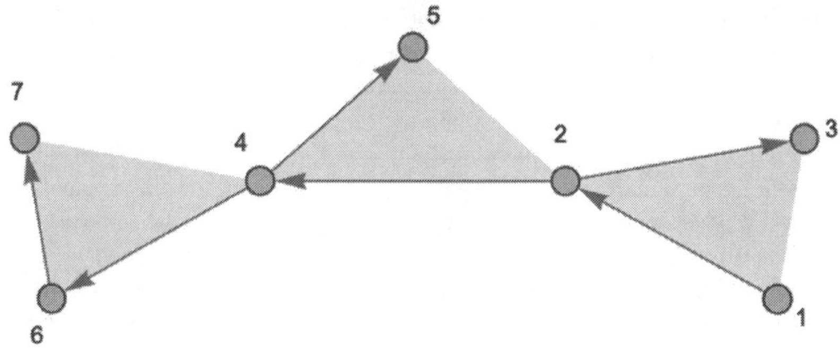

Figure 12.1 An example of a Wolfram model hypergraph with directed hyperedges: $\{\{1,2,3\},\{2,4,5\},\{4,6,7\}\}$.

set of V, i.e., $E \subset \mathcal{P}(V)\backslash\{\emptyset\}$. A hyperedge in E is an unordered collection of nodes from the vertex set V. Hyperedges in the Wolfram model, can be both, either directed or undirected.

One can then define the dynamics of a Wolfram model system in terms of hypergraph rewriting rules as follows:

Definition 12.2: A "Rewriting Rule", denoted as R, for a spatial hypergraph $H = (V, E)$, is an abstract rewriting rule expressed in the form $\mathrm{H}_1 \rightarrow \mathrm{H}_2$. In this rule, a subhypergraph that matches the pattern H_1 is replaced by a subhypergraph that matches the pattern H_2.

Definition 12.3: A Wolfram model is an abstract rewriting system founded on the principles outlined in definitions 12.1 and 12.2. It's worth noting that Wolfram models are not solely limited to hypergraph rewriting systems; they encompass a range of other rewriting systems, including but not limited to string rewriting systems, term rewriting systems (TRS), (hyper)graph rewriting systems, and cellular automata.

Every rewriting rule in this context can be formally mapped to a set-substitution system, where a specific subset of ordered (unordered) relations that matches a given pattern is replaced with another distinct subset of ordered (unordered) relations that also corresponds to a particular pattern, as shown in Fig. 12.2.
It is worth noting that the sequence in which transformation rules are applied is generally not predetermined. Even in the simplest scenario where the rule is applied to every matching and distinct subhypergraph (see Figs. 12.3 and 12.4), the initial selection of which subhypergraph to transform first remains open-ended. This multiplicity of choices typically leads to different, non-equivalent

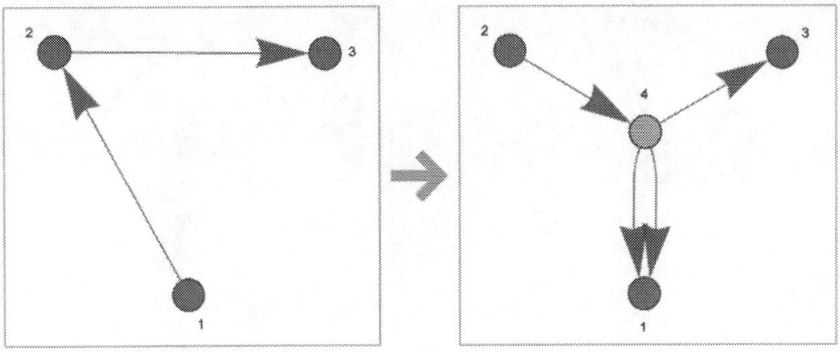

Figure 12.2 A hypergraph transformation rule corresponding to the set-substitution system $\{\{1,2\},\{2,3\}\} \rightarrow \{\{4,1\},\{4,1\},\{4,3\},\{2,4\}\}$.

sequences of evolving hypergraphs. Hence, the evolution of a given spatial hypergraph is inherently non-deterministic due to the absence of a fixed updating order (or rather the possibility of multiple possible updating orders). Therefore, we can treat the Wolfram model as a non-deterministic abstract rewriting system.

Furthermore, within the conventional computational paradigm, systems typically evolve through a series of sequential steps by applying specific rules. However, in the case of Wolfram models (for certain rules), determinism is not inherent. Multiple choices of substitutions are possible, resulting in diverse outcomes. Usually, we select one possibility (e.g. from the possibility space mentioned earlier) and disregard the others (as ways the system *might* have

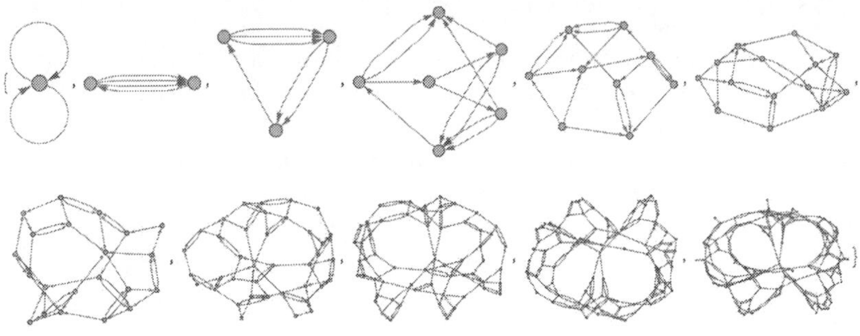

Figure 12.3 The results of the first 10 steps in the evolution history of the set-substitution system $\{\{1,2\},\{2,3\}\} \rightarrow \{\{4,1\},\{4,1\},\{4,3\},\{2,4\}\}$, starting from a double self-loop initial condition $\{\{1,1\},\{1,1\}\}$.

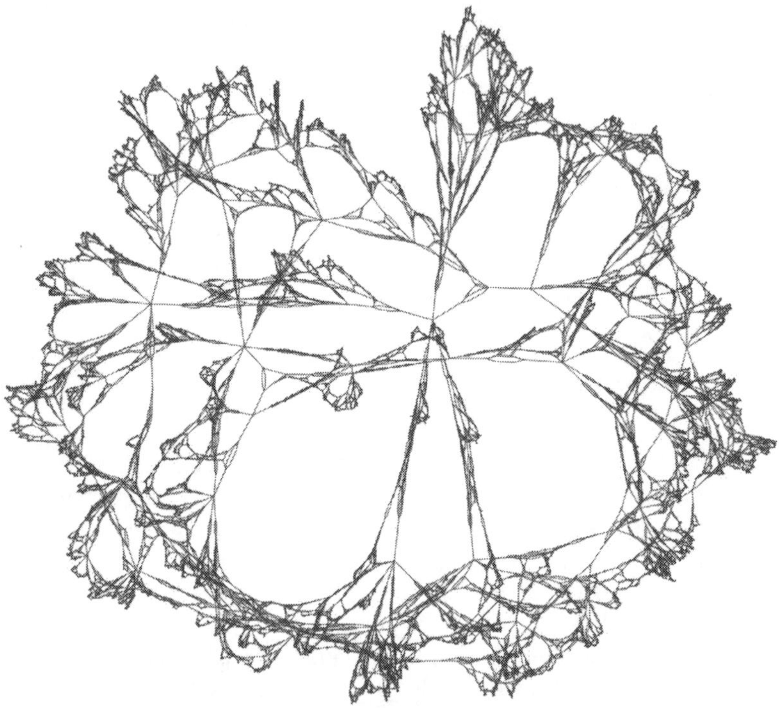

Figure 12.4 The result after 19 steps of evolution of the set-substitution system $\{\{1,2\},\{2,3\}\} \rightarrow \{\{4,1\},\{4,1\},\{4,3\},\{2,4\}\}$, starting from a double self-loop initial condition $\{\{1,1\},\{1,1\}\}$.

evolved), but the concept of a multiway system allows for simultaneous exploration of all potential choices. A key idea is to consider all those possible threads of history—and to represent these in a single object that we call a multiway graph.[12] Consider as an example (Fig. 12.5) a system defined by the string rewrite rules: $A \rightarrow BBB, BB \rightarrow A$. Starting from A, the next state has to be BBB. But now there are two possible ways to apply the rules, one generating AB and the other BA (thus forming a fork in the graph). And if we trace both possibilities, we get what we call a multiway system—whose behavior we can represent using a multiway graph. And it's not really difficult to construct multiway system models. There are multiway Turing machines. There are multiway systems based on rewriting not only strings, but also trees, graphs, or hypergraphs. There are also multiway systems based on numbers and all kinds of multiway systems. Combinatorially, a multiway system is simply a directed, acyclic graph of states, determined by abstract rewriting rules that inductively

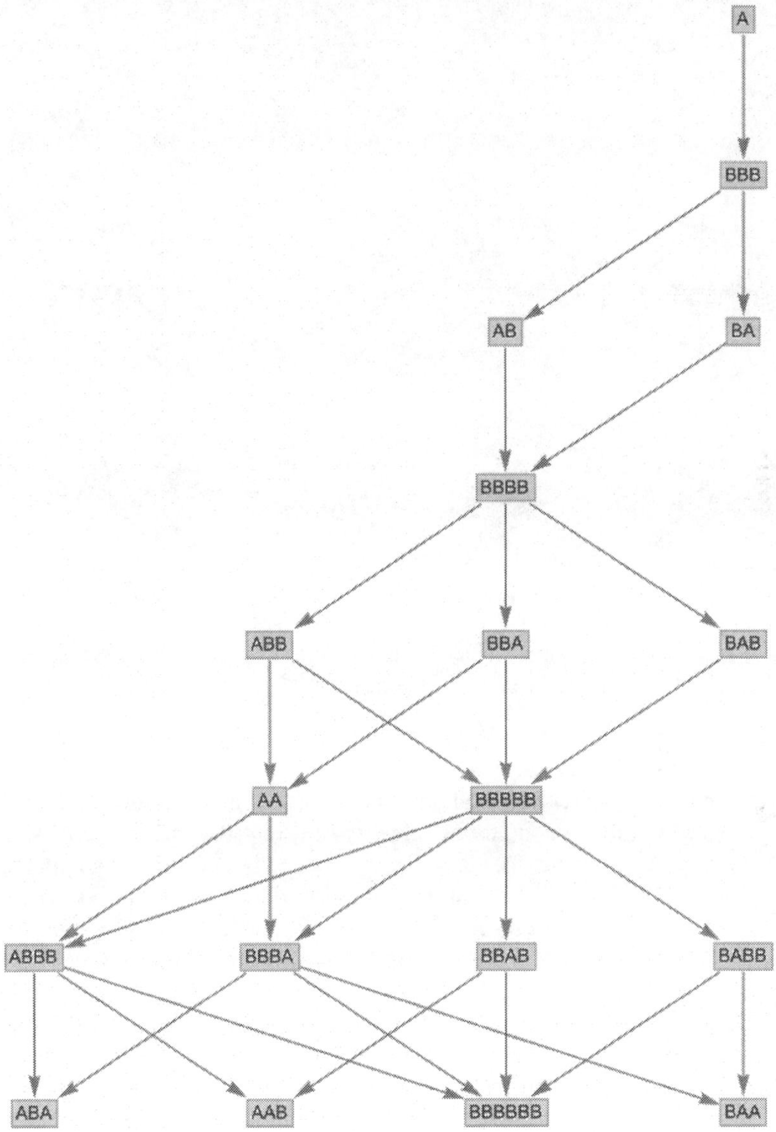

Figure 12.5 The multiway evolution graph corresponding to the first 7 steps in the non-deterministic evolution history of the string rewrite rules: $A \rightarrow BBB, BB \rightarrow A$ (cf. [7, p. 8]).

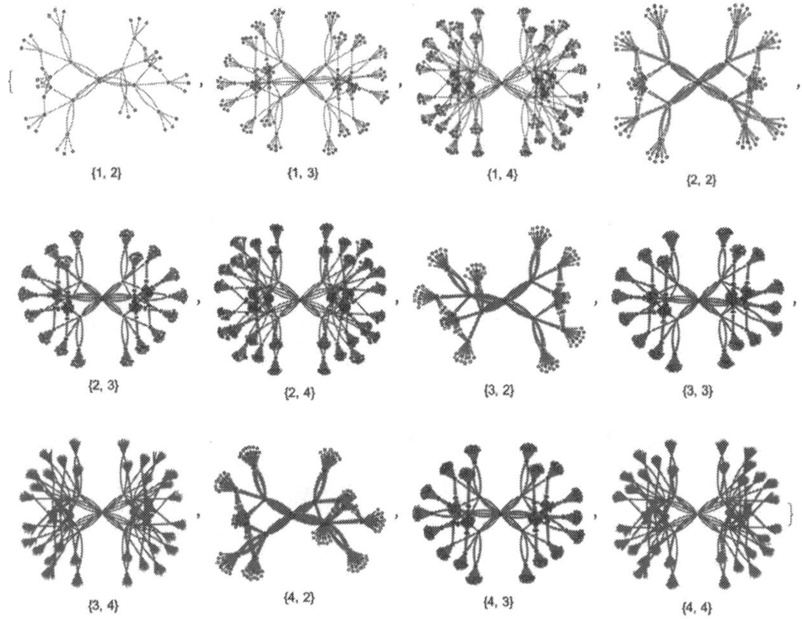

Figure 12.6 Rulial multiway graphs after 3 steps for Turing machines with various numbers of states and colors $\{s, k\}$ (Adapted from [65]).

generate a (potentially infinite) *multiway evolution graph*, together with a partial order on the rewrite rule applications, determined by their causal structure.

Now instead of looking at all possible ways a given rule can update these rewrite systems, imagine the structure of spaces created by applying all possible rules. Instead of just forming a multiway graph in which we do all possible updates with a given rule, we form a *rulial* multiway graph in which we follow not only all possible updates but also all possible rules (an illustrative figure of rulial multiway graphs for Turing machines with various numbers of states and colors $\{s, k\}$ is shown in Fig. 12.6) [64, 65]. This construction allows us to think about the notion of rulial space, i.e. the space of all possible rewriting rules of a given signature. By applying all possible rules in all possible updates, we get what we call the *Ruliad* the result of following all possible computational rules in all possible ways (a schematic depiction of a finite approximation of the Ruliad is shown in Fig. 12.7).[13] It's the ultimate limit of all rulial multiway systems. And as such, it traces out the entangled consequences of progressively applying all possible computational rules. The concept here is to use not just all rules of a given form, but all possible rule, and to apply these rules to all possible initial conditions, as well as to run the rules for an infinite number of

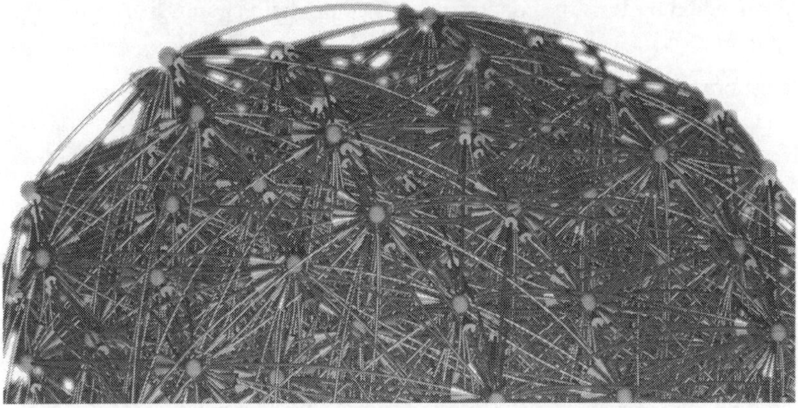

Figure 12.7 A schematic figure of the Ruliad (Adapted from [64]). The image presented can be considered as a rough, finite approximation of the Ruliad. The complete Ruliad encompasses the exploration of infinite limits across all conceivable rules, initial conditions, and steps. Here the nodes and edges are not specific to any single entity. Nodes can represent various entities such as hypergraphs, strings, or states, while the edges signify the myriad potential causal connections between those entities.

steps. Essentially, the Ruliad involves taking the infinite limits of all possible rules, all possible initial conditions, and all possible steps.[14] Consequently, the Ruliad is in effect a representation of all possible computations. A conceptual definition of the Ruliad is given below:

Definition 12.4: *Ruliad* is a meta-structural domain that encompasses every possible rule-based system, or computational eventuality, that can describe any universe or mathematical structure. It acts as a theoretical space wherein the boundaries between map and territory blur, pushing beyond mere perception and functioning as the ground for the possibility of multi-computation. Within the Ruliad, every conceivable physical and mathematical system can be situated, but their accessibility or meaningfulness is determined by the specific observer-related frames or constructs. The Ruliad is thus a pre-physical framework, and its utilization in physics is to pinpoint the exact rule-based system that corresponds to our observed reality.

A computational definition is as follows:

Let $\mathcal{R}$ be the space of all possible computational (or rewriting) rules. We refer to $\mathcal{R}$ as the "Rulial space". Consider $r \subseteq \mathcal{R}$. r may represent either a single rule or a collection of rules capable of generating a computational universe U_r (the outcome of all possible computations following a given rule set). The

Ruliad, denoted by **R**, is the collection of all computational universes. That is, $\mathbf{R} = \{U_r \mid r \in \mathcal{R}\}$.

Furthermore,

1 An observer O with a frame or construct F_O can interpret or access a subset of $\mathcal{R}$ based on their specific observational constraints. The observability of any U_r is contingent on F_O.

2 The entirety of **R** serves as the foundational ground for multicomputation, transcending the dichotomy between representation and reality.

Ruliology, intricately intersects with this multicomputational paradigm. This nexus between Ruliology and multicomputation sets a fresh foundation for understanding the vastness and versatility of rule space, where different rules can lead to multifaceted and complex behaviors. Thus, one can contend that the Wolfram model encompasses a novel paradigm that extends beyond traditional computation [66]. This is what Stephen Wolfram calls the multicomputational paradigm [66]. It not only traverses the boundaries of physics but also paves the way for a foundational and versatile methodology for crafting models in theoretical science. Historically, three paradigms have dominated theoretical science: mathematical equations, which rely on formulas to describe phenomena; mechanistic models, which provide detailed, step-by-step explanations of how systems function, likened to "machines" with distinct parts and processes; and computational models, which view systems as computational entities, allowing for the definition of rules and initial conditions, and then observing the resulting behaviors. However, the multicomputational paradigm goes a step further. It's not just about analyzing specific historical paths but delving into the evolution of all conceivable histories, epitomized by the Ruliad. In many instances, it may not offer insights into specific histories. Instead, what it will describe is what an observer sampling the whole multicomputational process will perceive. This pivotal intersection of Ruliology and multicomputation is where our exploration now focuses.

Definition 12.5: *Multicomputation*, also known as the multicomputational paradigm, is a generalization of the traditional computational paradigm to encompass multiple computational histories or threads of time. In the standard computational approach, time progresses in a linear fashion. This means that the next state of a system is computed successively from its previous state. In contrast, multicomputation allows for every possible path of computation to proceed through distinct, interwoven threads of time. Instead of a single linear progression, there are thus multiple threads of computational time that can be explored. In essence, multicomputation expands the scope of computational exploration by considering all possible computations simultaneously, rather than just one at a time.

The following provides a characterization of how computational steps are executed in a multicomputational paradigm:

Let S be a system defined by a set of computational rules R and initial conditions I. In the traditional computational paradigm, the evolution of S is represented by a sequence of states $s_1, s_2, \ldots, s_n$ such that each state s_i is derived from s_{i-1} using rules R. In the multicomputational paradigm, the evolution of S is represented by a network T of states, where each node represents a state of S and each branch represents a possible path of computational evolution based on rules R. Each node can have multiple child nodes, representing different possible next states. Furthermore, state equivalences between nodes in different branches allow for intersections of different evolution paths. In other words, for each state s_i in T, there exists a set of states $C(s_i)$ such that for each s_j in $C(s_i), s_j$ is a possible next state of s_i following rules R. The network T thus captures all possible computational trajectories of S starting from initial conditions I.

Definition 12.6: *Ruliology*, derived from the term "rule", is the systematic study and exploration of computational rules and their myriad manifestations within computational systems. It delves into the intricacies of rule space, examining how diverse collections of rules can give rise to complex behaviors and structures. Ruliology transcends traditional computational boundaries, aiming to comprehend the foundational principles behind all possible computations, and seeking to understand how distinct collections of rules can generate entire universes of computation. At its core, Ruliology is an attempt to map out and understand the vast, multifaceted landscape of the Ruliad, where every conceivable rule is executed in every possible way.

In terms of the Ruliad, Ruliology as the study of computational rules and universes can be characterized as follows:

Let R be the collection of all possible computational rules, S be the collection of all possible states, and $F : R \times S \to \mathcal{P}(S)$ be a map denoting the evolution or transformation of states dictated by a rule from R upon a state from S. The Ruliad $\mathbf{R}$ (defined above) includes all such computational evolutions. Ruliology is then the study of the properties, structure, and implications of the Ruliad $\mathbf{R}$, as well as the exploration of individual and collective behaviors arising from elements of R when acted upon S.

12.3 Observers, sampling, and the physical world

> O God! I could be bounded in a nutshell, and count myself a King of infinite space...
>
> William Shakespeare, *Hamlet*, II, 2

A potentially serious stumbling block with the Ruliad idea as it stands is what we might call the "realization problem": how does an abstract rule get turned into physical reality? How do we end up with a particular history, to use the previous section's terminology? If this reality is the result of the computation of rules, then what is *doing* the computation? Is it a computer of some kind? But then if this is a fundamental theory, should this computer not itself be a part of the Ruliad? We find ourselves in this way in a loop which cannot possibly be a virtuous circle of reasoning. It is more akin to that famous adventurer Baron von Munchausen rescuing himself and his horse from a quagmire by lifting himself up by his own hair. We can expect something like this problem to face any pre-geometry-type proposal that intends to dig beneath the spatiotemporal world populated with matter to something more abstract lying beneath—in several quantum gravity proposals, the task of getting the world we are acquainted with, with its description in terms of fields on differentiable manifolds, from a deeper discrete theory, is known as the "reconstruction problem". Part of the problem is that the deeper theories do not involve things spatially located, and evolving dynamically, but an abstract and far more primitive structure, often based on more relational concepts such as graphs and networks. It is, of course, a general and well-known problem to explain how we move from abstract formalism to physical reality. Usually, we start from the physical reality as a foundation, and then develop an abstract representation of it. In the case of pre-geometrical approaches (such as the Wolfram model), one makes no initial theoretical assumptions about the nature of physical reality, but starts instead from an abstract domain, with the hope of then recovering the physical aspects from this.

A related problem, of moving from abstract to concrete, is well expressed by John Wheeler [40, p. 1208]:

> Paper in white the floor of the room, and rule it off in one-foot squares. Down on one's hands and knees, write in the first square a set of equations conceived as able to govern the physics of the universe. Think more overnight. Next day put a better set of equations into square two. Invite one's most respected colleagues to contribute to other squares. At the end of these labors, one has worked oneself out into the doorway. Stand up, look back on all those equations, some perhaps more hopeful than others, raise one's finger commandingly, and give the order "Fly!" Not one of those equations will put on wings, take off, or fly. Yet the universe "flies".

A well-known related sentiment was expressed through Stephen Hawking's question "What breathes fire into the equations?" in his book, *A Brief History of Time* [35]. In other words, what makes an (abstract) equation or generalization (which "oversees" a set of possibilities) a physical reality? Hawking elaborates:

> Even if there is only one possible unified theory, it is just a set of rules and equations. What is it that breathes fire into the equations and makes a universe for them to describe? The usual approach of science of constructing a mathematical model cannot answer the questions of why there should be a universe for the model to describe. Why does the universe go to all the bother of existing?

This is an issue concerning the metaphysics of the laws of nature. We can give much the same response for the Ruliad as Wheeler did here: observership, or rather participation, provides the necessary engine that powers the creation of the physical universe. Hence, the Ruliad will need to include a "theory of the observer" such that "the universe as we know it" seen from the vantage point of "observers like us" realizes observable physical attributes of the universe. However, it would be a mistake to view the physical universe as *unique* and *absolute*. Moreover, any computation is a result of the perspective of a computationally-bounded observer, rather than a fundamental feature of the universe itself: we are now in the realm of epistemology (i.e. description or representation) rather than ontology (how things are in a fundamental sense). The observer acts as a kind of transducer for the Ruliad, converting the abstract computations into (apparently) physical form.[15]

There is an interesting philosophical relation of the Ruliad to Leibniz's system of monads. Leibniz's *monadology*, his last attempt to codify his philosophical system, can certainly rival Wolfram's Ruliad for all encompassing majesty, despite its extreme brevity. Each monad is an individual that reflects the rest of the universe from its own unique point of view. The parts shape the whole and in turn, the whole back-reacts on the parts. Likewise, the Ruliad has similarity to *Indra's Net* from *The Flower Garland Sutra*—a kind of representation of a totality in terms of bejeweled vertices which encode the whole. Each is a vista of the whole. Every possible view is present in the whole. It is interesting to see how this basic idea, in which a totality is decomposed into interdependent parts, repeats.[16]

While monads collectively provide all possible perspectives of a world, as tiny independent mirrors (or points of view), Wolfram's Ruliad deals with all possible rules applied to some initial collection of abstract relations. However, there is also a sense in which monads are carrying out a pre-set program (or entelechy), coordinated with all other monads, in a pre-established and divinely choreographed dance determined to generate (i.e. construct) the best of all possible worlds. While there are the well-known principles of sufficient reason and identity of indiscernibles providing basic constraints on this construction, the principles themselves do not directly determine what is constructed. Rather, they inform the composition of monads into complex structures, which is then carried out through the pre-established harmony. A major reason for the introduction of pre-established harmony was to explain the mind-body (or soul-body) correlations. For Leibniz there was no causal link and the correlation stemmed from the common cause in which both were set on their way like a pair of perfectly synchronized watches. Interestingly, as we will develop further in another paper [6], Wolfram model has a remarkably similar explanation for the correspondence of the world to the mind in that they both emerge from the same initial rules for construction and emerge in parallel with the mind (or observer) simply sampling the world and providing a perspective, much like a monad, where different observers represent the whole universe from different points of view. Likewise, one can find a similar generation of variety in the Wolfram model through this dislocation of a single, unified structure into many of points of view.

Of course, Leibniz's theory, as it stands, is not of much practical value in terms of showing how our present theories and phenomena can be *constructed*. The approach of the Wolfram model, involving hypergraph rewriting systems, places the ontological weight on the very rules of construction themselves. By contrast with Leibniz's "God as architect" (as he puts it in S.89 of his *Monadology*), here the metaphor is better expressed, following Chaitin [17], as "God as programmer", though here employing a multiway approach rather than a single-track, Turing machine approach. A more crucial distinction, related to the constructive approach, is that physics (and mathematics) emerges from the interplay of computationally-bounded, embedded systems (observers) with the structure in which they are embedded (and therefore sampling), namely the Ruliad.

What an observer ultimately does is to take in an input from a large set and return an output from a much smaller set thus acting as a kind of idempotent filter. It's a concept that's appeared in many fields under many different names. It can be called a contractive mapping, a reduction to canonical form, a classifier, a forgetful functor, lossy compression, projection, renormalization group transformation, and so on. It's what's fundamentally going on whenever we use a sensor or a measuring device, or for that matter, our human senses: we extract statistics, fit to models, and describe things symbolically. As a basic physical example, consider a gas pressure sensor based on a piston. Within the gas, individual molecules move around in complicated and seemingly random ways, hitting the piston in all kinds of configurations. But the piston "reduces out" all those details, responding just to the aggregate force of all the molecules, the same one of which can be realized in potentially infinitely many ways. The main point is that we can describe what's going on more formally by saying that "observations by the piston" identify all the different detailed configurations of molecules, preserving only information about their aggregate force, forgetting the finer details.

This same idea can then be given slightly different interpretations, revealing how observers influence various branches of science. In statistical physics, for example, observers have, as just described, the effect of averaging over many particles or other degrees of freedom [68]. In general relativity, they are averaging over spacetime regions, and forgetting those details having to do with coordinate transformations. In quantum physics, they are basically averaging over many quantum histories. In mathematics the "same" statements are stated differently in terms of underlying axioms [63]. Gauge theories can be understood in the same way: the equivalence classes will in this case be generated by the gauge transformations which will be identified relative to an observer (though perhaps some other observer could view absolute structures such as the individual gauge potentials, much as a skilled musician with absolute pitch can hear differences that most others identify). In high energy physics, black holes in various dimensions have an associated thermodynamic description, such that the physical charges of the black hole or black ring depend on whether the object is viewed from a 4D perspective or a 5D perspective [2, 3]. Further afield, in economics, the focus might be on certain indices generated by the behaviors of a country's people. Even in linguistics we have identifications (between systems that differ in details) given by the equivalence relation defining concepts, such as what counts as a chair, coat hook, or cabbage. This can then be applied to all scientific areas (and beyond) in which observers are involved.[17]

Let us now consider some further conceptual implications of this overall structure or observers sampling the Ruliad, along with the notion of updating/rewrite rules. We start with the status of the approach *vis-à-vis* determinism and modality.

12.4 Computation, determinism, and free will

The question of determinism versus indeterminism is not so clear-cut in the Ruliad model and intersects with the issue of descriptions from the inside versus outside [i.e. computationally unbounded versus computationally bounded respectively]. One reason is that the inclusion of quantum mechanics into this picture is achieved through a notion of a branching (i.e. multiway), rather than linear, structure connecting the states in a process. This is supposed to represent the multiplicity of possible paths that quantum, but not classical, mechanics entails. In much the same way that the universal wave function, while itself deterministic (evolving according to the linear Schrodinger equation), nonetheless contains a kind of *local* indeterminism if one follows specific paths through the space defined by the universal wave function.[18] In other words, our answer to the question "Is this theory deterministic or indeterministic?" hinges on whether we are viewing things from an embedded perspective or from a God's eye perspective.

However, to view this branching itself as parallel to the indeterminism of quantum mechanics is to deny the possibility of classical, deterministic physics. And, of course, the Wolfram model should contain both classical and quantum theories according to its role as a theory of all theories. Wolfram has a way of explaining this, of course, by pointing to a coarse-graining effect mentioned in the previous section. The explanation of definite happenings (phenomena) despite the multiplicity of paths then comes about through the observer-participator as embedded in the multiway system. However, as with many-worlds interpretations, from the perspective of the totality (the Ruliad), everything that can happen (from the point of view of rules) does happen. However, the approach here goes beyond many-worlds—or is perhaps more in keeping with Hugh Everett's original ideas—since it incorporates the observer in the model itself (something we explore more in the next section).

Given the branched-system understanding of quantum mechanics, it is clear that the model can support a grounding of counterfactuals along the same lines as constructor theory, in which the Everettian approach is adopted (at least for counterfactuals that occur in some branch). One of the problems faced in that approach is precisely that there are modal claims (e.g. those involving the very laws of physics themselves) that do *not* occur in a branch of the multiverse. In this sense, the Ruliad has more modal breadth, since it encompasses the entire theoretical structures.

Does this have anything to tell us about free will? One might, for example, take the idea of computational irreducibility, in which one must let a system evolve to know its evolution, as some kind of free will proxy. However, this is simply unpredictability, not indeterminism. We find the same behavior in chaotic systems, of course, where it is understood that the systems are perfectly deterministic albeit with agent-relative uncertainty about the development. It is why we simulate some systems, and in which the best we can do to know a future state is simply increase the performance power of the computer cranking through each iteration. So as observers watching a computationally irreducible process, we have uncertainty about the future, but the future is not uncertain *simpliciter*. This must be the case if the Ruliad is seen to exist eternally, so that it is not a growing block structure. Such a degree of uncertainty does not imply free will. Even if there was pure randomness, it would not imply free will. Free will is the ability for the evolution of a system to fork in such a way that an agent has the ability to decide which path is chosen. That is, one could have isomorphic histories up to a point t which diverge thereafter. There seems to be a confusion occurring between one's actions not being pre-determined and free will. However, with the Ruliad we have neither of these situations, but simply an epistemic uncertainty about the future based on embedding an observer in a system with a kind of temporal gradient that emerges from the hypergraph rewriting process that the selfsame observer will be a part of.

Wheeler famously asked "Why the quantum?". We have an answer here: the Ruliad (and its included samplers). And if we are then pushed to ask "Why the Ruliad?", then we have an answer there too: there could not *not* be the Ruliad, since it is a mathematically necessary object. But we can say a little more here. The quantum is not only a matter of the multiway picture. It is also a demand that the observer be included. Without this we have something akin to a space of pure potentiality, but in which nothing is made concrete or actual. There are no happenings without the consideration of a frame with respect to which something happens. This is very similar to the way in which there is no moon in quantum mechanics when no one is looking: there simply is no objective *way* the world is in quantum mechanics, and likewise not in the Wolfram model either. Rather, there are all ways, which implies no way. Hence the need to introduce something like a Wheelerian observer-participator to select one such way the world can be, though with uncertainty as to which way that is. This notion of including the observer in so central a way suggests that the Wolfram model would benefit from ideas originating in second-order cybernetics. Indeed, the Wolfram model might be a fine example of a naturally second-order cybernetic system.

12.5 The algorithmic nature of observers

Circling back to the dual relation between computation and measurement of an observable, one may think of measurement as an inference process under conditions of uncertainty. In other words, the process of observation (at least in the sense of making measurements of physical quantities) is the act of inferring approximate or coarse-grained causes of the outcomes that computations (around the observer) generate. Given the observer's own computational boundedness, it has to base these estimates on limited samples and restrict to causes that provide description only up to those levels of complexity that its own inferential engine can handle. This is where making equivalences and coarse-graining states of high complexity become relevant for any observer theory. This suggests an algorithmic nature of the observer as an inference engine within the Ruliad seeking states or computations with lowest complexity.

Apart from the setting of the Ruliad, a theory of inference engines has been described in the context of cognitive neuroscience in terms of Karl Friston's "Free Energy Principle," where a cognitive agent seeks to minimize its free energy either by performing actions upon the world or by changing its perceptions/representations of the state of the world based on new incoming data [28].

The free energy here is a complexity measure[19] whose minimization is associated to minimizing the "surprise" or uncertainty in the agent's representation of the world around it. More generally, for cognitive agents (both, biological and artificial), this minimization is achieved algorithmically using a hierarchical inference scheme based on feedback loops involving predictions and errors concerning the states of the world in comparison to the agent's own prior expectation [10, 12].

Coming back to the Ruliad, the idea of an observer as an inference engine may be abstracted as a theory of sampling and measurement of low-complexity (or at least comparable to the observer itself) states within the Ruliad. Any complexity minimization principle akin to the free energy principle is in fact a second-order cybernetic construct. Presumably, this has to be included as a meta-rule upon the Ruliad.

In contrast, Roger Penrose has famously defended the view that human consciousness is non-algorithmic [41] (see also [9, 11]). Prima facie, if we are treating observers as an emergent feature of the Ruliad, then we must respond to Penrose's challenge. If we take Penrose's view of consciousness, as developed with Stuart Hameroff [34], then we can see how this can be accommodated by accommodating microtubles within the Ruliad, and having an account of the coherence they exhibit.

Penrose and Hameroff posit that orchestrated objective reduction of the wave function is associated to proto-consciousness, and this is non-computable. When Wolfram speaks of computation as omnipresent, he refers to a general use of the term that includes both computations that are reducible and irreducible. It is the irreducible ones that correspond with what Penrose refers to as non-computable. From the point of view of microtubuli represented within the Ruliad, they are running some irreducible rules (analogous to CA rule 30 [59]) which to another observer (within the Ruliad) does not lend itself to full predictability. When an observer conflates histories of the multiway, that constitutes a measurement upon the external world. This measurement itself may be computationally reducible. But the observer also needs a higher-order computation which determines which measurements to make and which histories of the multiway it should conflate—that higher-order process may be irreducible (one may call that meta-cognition). If these higher-order processes are required for consciousness, then the conscious observer is not just a program, but a meta-program (and an irreducible one).

Ultimately, what computational irreducibility means is, as the name suggests, that there is no redundancy in the process that can be eliminated to shorten or compress the process. This means that there can be no "short-cuts" in which features of the process can be ignored, made equivalent, or in some other such way utilized to jump to the end. The best one can do is to run the process or simulate it. Again, one might be able to throw more performance power at it, but still it must run through step-by-step. In this sense we see that predictability is bound up with the notion of reducibility, and we have something like an open future if not quite full-blown free will. We are thus left with incompleteness, however, which is related to the so-called hard problem of consciousness. We are giving a model of an observer "from the outside" as it were. Yet how do we find a place for subjectivity (the inside view) here?

12.6 Seeing the Ruliad from the inside: Second-order cybernetics

> Space and time, defining everything we cognize by sensuous means, are in themselves just forms of our receptivity, categories of our intellect, the prism through which we regard the world - or in other words, space and time do not represent properties of the world, but just properties of our knowledge of the world gained through our sensuous organism. From this it follows that the world apart from our knowledge of it, has neither extension in space nor existence in time; these are properties which we add to it.
>
> (P. D. Ouspensky, *Tertium Organum*, 4)

The notion that the world we experience (the phenomenal world or manifest reality) is *conditioned* by our faculties as observers, including the notion of computational boundedness or limitation, can be traced to Kant's theory of the categories. This traces many features that we might naively impute to the world itself back to features of the observer. A natural question, and one considered by Kant, is what happens when different observers are considered. The Wolfram model also involves the idea that different observers might generate very different descriptions of the Ruliad, and so would discover different laws in their world. It is, in other words, vital that the specifics of observers be provided, in order to get a world-description out, and as such the former is the *sine qua non* of the latter.

Several examples of such "alien" scenarios were presented in the early flatland ideas. While Edwin Abott's approach is the best known, the most useful for our purposes is Charles Howard Hinton's (grandson of George Boole), who writes:

> Thus if we make up the appearances which would present themselves to a being subject to a limitation or condition, we shall find that this limitation or condition, when unrecognized by him, presents itself as a general law of his outward world, or as properties and qualities of the objects external to him. He will, moreover, find certain operations possible, others impossible, and the boundary line between the possible and impossible will depend quite as much on the conditions under which he is as on the nature of the operations.
>
> [37, p. 40]

Our epistemological equipment allows us to generate a kind of screen on which reality can display itself. But, of course, what is manifest is only a relative appearance and has much to do with the equipment (including any necessary factors that enable it to exist in the first place). Thus, the observer (human or otherwise) acts as a kind of prism, or transducer, converting a potentially infinite spectrum of data into a finite package capable of being processed. The prism is a good analogy because without it, there would be no such phenomena. And had we placed a distinct observer where the prism is, perhaps a mirror, then we would generate a very different kind of display of the *same* region.[20] But, to repeat, without some means of displaying the world, there is nothing other than a kind of potentiality to display.

Second-order cybernetics is based on the idea that no science is possible from a "view from nowhere" in which one can view reality unveiled as it were [30, 36, 50, 54]. One has to consider a standpoint, or perspective, or frame from which the universe is viewed. Without it (i.e. the viewer), there is no view. The Ruliad as it stands, is abstractly defined as a view from nowhere: a totality. Wolfram himself speaks explicitly of the Ruliad "viewed from the outside" [60, p. 235]. To carry out scientific exploration in the Ruliad, we must include a system, an observer, capable of sampling the space.[21]

Wolfram elsewhere defines the Ruliad as "result of following all possible computational rules in all possible ways" [63, 64]. This is more in line with the second-order cybernetics approach, but we must ask: *who* is following the results? Who is the observer in this case? And who models that observer? From the outside, the Ruliad is simply understood as the totality of all possible computations. However, from the point of view of any of its parts (which satisfy criteria of computational boundedness and persistence), any part of the Ruliad, qualified by boundedness and persistence, is potentially an observer of its complement (within a specified horizon, which would again depend on its computational boundedness). The computations performed by this localized observer realize measurements in the universe. Hence, from the perspective of the second-order cybernetics, the object of interest may not be the Ruliad by itself, but rather something like the power set (or appropriate categorical generalization of a power set, and appropriately restricted by the conditions for observers) of the Ruliad that encapsulates observers, the observed, and their interactions. Teleology and mereology will both be relevant to this power object (of the Ruliad).

In the past, it was usually possible to do theoretical science without explicitly discussing the observer. But it turns out that to say anything about "what happens" requires knowing about the observer. In general, what an observer does is take the raw complexity of the world, and reduce it in such a way that conclusions can be made from it. And here one can think of a certain fundamental duality between computation and observation: the process of computation has the effect of generating new outcomes; the process of observation has the effect of reducing outcomes by"equivalencing" different ones together, as discussed earlier. Computation theory gives us a way to describe possible processes of computation. And the goal of what we're calling "observer theory" is to give us an analogous way to describe possible processes of observation. Because it turns out that our limitations as observers are in a sense what gives us many of the most fundamental scientific laws that we perceive. And it's really all about the interplay between the underlying computational irreducibility and our nature as computationally bounded observers.

The crucial feature of observers seems to be that the observer is always ultimately some kind of "finite mind" that takes all the complexity of the world and extracts from it just certain "summary features" that are relevant to the "decisions" it has to make [67]. Observers like us have two basic characteristics: first, that they are computationally bounded, and second, that they are persistent in time. Computational boundedness is essentially the statement that the region of space observers occupy is limited, i.e. we can't expect to "reverse engineer" computationally irreducible processes that are going on "underneath" [67].

The Wolfram model takes the perspective that an observer has to be a part of the underlying multiway system (possibly as a subgraph spread across branches). In this view, measurement is consequently the process of the observer conflating parallel threads of multiway history with a single evolution leading to the illusion of a unique sequential thread of time. Furthermore, the notion of causal invariance, which can be thought of as being associated with paths of history that diverge eventually converging again, is what guarantees a coherent eventual consistency. And since the Ruliad contains paths corresponding to all possible rules, it's basically inevitable that it will contain what's needed to undo whatever divergence occurs—because of causal invariance, the laws of physics are invariant in any frame of reference (though to realize the laws, one has to set up a system within a given reference frame). Hence, anything physically observable is going to be in a subjective setting, and from this subjective setting, when we infer anything objective, it is relative-objectivism.

So any given observer interprets what one sees in terms of a description language, which causes one to attribute certain rules to be "the rules of the universe"—one has to choose what kind of system one is working with, and it is almost impossible to state a law without those choices. Once we make those choices, we are already in a constructive domain. Hence, if one sets up some particular computational system or mathematical theory, there will always be choices to be made, and our most important feature as observers is that we're computationally bounded, i.e. the way we parse the universe involves doing an amount of computation that's absolutely tiny compared to all the computations going on in the universe. We sample only a tiny part of what's really going on underneath, and we aggregate many details to get the summary that represents our perception of the universe. Recall our earlier example of the molecules in a gas. The molecules bounce around in a complicated pattern that depends on their detailed properties, but an observer like us doesn't trace this whole pattern. Instead, we only observe certain "coarse-grained" features (e.g. pressure and temperature). In this sense everything then boils down to how an observer chooses/samples the space in which they are located, so that their properties are of the essence, which reveals the Wolfram framework as an already second-order cybernetic system.

12.7 The limits of Ruliology: The impossibility of seeing the Ruliad from the outside

> Philosophy is an attempt to express the infinity of the universe in terms of the limitations of language.
>
> A. N. Whitehead

If, as Whitehead put it, philosophy is an attempt to express the infinity of the universe in terms of limitations of language, Ruliology is likewise an attempt to express that infinity in the somewhat less limited framework of representations of computationally-bounded observers that are embedded within it. The Wolfram model depends on coarse-graining over paths in order to model the observed physics. The coarse-graining is, of course, relationally linked to specific observers (or classes of observers). This introduces a limit, since any ruliadic properties that we can speak about are, of course, from the point of view of a member of such a class of observers. We can model other observers by changing the properties defining the observer-class, but even this is itself generated from our own perspective and so will inherit any associated limitations.

Following on from the two ways of thinking about the Ruliad, from inside versus outside, we can see that irreducibility is a feature of the embedded view: it is a feature of the relationship between observers and the Ruliad of which they are a part. We can link this to the two broad approaches to the ontology of mathematics, and note the role they play in physics. From the inside view, in this case, the appropriate ontological picture is that of constructivism, with intuitionistic logic playing the role, and in someway paralleling the computational irreducibility that the observers face in their knowledge claims. But taken as a completed object, where all processes have been carried out to their infinite limits, there is, of course, no computational irreducibility because the object is eternally given and we view it *sub specie aeternitatis*.

It seems that Wolfram is acutely aware of the necessity to include the observer in the description itself [61]:

> It's a typical first instinct in thinking about doing science: you imagine doing an experiment on a system, but you—as the

> "observer"—are outside the system. Of course if you're thinking about modeling the whole universe and everything in it, this isn't ultimately a reasonable way to think about things. Because the "observer" is inevitably part of the universe, and so has to be modeled just like everything else.

Yet Wolfram also writes that [63]:

> [T]he Ruliad is not just a representation. It's in some way something lower level. It's the "actual stuff" that everything is made of. And what defines our particular experience of physics or of mathematics is the particular samples we as observers take of what's in the Ruliad.

This stuff is made of "emes", which function as the most fundamental layer [63]. It is supposed to transcend observers and goes beyond representation. However, given our discussion of the second-order cybernetics, can this be right? How can we possibly make any statements about reality that do not carry with them their source from us *qua* observers? To say it is not just representation implies that we can somehow step outside of all representations, and step outside of our position as observers, to see that more lies beyond: *plus ultra*. In doing so we have stepped outside of Ruliology proper, and entered speculative metaphysics. Rather, as Heinz von Foerster puts it:

> [A] brain is required to write a theory of a brain. From this follows that a theory of the brain, that has any aspirations for completeness, has to account for the writing of this theory. And even more fascinating, the writer of this theory has to account for her or himself. Translated into the domain of cybernetics; the cybernetician, by entering his own domain, has to account for his or her own activity.
>
> [54, p. 289]

The observer in the Wolfram model must ultimately also be ruliadic if this theory is to be truly fundamental. Indeed, the necessity for the second-order cybernetics suggests the need for *meta-rules* upon the Ruliad itself. These rules when instantiated locally within computationally bounded patches of the Ruliad operationalize an abstract notion of observers.

In this case, when we speak of ourselves sampling the Ruliad to generate particular systems of mathematics and physics, we are really speaking about the Ruliad *self*-sampling.[22] In this way we can compare the role and status of observers in the world to the role of humans in such religious systems as Sufism. A totality has splintered into many (relative) perspectives, each with the ability to explore a particular part of that totality. The perspectives are transducing the ineffable Ruliad (akin to an *absolute* unconditioned reality) into something "effable". The self-sampling naturally leads to reflexivity and looping elements, linking the observer and the observed. As Kauffman explains:

> In an observing system, what is observed is not distinct from the system itself, nor can one make a separation between the observer and the observed. These stand together in a coalescence of perception. From the stance of the observing system all objects are non-local, depending upon the presence of the system as a whole. It is within that paradigm that these models begin to live, act and converse with us. We are the models. Map and territory are conjoined.
>
> [38, p. 1]

This is not a flaw with such a model, but rather a virtue. It is quite clear that if we consider some global system (a universe or Ruliad), then we can see that it is trivially the case that some sub-system (the observer) of that system cannot observe the whole system. In other words, the system as a whole is not an observable in the strict sense: there is no operation that we can envisage to measure it.[23] That the Wolfram model contains observers and their viewpoints and generates physics through sampling (in a consistent loop), is the most fundamental model one can manage if one does indeed take seriously the fact that we ourselves must be such observers. It is a virtue for a theory to describe its limitations; in this case containing its limits as theoretical outcomes.

Of course, the Ruliad is not something that we could ever directly observe, and nor is it presented as such. It is an abstract entity that if picturable in any way would be akin to a kind of hyper tensorial object. However, inasmuch as it is abstract, it exists as a representation in the mind of an agent and so inherits limitations. In this case we cannot quite speak of the distinction between map and territory blurring, as with perceptions of the world, because the Ruliad is supposed to go beyond any possible perception. In this sense it stands more in the position of an unknowable God, and the evidence we have is more of the form of a transcendental argument, such that it functions as the *ground* required for having the kinds of experiences we do have and for there being an apparently existing universe in the first place. It is the ground of the possibility of multicomputation.

As we are familiar with general relativity, there are many systems of coordinates that can serve to fix a gauge on the universe, each providing a foliation of spacetime that is invariant when completed. There is also the notion of the quotient space that, in a sense, averages over all the gauge freedom, spitting out the invariant structure.[24] Of course, the resulting entity is harder to deal with from a physical point of view since the gauge (the coordinate frame) is what allows us to epistemically access the universe. Of course, this also tells us that the epistemic access is also partly one of the constructions of what is observed. The frames are observer-constructs and must be purged in any consideration of what is the "true picture" of reality. Of course, one can also simply speak of the frame-relative picture in such a way that so long as the frame is considered in the evaluation of some physical quantity, there is a kind of observer-invariance of the quantities by virtue of observers being able to change their own coordinates into new frames.

This issue of attempting to describe a reality beyond our selective, descriptive capacities is a common problem facing those dealing with apophatic theology, in which one cannot speak of the thing in question with positive characteristics because they thereby bound it, and yet the very bounds come from us. But without such an extremely deep level of probing, we cannot be said to have a fundamental theory. By focusing on the building rules themselves, we find both physics and mathematics emerging, which is same as we should expect. Along these lines, a complaint that has been leveled against the Wolfram model described here is that it is incapable of making predictions. The same complaint was lodged against Eddington's Fundamental Theory. But it entirely misses the point, which is that the Ruliad is a home for physical theories. It is the ground.

A fundamental theory, that is this fundamental, cannot possibly be expected to make direct predictions of the sort that the critics clearly desire. But what it can do is locate them in a web of theories, and moreover it can suggest entirely new kinds of theory that would then themselves make predictions when properly worked out in the manner appropriate for less fundamental approaches. The task of physics, indeed, is to figure out where in the Ruliad we are located. In this sense physics (and mathematics) amounts to the task of a librarian working in Borges' Total Library. Every possible book is contained within that library, much the same as every possible theory is contained in the Ruliad. While all possible books are therein, one needs the right indexing system to locate the correct book, so as not to grab some book of gibberish. The observer is the linchpin that connects the Ruliad with scientific theories, since it is the locus of indexing. Ruliology is not then a replacement of physics, but a way of making sense of it. Moreover, what might appear to be physically nonsensical sectors of rulial space to us, might be perfectly experienciable (as a quite different kind of universe) to other kinds of observers. What we have described is, then, not a theory of physics in the ordinary sense at all. It is a pre-physical framework for any possible theory of physics and should not be analyzed (or critiqued) in the same terms as orthodox physical theories.

Acknowledgments

We thank Stephen Wolfram for engaging discussions on the Ruliad. This work is supported by the following grants: The Foundational Questions Institute and Fetzer-Franklin Fund, a donor advised fund of Silicon Valley Community Foundation [FQXi-RFP-1817]; The John Templeton Foundation [Grant ID 62106]; and the Australian Research Council [Grant DP210100919].

Notes

1. *Dimensions: A Casebook of Alien Contact* (Anomalist Books, 2008, p. 287).
2. Thanks to Samuel Zinner for bringing this quote to our attention.
3. This is formally akin to the notion of moduli spaces in algebraic geometry, where points in the moduli space correspond to isomorphism classes of algebro-geometric objects and trajectories yield a formal notion of dynamics on the space. Moduli spaces are useful for classification problems, where coordinatizing the space is useful for studying various classes of deformations (of the moduli parameters) corresponding to the objects in question (see [46] for a useful introduction).
4. This is the infamous Wheeler-DeWitt equation with its problematic interpretation in terms of dynamical evolution resulting from the absence of a time-parameter t, itself stemming from the fact that time evolution is a kind of diffeomorphism in general relativity and so belonging to the category of gauge degrees of freedom rather than the structure to be assigned a physical interpretation (see [42]).
5. This holds also for the "upgrade" of quantum geometrodynamics from 3-geometries and their histories to spin-networks and their spin-foams, though the latter formalism features a slightly more primitive structure in that it involves abstract graphs as its fundamental objects and whose relations build up the various layers of structure we associate with out physical theories.
6. Indeed, Deutsch has compared his own approach to cybernetics (which involves steering systems to pre-defined goals), which he describes as a possible "early avatar" (`https://www.edge.org/conversation/constructor-theory`). But while Deutsch believes he has provided a perfectly non-abstract, physical description of the world (e.g. talks of abstract computers not making sense), he then slips to mentioning information as if it itself were concrete. It is not, and this is where Wolfram's approach has the advantage, since it directly brings in the additional element that allows for the inclusion of information into the model. If constructor theory has cybernetics as an early avatar, then second-order cybernetics is an early avatar of the Wolfram model. Indeed, Deutsch's claim about materiality being necessary for realization is a tautology since by "realization" he *means* within a material system. This ignores the fact that there is clearly an information template beyond that realization which is what such realizations in matter are realizing. There is quite simply no getting around the fact that if one adopts an information-based ontology, then one is stepping somewhat outside of orthodox materialism.
7. But note, again, that on Deutsch's own interpretation, *some* physical substrate or other is required. The substrate would ground particular instances, allowing for task realization.
8. A recent formalization of pregeometry based on homotopy type theory, as well as one based on pre-quantum structures from noncommutative operator algebras can be found in these works: [4–8]. These studies borrow from ongoing advances at the foundations of mathematics [43, 51–53] as well as applications of higher category theory to physics [16, 47–49].

9. A more detailed version of this material can be found in [31, 60, 64, 66], from which portions of the text in this section are taken (as indicated by specific citations).
10. See [15, 45] for a background overview on rewriting systems; and [69–71] for an overview on hypergraph algebras.
11. For more detail, see [31], section 2, from which some part of the material that follows, including Definitions 1 and 2, has been taken.
12. In some ways this is similar to the universal wave function of Everettian quantum mechanics [57], though it sits at a far lower-level of structure. Indeed, it sits at what might be called a "sub-structural" level (see [6] for more details on this, including the idea of a"structureless structure" from which structure is generated).
13. Here we see another crucial difference to other ensemble theories, such as the Everettian multiverse, which means that such approaches will be automatically subsumed in the Ruliad, as the exhaustive application of just one rule (or category of rules). The Ruliad is instead the ultimate ensemble theory, or the ultimate multiverse.
14. A category theoretic description of the limiting rulial multiway system in terms of infinity-categories can be found in [7].
15. Though ultimately everything (observer and observed) is supposed to remain part of the Ruliad of course, and so will remain abstract when conceived from a third-person perspective. However, if given a complete treatment, then the notion of the Ruliad is also a representation via an observer.
16. See [14] for a treatment of such a decompositional metaphysical position (decompositional dual-aspect monism) as elucidated in several case studies from physics, of which the Wolfram model appears to be another convincing instance.
17. Note that we are able to see a clear explanation of the so-called "unreasonable effectiveness of mathematics" here [58]. The mathematical and physical structure emerge from the selfsame source, namely ruliadic sampling. Since they are constrained in the same way, a specific observer's sampling system will pick up correlations between the systems and properties it generates, and the way they are encoded in mathematics.
18. We have in mind the Deutsch-Wallace [56] approach to the interpretation of probabilities in quantum mechanics according to the Everet's, interpretation in which one also has to square an ultimately deterministic, branching process with the apparent indeterminism in measurement results. Indeed, there is no reason why one might not adopt the self same decision-theoretic approach in the Wolfram model in which "things occur" (i.e. outcomes) only relative to the embedded observers. The idea would, in this case, be to view the probabilities in terms of rational decision under uncertainty about one's location in the Ruliad.
19. See [13] for an overview of complexity measures related to cognition and consciousness.
20. Note also that some of these "displays" might be mutually incompatible leading to the kinds of complementarity one finds in quantum mechanics, e.g. with the inability to place equipment capable of both position and momentum measurements.
21. We don't go into any details here, beyond simply noting that the Wolfram model fits the basic mould of the second-order cybernetic framework. A future paper will consider pairing in more detail. See [55] for a superb review of the basic ideas of the second-order cybernetics.
22. This is also known as endophysics, or the physics from the inside (see e.g. [44]). Rössler argues that the world is always relative to an observer-perspective, so that an "interface" is involved in which the world appears as a kind of screen to the observer, though not in any fixed way. Rather, the cut between self and world (or observer and observed/environment) is variable. In the case of the Observer-Ruliad system the observer is a kind of bounded foliation of the whole.
23. Here we are assuming that the system is closed, of course, so that there is no external interaction and no sense in which the system is itself a sub-system of some larger system. In the case in which, e.g. the system is a simulated universe, then of course one can well imagine an external observer being able to make appropriate measurements on the system, though

they would then face the same problem in their own universe (cf. [26, section 2.1], [25]) and the description in question would no longer be fundamental.

24. In a similar way, Wolfram speaks of "bulk" features, in the sense of something like the quotient object, i.e. which transcends the baggage brought by observers and their gauges or frames.

References

[1] Abramsky, S., Coecke, B. (2009). Categorical Quantum Mechanics. *Handbook of Quantum Logic and Quantum Structures*, 2, 261–325.

[2] Arsiwalla, X. D. (2008). More Rings to Rule Them All: Fragmentation, 4D $\leftrightarrow$ 5D and Split-Spectral Flows. *Journal of High Energy Physics*, 2008(02), 066.

[3] Arsiwalla, X. D. (2009). Entropy Functions with 5D Chern-Simons Terms. *Journal of High Energy Physics*, 2009(09), 059.

[4] Arsiwalla, X. D. (2020). Homotopic Foundations of Wolfram Models. *Wolfram Community*. `https://community.wolfram.com/groups/-/m,2032113`.

[5] Arsiwalla, X. D., Chester, D., Kauffman, L. H. (2022). On the Operator Origins of Classical and Quantum Wave Functions. *ArXiv:2211.01838.*

[6] Arsiwalla, X. D., Elshatlawy, H., Rickles, D. (forthcoming). Pregeometry from Formal Language. *In preparation.*

[7] Arsiwalla, X. D., Gorard, J. (2021). Pregeometric Spaces from Wolfram Model Rewriting Systems as Homotopy Types. *ArXiv:2111.03460.*

[8] Arsiwalla, X. D., Gorard, J., Elshatlawy, H. (2022). Homotopies in Multiway (Nondeterministic) Rewriting Systems as n-Fold Categories. *Complex Systems*, 31(3), 261–277, *ArXiv:2105.10822.*

[9] Arsiwalla, X. D., Herreros, I., Moulin-Frier, C., Sanchez, M., Verschure, P. (2016). Is Consciousness a Control Process?. In *Artificial Intelligence Research and Development* (pp. 233–238). IOS Press.

[10] Arsiwalla, X. D., Herreros, I., Verschure, P. (2016). On Three Categories of Conscious Machines. In *Biomimetic and Biohybrid Systems: 5th International Conference, Living Machines 2016, Edinburgh, UK*. Proceedings 5 (pp. 389–392). Springer International Publishing.

[11] Arsiwalla, X. D., Signorelli, C. M., Puigbo, J. Y., Freire, I. T., Verschure, P. F. (2018). Are Brains Computers, Emulators or Simulator?. In *Biomimetic and Biohybrid Systems: 7th International Conference, Living Machines 2018, Paris, France*, Proceedings 7 (pp. 11–15). Springer International Publishing.

[12] Arsiwalla, X. D., Solé, R., Moulin-Frier, C., Herreros, I., Sánchez-Fibla, M., Verschure, P. (2023). The Morphospace of Consciousness: Three Kinds of Complexity for Minds and Machines. *NeuroSci*, 4(2), 79–102.

[13] Arsiwalla, X. D., Verschure, P. (2018). Measuring the Complexity of Consciousness. *Frontiers in Neuroscience*, 12, 424.

[14] Atmanspacher, H., Rickles, D. (2022), *Dual-Aspect Monism and the Deep Structure of Meaning*. Routledge.

[15] Baader, F., Nipkow, T. (1998). *Term Rewriting and All That*. Cambridge University Press.

[16] Baez, J. (2007). The Homotopy Hypothesis. Available online at `https://math.ucr.edu/home/baez/homotopy/homotopy.pdf`.

[17] Chaitin, G. (2018). Building the World Out of Information and Computation: Is God a Programmer, Not a Mathematician? In Wuppuluri Shyam & Francisco Antonio Dorio (eds.), *The Map and the Territory: Exploring the Foundations of Science, Thought and Reality* (pp. 431–438). Springer.

[18] Coecke, B., ed. (2011). *New Structures for Physics*. Springer.

[19] Coecke, B., Duncan, R. (2011). Interacting Quantum Observables: Categorical Algebra and Diagrammatics. *New Journal of Physics*, 13(4), 043016.

[20] Cronin, L., Imari Walker, S. (2016). Beyond Rebiotic Chemistry: What Dynamic Network Properties Allow the Emergence of Life? *Science*, 352(6290), 1174–1175.

[21] Deutsch, D., Marletto, C. (2015). Constructor Theory of Information. Proceedings of the Royal Society A, 471.

[22] Döring, A., Isham, C. (2010). "What Is a Thing?": Topos Theory in the Foundations of Physics. In *New Structures for Physics* (pp. 753–937). Springer Berlin Heidelberg.

[23] Du Plessis, J. F., Arsiwalla, X. D. (2023). A Cosine Rule-Based Discrete Sectional Curvature for Graphs. *Journal of Complex Networks*, **11**(4), cnad022.

[24] Einstein, A. (1936). Physics and Reality. In his *Ideas and Opinions* (pp. 290–323). Bonanza.

[25] Fields, C. (2012). If Physics Is an Information Science, What Is an Observer?. *Information*, 3(1), 92–123.

[26] Fields, C. (2016). Building the Observer into the System: Toward a Realistic Description of Human Interaction with the World. *Systems*, 4(4), 32.

[27] Frauchiger, D., Renner, R. (2018). Quantum Theory Cannot Consistently Describe the Use of itself. *Nat Commun* 9, 3711.

[28] Friston, K. (2010). The Free-Energy Principle: A Unified Brain Theory? *Nature Reviews Neuroscience*, 11, 127–138.

[29] Gisin, N. (2021). Indeterminism in Physics and Intuitionistic Mathematics. *Synthese*, 199, 13345–13371.

[30] Glanville, R. (2002). Second Order Cybernetics. *Systems Science and Cybernetics*, 3, 59–85.

[31] Gorard, J., Namuduri, M., Arsiwalla, X. D. (2020). ZX-Calculus and Extended Hypergraph Rewriting Systems I: A Multiway Approach to Categorical Quantum Information Theory. *ArXiv:2010.02752.*

[32] Gorard, J., Namuduri, M., Arsiwalla, X. D. (2021). ZX-Calculus and Extended Wolfram Model Systems II: Fast Diagrammatic Reasoning with an Application to Quantum Circuit Simplification. *ArXiv:2103.15820.*

[33] Gorard, J., Namuduri, M., Arsiwalla, X. D. (2021). Fast Automated Reasoning over String Diagrams Using Multiway Causal Structure. *ArXiv:2105.04057.*

[34] Hameroff S., Penrose, R. (2014) Consciousness in the Universe: A Review of the 'Orch OR' Theory. *Physics of Life Reviews*, 11(1), 39–78.

[35] Hawking, S. (2009). *A Brief History of Time: From Big Bang to Black Holes*. Random House.

[36] Heylighen, F., Joslyn, C. (2001). Cybernetics and Second-Order Cybernetics. *Encyclopedia of Physical Science & Technology*, 4, 155–170.

[37] Hinton, C. H. (1900). *A New Era of Thought*. Swan Sonnenschein & Co.

[38] Kauffman, L. H. (2003). Eigenforms — Objects as Tokens for Eigenbehaviors. *Cybernetics And Human Knowing*, 10(3–4).

[39] Marletto, C. (2021). *The Science of Can and Can't: A Physicist's Journey through the Land of Counterfactuals*. Penguin.

[40] Misner, C., Thorne, K., Wheeler, J. (1973). *Gravitation*. Princeton University Press.

[41] Penrose, R. (1989). *The Emperor's New Mind*. Oxford: Oxford University Press.

[42] Rickles, D. (2006). Time and Structure in Canonical Gravity. In D. Rickles, S. French & J. Saatsi (eds.), *The Structural Foundations of Quantum Gravity* (pp. 152–191). Oxford University Press.

[43] Riehl, E., Shulman, M. (2017). A Type Theory for Synthetic ∞-Categories. *ArXiv:1705.07442.*

[44] Rössler, O. E. (1987). Endophysics. In J. L. Casti & A. Karlquist (eds.), *Real Brains, Artificial Minds*. North-Holland, New York.

[45] Rozenberg, G. (ed.). (1997). *Handbook of Graph Grammars and Computing by Graph Transformation* (Vol. 1). World Scientific.

[46] Schlichenmaier, M. (2007). *An Introduction to Riemann Surfaces, Algebraic Curves and Moduli Spaces*. Springer.

[47] Schreiber, U. (2013). Differential Cohomology in a Cohesive Infinity-Topos. *ArXiv:1310.7930.*

[48] Schreiber, U. (2021). Higher Prequantum Geometry. *New Spaces in Physics: Formal and Conceptual Reflections*, 2, 202.

[49] Schreiber, U., Shulman, M. (2014). Quantum Gauge Field Theory in Cohesive Homotopy Type Theory. *ArXiv:1408.0054.*

[50] Scott, B. (2004). Second-Order Cybernetics: An Historical Introduction. *Kybernetes*, 33(9/10), 1365–1378.

[51] Shulman, M. (2017). Homotopy Type Theory: A Synthetic Approach to Higher Equalities. In *Categories for the Working Philosopher*, (PP. 36–57).

[52] Shulman, M. (2021). Homotopy Type Theory: The Logic of Space. In *New Spaces in Mathematics* (Vol. 1, pp. 322–404). Cambridge University Press.

[53] The Univalent Foundations Program. (2013). Homotopy Type Theory: Univalent Foundations of Mathematics. *ArXiv:1308.0729.*

[54] von Foerster, H. (2003). *Understanding Understanding: Essays on Cybernetics and Cognition*. Springer.

[55] von Foerster, H. (2014). *The Beginning of Heaven and Earth Has No Name: Seven Days with Second-Order Cybernetics*. Fordham University Press.

[56] Wallace, D. (2003). Everettian Rationality: Defending Deutsch's Approach to Probability in the Everett Interpretation. *Studies in the History and Philosophy of Modern Physics*, 34, 415–439.

[57] Wallace, D. (2012). *The Emergent Multiverse: Quantum Theory According to the Everett Interpretation*. Oxford University Press.

[58] Wigner, E. P. (1960). The Unreasonable Effectiveness of Mathematics in the Natural Sciences. *Commun. Pure Appl. Math, XIII*, 1–14.

[59] Wolfram, S. (2002). *A New Kind of Science*. Wolfram Media.

[60] Wolfram, S. (2020). A Class of Models with the Potential to Represent Fundamental Physics. *Complex Systems*, 29(2).

[61] Wolfram, S. (2020). Finally We May Have a Path to the Fundamental Theory of Physics... and It's Beautiful. `http://writings.stephenwolfram.com/2020/04/`.

[62] Wolfram, S. (2021). *Combinators: A Centennial View*. Wolfram Media.

[63] Wolfram, S. (2022). *Metamathematics: Foundations & Physicalization*. Wolfram Media.

[64] Wolfram, S. (2021). The Concept of the Ruliad. `http://writings.stephenwolfram.com/2021/11/the-concept-of-the-ruliad/`.

[65] Wolfram, S. (2021). Exploring Rulial Space: The Case of Turing Machines. *ArXiv:2101.10907.*

[66] Wolfram, S. (2021). Multicomputation: A Fourth Paradigm for Theoretical Science. `https://writings.stephenwolfram.com/2021/09/multicomputation-a-fourth-paradigm-for-theoretical-science/`.

[67] Wolfram, S. (2021). What Is Consciousness? Some New Perspectives from Our Physics Project. `https://writings.stephenwolfram.com/2021/03/what-is-consciousness-some-new-perspectives-from-our-physics-project/`.

[68] Wolfram, S. (2023). Computational Foundations for the Second Law of Thermodynamics. `https://writings.stephenwolfram.com/2023/ 02/computational-foundations-for-the-second-law-of-thermodynamics/`.

[69] Zapata-Carratalá, C., Arsiwalla, X. D. (2022). An Invitation to Higher Arity Science. *ArXiv:2201.09738*. In press *Complex Systems*.

[70] Zapata-Carratalá, C., Arsiwalla, X. D., Beynon, T. (2022). Heaps of Fish: Arrays, Generalized Associativity and Heapoids. *ArXiv:2205.05456*.

[71] Zapata-Carratalá, C., Schich, M., Beynon, T., Arsiwalla, X. D. (2023). Beyond Binary: Hypermatrix Algebra and Irreducible Arity in Higher-Order Systems. *ArXiv:2301.07494*.

Part 5

Lattice and Hybrid Theories

13 Nonperturbative Quantum Gravity, Unlocked through Computation

Renate Loll

This contribution is dedicated to Chris Isham, my former teacher and our all quantum gravity guru of old.

13.1 Quantum gravity and the lack of computability

Understanding quantum gravity, the elusive fundamental quantum (field) theory underlying the classical theory of general relativity, obviously includes the ability to perform computations that quantify its physical properties, for example, in terms of the spectra of suitable quantum observables. Since gravity is perturbatively nonrenormalizable, neither perturbative nor effective quantum field theoretic methods are sufficient to reach such an understanding, and an approach beyond perturbation theory is called for.

Historically, the failure of standard, perturbative[1] tools, encapsulated in the famous two-loop computations of Goroff, Sagnotti [1] and van de Ven [2], and the failure of standard, nonperturbative tools, in the attempt to emulate the successes of lattice quantum chromodynamics (QCD),[2] led to a search for *non*standard solutions of quantum gravity. The 1980s saw the beginnings of what many popular science publications continue to call "the two leading approaches to quantum gravity", superstring theory and loop quantum gravity. One of the few features these two formulations have in common is a reliance on

DOI: 10.4324/9781003546139-18

fundamental, one-dimensional excitations, which places them outside the conventional framework of quantum field theory. Whether the assumption of their strings or loops is correct or indeed testable remains unclear, but, more to the point, in either approach one is still far from being able to perform any meaningful computations on distance scales near the Planck length (1.6×10^{-35} m), let alone make new physical predictions based on such computations.

Because of this lack of nonperturbative computational tools, many discussions and disputes about quantum gravity in recent decades have focused on formalisms rather than results and on general principles and concepts rather than concrete calculations of local dynamics. Long wishlists of problems that quantum gravity should solve – if only we knew what it was – have been compiled [4], largely without the benefit of numerical or other reality checks. Moreover, the weakness of the gravitational interactions makes it unlikely that we will accidentally encounter a new physical phenomenon attributable to quantum gravity that could provide guidance on building a fundamental theory.

Long before addressing the challenge of experimental verification, which is due to the extreme scales involved, we therefore seem to face a very unsatisfactory situation: without a sufficiently stringent computational framework, it is difficult to formulate objective criteria for the validity and correctness of candidate theories, and even a requirement like internal consistency becomes a hazy notion. Invoking qualitative, "intuitive" criteria like simplicity or beauty instead may be outright misleading, given that the searched-for theory describes an unknown physical regime far beyond classicality and the validity of perturbation theory, where many standard concepts of classical spacetime are not expected to apply.

Another elephant in the room that deserves attention is a high degree of nonuniqueness, which is associated with a large number of free parameters and other free choices that come with a particular candidate theory and imply a lack of physical predictivity. As we have learned from string theory, the prime example of a grand unified theory of all the interactions, including gravity, this correlates with the theory's richness of ingredients, in this case, many unobserved fundamental excitations, supersymmetry and extra dimensions. It raises the interesting question of how little in terms of ingredients we can get away with when constructing a theory of quantum gravity, to avoid such a scenario.

13.2 Quantum gravity is not simple

However, it is worth emphasizing that there is absolutely no reason to expect quantum gravity to be a *simple* theory, even in the absence of exotic ingredients. Let us begin by recalling the complicated structure of the *classical* field theory of general relativity, whose basic field is a Lorentzian metric $g_{\mu\nu}(x)$. The local curvature properties of such a spacetime are encoded in its Riemann tensor, a quantity with 256 components, and its dynamics are described by a coupled set of nonlinear partial differential equations, whose exact solution is only known in very special cases [5]. Highly refined and dedicated numerical methods are needed to extract the physical content of Einstein's equations whenever gravity is strong, like in the collision of black holes [6]. As already stressed above, the quantum theory has an analogous need for numerical tools.

Another useful reference for estimating complexity and computability are the quantum field theories of the standard model of particle physics, and QCD in particular that, like gravity, has nonlinear classical field equations and a complicated gauge group action. Most relevant to this comparison is its nonperturbative sector, which has been investigated extensively with the help of powerful lattice methods, yielding quantitative results about the QCD spectrum not obtainable by other means [7]. Likewise, quantum gravity is an interacting, nonperturbative quantum field theory, but with an arguably even more complicated field content, dynamics and symmetry structure. This strongly suggests that quantum gravity will *not* be simple, computationally or otherwise, and certainly not simpler than QCD. In other words, hoping for a magical insight that dissolves the known structural features and difficulties of gravity to yield a quantum theory governed by simple relations and dynamical outcomes seems highly unrealistic.

In further assessing what is feasible and what is not in quantum gravity, we also need to examine the power and limitations of our most advanced theoretical and computational tools, and to what extent they have enabled us to quantitatively understand the nongravitational fundamental interactions and QCD in particular. This provides useful benchmarks for what can realistically be achieved in terms of "solving" quantum gravity.[3] On top of this, we still need to take into account that these tools must be adapted to the gravitational case, where spacetime is dynamical and not part of the fixed background structure.

Lastly, exact mathematical methods are unlikely to solve quantum gravity because they are not even able to describe the renormalizable quantum field theories of particle physics. This message may be underappreciated, since there have been many studies of quantum gravity-inspired toy models that are sufficiently simple to allow for an exact treatment beyond perturbation theory (see e.g. [8]). However, they are based on unphysical, simplifying assumptions, like reducing the spacetime dimension from four to two or three, or postulating additional spacetime symmetries.[4] Since this removes exactly the features that make physical quantum gravity in four dimensions interesting and difficult, it is not surprising that these models teach us very little about the full theory (see also [9], Sec. Q17 for further discussion and references).

13.3 Lattice quantum gravity reloaded

Viewing quantum gravity through the lens of computability and the availability of suitable tools, and combining this with some of the lessons of the past

40 years of quantum gravity research strongly suggest a refocusing on *nonperturbative computation as the key to progress*. Since technical and conceptual issues in quantum gravity tend to be closely intertwined, this will help to inform expectations of what the theory can deliver. Recognizing the nature and magnitude of the challenge of making quantum gravity computable should not be a deterrent, but should allow us to take a realistic perspective on the effort and time frame needed.

Fortunately this effort does not have to start from scratch, since a quantum field theoretic formulation of quantum gravity with a functioning, well-tested and nonperturbative computational framework is already available. This "lattice quantum gravity 2.0" is formulated in terms of causal dynamical triangulations (CDT) and has been developed over the past 25 years [10–12], building on previous developments (see [3] for a review). Unlike its lattice predecessors "1.0", it has the dynamical and Lorentzian nature of spacetime built into its construction from the outset. In a nutshell, this formulation has opened a computational window near the Planck scale where certain "numerical experiments" a.k.a. Monte Carlo simulations of quantum gravity can be performed with the help of dynamical lattice methods. Geometric quantum observables can be measured, giving us for the first time a quantitative insight into the nature and properties of quantum spacetime and its dynamics in this nonperturbative regime. Numerous nontrivial lessons have been learned in the course of these developments, and some of the results already obtained have been totally unexpected from the point of view of the classical and perturbative theories.

The research programme of CDT quantum gravity is very much ongoing, and its achievements and future perspectives will be sketched below. How does it fit into the larger quantum gravity landscape? Alongside the dominant superstring and subdominant loop paradigms (when measured in terms of publications, grant moneys and media attention), research on alternative approaches beyond perturbation theory has always continued.[5] In broad brushstrokes, one can distinguish between more or less conventional quantum field theoretic formulations, where also lattice quantum gravity belongs, and other approaches with a looser or unclear relation to the concepts of quantum field theory. In the latter category are also formulations that posit some form of fundamental discreteness of spacetime, usually at the Planck scale, like the causal set approach [13].

When it comes to comparing these different candidate theories, one is faced with the usual conundrum that there are no quantities that can be meaningfully compared, because of the theories' incompleteness and a lack of effective computational tools in many of them. Comparing the various formalisms instead is not a particularly fruitful exercise because of their vastly different starting points and choices of ingredients, which do not have a direct physical interpretation in themselves. Examples of a successful comparison are the spectral properties of selected observables measured in CDT[6] and aspects of renormalization group flows [14], which can be reproduced by using functional renormalization group methods in so-called asymptotic safety [15], an approach that combines perturbative and nonperturbative elements of quantum field theory. If and when other nonperturbative computational schemes become operative in the future, we will be able to formulate further quantitative criteria to assess the equivalence or otherwise of the corresponding candidate theories of quantum gravity.

13.4 Lattice quantum gravity is *not* discrete quantum gravity

In the search for a theory of nonperturbative quantum gravity, so-called fundamentally discrete approaches have had an enduring popularity among practitioners. Loosely speaking, the underlying idea is that spacetime should come in discrete units or "bits", just like matter is composed of elementary quantum particles. These building blocks are usually assumed to be Planck length-sized, with or without individual shapes or other properties that are usually guessed, either in analogy with systems on much larger scales or because of some other expediency. After adding a prescription for how the microscopic bits can interact or relate to each other, one then envisages that a large number of them will coalesce or self-organize dynamically in such a way that quantum gravity and spacetime "emerge".

If it could be realized, such a picture seems attractively simple: all we have to do is consider finite arrangements of some "Lego blocks" of finite size, relying on our everyday intuition, which is much more attuned to working with natural numbers than with real ones. As an added bonus, if all physical quantities come in terms of some minimal, fundamental length unit, the infinities characteristic of quantum field theory and the ensuing need to renormalize them will simply disappear. Everyone can do quantum gravity. Not surprisingly, such a scenario is far too simple to be true or have anything much to do with gravity, for reasons that will be explained below.

For simplicity, let us ignore that any notion of fundamental discreteness requires an operational definition. Intuitive, classical ideas will be meaningless in a Planckian quantum regime that lacks a pre-existing spacetime, which in this scenario is expected to be generated dynamically and not put in by hand. Even assuming such a definition, we run into the problem that for any choice of building blocks and interaction rules at the Planck scale, there will be an infinity of other choices that are equally well motivated. Imagine that a particular choice could be shown to lead to a viable candidate theory of quantum gravity, in the sense of reproducing one or more known features of general relativity. (This is a necessary but not sufficient condition, and has not actually happened yet.)

Then there will be many other choices that are equally viable in this sense, but which by construction are *different* theories at the same, Planckian scale, which after all is the primary habitat of quantum gravity. We conclude that formulations of quantum gravity based on the assumption of fundamental discreteness at the Planck scale have a structural problem because of their high degree of nonuniqueness, with a corresponding lack of predictivity.

Although the nonperturbative lattice formulation of quantum gravity also has some discrete features, they are not fundamental in nature, but part of a regularization, where an unphysical lower ("ultraviolet") cutoff on the length of the lattice edges is employed at an intermediate stage of the calculation to "tame" infinities. Subsequently, one takes a scaling limit by sending this cutoff to zero while renormalizing coupling constants appropriately, a process which under favourable circumstances leads to an essentially unique[7] continuum quantum theory without infinities [16]. Importantly, by a mechanism called universality [17], the final theory does not depend on the details of how the regularization was set up, like the shape of the building blocks or the detailed manner of their interaction. This provides the uniqueness mechanism that is missing in the fundamental discreteness scenario.

Since lattice investigations are run on computers with finite processing power and storage capacity, the continuum limit of vanishing cutoff and infinitely fine lattices cannot be reached in practice, but is extrapolated systematically from sequences of ever finer lattices. From a practical point of view, it implies that to extract universal results at a given scale, say, the Planck length, one must use a lattice resolution that is significantly smaller than this scale to avoid that the measurements are dominated by discretization artefacts.

13.5 CDT and the challenges of lattice quantum gravity

Let us introduce the basic principles and structural features that enter into the construction of modern lattice quantum gravity based on causal dynamical triangulations. Rather than an *approach* to quantum gravity, distinguished by a specific choice of nonstandard ingredients (loops, strings, spin foams, causal sets or others [18]), it is a minimal nonperturbative quantum extension of general relativity, using only standard principles from (lattice) quantum field theory *adapted to accommodate the dynamical nature of spacetime*.

This adaptation, beyond the framework of relativistic quantum field theory on a fixed Minkowski space, did not just involve a few minor tweaks but required solutions to long-standing problems that have hampered many approaches to quantum gravity: how to regularize and renormalize in a way that is compatible with diffeomorphism invariance, how to analytically continue ("Wick rotate") the path integral to make it amenable to computation, how to deal with the conformal divergence of the resulting path integral [19] and how to achieve unitarity [10, 11].

The dynamical principle at the heart of the quantum theory is the usual Feynman path integral, which in the gravitational case implements the quantum superposition of curved spacetime geometries g, schematically written as the functional integral

$$Z = \int_{\mathcal{G}} \mathcal{D}g \, \mathrm{e}^{iS[g]}, \tag{13.1}$$

where each geometry $g \in \mathcal{G}$ is weighed by a complex phase factor depending on the gravitational action $S[g]$. The expression (13.1) is entirely formal and needs to be accompanied by explicit definitions of the nonlinear configuration space $\mathcal{G}$ and its parametrization, the measure $\mathcal{D}g$ and a prescription for how to compute Z nonperturbatively, without resorting to a perturbative linearization of $\mathcal{G}$ around a solution of the classical Einstein equations.

However, even after making these specifications, Z will be ill-defined and infinite, due to quantum field theoretic divergences that need to be renormalized. This has nothing to do with gravity but happens just as well for a scalar field theory, say. At this point, it is natural to invoke a lattice regularization and renormalization to evaluate the path integral nonperturbatively, following the highly successful example of lattice QCD. Such a strategy was suggested early on in the history of lattice gravity, using various classical gauge-theoretic (re-formulations) of gravity as a starting point (see e.g. [20]). It was pursued for a number of years, including numerical lattice implementations, but remained unsuccessful and inconclusive [3], as already mentioned earlier.

Even in hindsight it is difficult to pinpoint which of the shortcomings of "lattice gravity 1.0" contributed most to this negative outcome, but a prime culprit was the use of fixed, hypercubic lattices, on which the gravitational holonomy variables were placed. The problem is that the diffeomorphisms,[8] which form the invariance group of general relativity, do not act on such lattices and the naively discretized continuum fields defined on them. It implies that the corresponding gauge group action cannot be "factored out" in a controlled way and the lattice fields do not properly represent the physical gravitational degrees of freedom.

Another major problem is that the Monte Carlo (MC) techniques used to evaluate lattice-regularized path integrals require a *Euclidean* quantum field theory. For a theory on Minkowski space, this can be obtained by an analytic continuation from real to imaginary time, which under suitable conditions converts the complex phases $\exp(iS)$ in the path integral (13.1) to real Boltzmann factors $\exp(-S^{\mathrm{e}u})$, as needed in the MC simulations,[9] where $S^{\mathrm{e}u}$ denotes the action of the Euclidean theory. The problem in quantum gravity beyond perturbation theory is that spacetime and, therefore, time are dynamical. For arbitrary curved spacetimes, there is no distinguished choice of time, and moreover the time dependence of the metric field tensor can be arbitrarily complicated. As a consequence, no Wick rotation for metrics $g_{\mu\nu}(x)$ is known that achieves the required conversion of the phase factors.

This has motivated many researchers to consider a different and a priori unrelated theory, so-called Euclidean quantum gravity, which is defined by a real "path integral" (a.k.a. a partition function) over Riemannian instead of Lorentzian geometries, possessing no notion of time or causality. Prior to CDT, all attempts at lattice quantum gravity were of this type, including gauge-theoretic approaches, quantum Regge calculus [21] and Euclidean dynamical triangulations [22]. Even if one could make sense of these path integrals, which has proven very challenging, it is unclear what, if anything, they have to do with the physical, Lorentzian theory of quantum gravity.

13.6 CDT: Lattices going dynamical

To cut a long story short, since its inception [23] until today, CDT quantum gravity takes both the dynamical *and* the causal Lorentzian nature of spacetime into account by building them into the structure of lattices from the outset. In other words, the functionalities of the lattice as a tool for regularizing the infinities of the quantum field theory have been adapted to match the physical content and symmetry structure of gravity, which are significantly different from those of gauge field theories. The success of this ansatz until now shows that this is a fruitful strategy.

The dynamical lattices of CDT, which represent distinct curved spacetimes in a regularized version of the continuum path integral [Eq. (13.1)], play the same role for gravity as the lattice representation of QCD field configurations in terms of holonomy variables due to Wilson [24] does for the strong interactions. The beauty and power of the latter lies in the fact that the $SU(3)$-gauge transformations placed at the lattice vertices have a well-defined action on the group-valued holonomy variables on lattice edges, which yields an *exact* notion of gauge-invariant lattice field configurations and observables.

The curved geometry of CDT lattices is defined by the geodesic edge lengths of their simplicial, four-dimensional Minkowskian building blocks[10] and the way in which these four-simplices[11] are glued together pairwise to obtain a triangulation, i.e. a piecewise flat spacetime manifold [25]. The length and gluing data are geometric in nature, but in order to locate the corresponding four-simplices inside a triangulation, the simplices need to be numbered or "labelled". This discrete labelling is arbitrary and unphysical in the sense that no observables can depend on it. The associated relabelling invariance may be thought of as an analogue of the coordinate- or diffeomorphism-invariance of general relativity, but unlike the latter is easily taken into account when evaluating the path integral.

Similar to what happens in Wilson's formulation of lattice QCD, CDT quantum gravity, therefore, has an exact notion of gauge-invariance, despite the presence of a lattice cutoff. In this formulation, unlabelled triangulations represent manifestly coordinate-invariant spacetime geometries. This remarkable property is due to how the discrete gluings capture the local curvature degrees of freedom of the regularized spacetimes, without referring to any continuously varying metric variables. Compared to what is done in standard lattice field theory, making the lattice itself dynamical is key to intrinsically combining the powerful idea of approximating spacetime by a lattice with the dynamical character of spacetime in gravity.

13.7 CDT: Lattices going causal

The idea of using dynamical, triangulated lattices has its origin in two dimensions [26], more precisely, the search for a nonperturbative description of the dynamics of two-dimensional world sheets in bosonic string theory. In due course, its application to intrinsic, embedding-independent curved geometries in four dimensions was considered, in an attempt to find a theory of Euclidean quantum gravity from a nonperturbative, regularized Euclidean path integral, of the kind already mentioned above. Claims of the presence of a second-order phase transition [27] in the corresponding lattice quantum gravity model, so-called Euclidean dynamical triangulations (EDT) or DT for short, signalling the possible existence of a continuum limit, generated much attention at that time, but were later shown to be erroneous.[12]

By contrast, CDT quantum gravity has two decisive new elements, which lead to much more interesting outcomes, including the presence of second-order phase transitions [29, 30] and the emergence of a macroscopic quantum spacetime with de Sitter properties [31, 32], to be discussed further below. The first novel feature, compared to EDT, is the use of a path integral over triangulated lattices that represent Lorentzian spacetimes rather than Riemannian spaces. Accordingly, one chooses Minkowskian instead of Euclidean four-simplices as elementary lattice building blocks and associated gluing rules, which ensure that each triangulation contributing to the path integral has a well-defined causal (or lightcone) structure globally [25]. Like in general relativity, each such spacetime consists of an ordered sequence of spatial slices representing moments in time. The letter "C" in "CDT" stands exactly for this causal ordering.

However, formulating a lattice version of the physical, Lorentzian path integral [Eq. (13.1)] is by itself not enough to achieve a breakthrough, since this form cannot be used as a direct input for MC simulations, as pointed out earlier. The crucial missing element is an analytic continuation of the complex path integral to a real partition function. Remarkably, this is also available in CDT and defines its second novel feature as follows. It was already mentioned that there are two different length assignments to the lattice edges, depending on whether they are space- or timelike. More precisely, spacelike edges have a squared[13] length $\ell_s^2 = a^2$ and timelike edges have $\ell_t^2 = -\alpha a^2$, where a is the so-called lattice spacing, i.e. the ultraviolet length cutoff that will be sent to zero eventually, and $\alpha > 0$ is a fixed positive constant, which from a classical point of view can be chosen arbitrarily. It has been shown that an analytic continuation of α to $-\alpha$ through the lower-half complex plane converts the path integral to a real partition function of the correct functional form [25], and makes it amenable to MC simulations. This is highly significant, since no analogous prescription is known in a continuum formulation based on metric fields $g_{\mu\nu}$. It opens the door to a quantitative exploration of the nonperturbative gravitational path integral.

13.8 Emergence: Aspirations and reality

Independent of whether one follows the path of lattice quantum gravity, of a fundamentally discrete model or of "something totally different", i.e. a conjectural theory with no resemblance to gravity as we know it at (sub-)Planckian scales,[14] one needs to show that its predictions are compatible with those of general relativity in physical situations where quantum effects are negligible, usually at large length scales and/or low energies. This turns out to be a very challenging task.

By construction, the formulations and variables used to describe the classical and nonperturbative quantum regimes are very different, with smooth classical tensor fields like the metric $g_{\mu\nu}(x)$ playing no role in the latter. More importantly, this will also be reflected in different observables and different ways in which diffeomorphism invariance is implemented, partly because of the absence of an a priori background geometry at the Planck scale.

The recovery of classical properties of gravity is a known difficulty of nonperturbative formulations. It is sometimes called the problem of the classical limit, but especially for the recovery of a classical spacetime, the term "emergence" is invoked frequently. For clarification, since this notion is sometimes used in a very loose sense, emergence here refers to (new) macroscopic properties arising from the collective behaviour of a large number of microscopic constituents. In the case of quantum gravity, the question is whether and how (sub-)Planckian building blocks or other ingredients and their interactions can on much larger scales give rise to an entity that resembles a four-dimensional extended spacetime, as well as to gravitational dynamics as we know it from general relativity.[15]

Emergence as an aspirational concept, without asking for a quantitative correlate, is seemingly straightforward: if we assume the presence of some incarnation of a Planckian "quantum spacetime foam", a process of emergence arguably *must* exist to take us from there to the classical theory. One might even dream of a universal emergence mechanism, whereby a wide range of microscopic ingredients ("at the Planck scale, anything goes") inevitably leads back to a nice classical spacetime like a Minkowski or de Sitter space.[16]

A majority of formulations uses some variant of the path integral as their dynamical principle, where "emergence" comes about through a superposition of amplitudes. Since there are currently no efficient computational methods to evaluate the complex path integral [Eq. (13.1)], one must rely on a suitable analytic continuation like in CDT, or work with a real, Euclidean partition function or state sum. Fortunately, there is already a significant body of work where such systems have been analyzed quantitatively, providing a much-needed reality check on the role of emergence. They include lattice quantum gravity, statistical models of random geometry, toy models in lower dimensions and some discrete quantum gravity formulations like causal sets, to the extent they allow a modicum of computational control. The overall conclusion is simple and largely independent of the details of the individual models: generically, *nothing emerges*, or at least nothing that has any obvious relation to general relativity or spacetime.

This negative outcome can have various origins. (i) Not enough is put in. Recall that classical gravity has a very complex local curvature structure and dynamics; if the choice of microscopic ingredients and interactions is too minimalist and does not capture the potentiality of these rich classical structures, they simply will not emerge – one cannot get something for nothing. (ii) Too much is put in. If the set of configurations that is summed or integrated over in the path integral is too large, the resulting infinities cannot be renormalized with standard methods, and nothing emerges either. The folklore that "one should sum over everything in the path integral" simply does not make sense in nonperturbative quantum gravity.

(iii) Even if (i) and (ii) are evaded, there are generic mechanisms which lead to a domination of the superposition by configurations that have nothing to do with any recognizable space or spacetime macroscopically, no matter how they are weighed or coarse-grained. The point to appreciate here is that very large quantum fluctuations are present in the nonperturbative Planckian regime, which generically do not cancel each other out to lead to a quasi-classical space "on average", i.e. in the sense of expectation values. Instead, even the dimensionality of space can become dynamical, as is illustrated by the spectral dimension [34].

An infamous pathological mechanism of this type found in Euclidean DT quantum gravity is *polymerization*, whereby building blocks preferably arrange themselves[17] into a so-called branched polymer, with Hausdorff dimension 2 and spectral dimension 4/3 in the limit of vanishing UV-cutoff, independent of the microscopic dimensionality d of the building blocks(!), as long as $d > 2$. A similar effect is also present in models other than DT (see [9], Sec. Q28 for further discussion and references).

In spite of these difficulties, the idea of emergence is not doomed, since there is a known cure for both polymerization and crumpling [22], another generic pathology. Judging by the results obtained, the crucial insight is to require path integral configurations to carry a well-defined causal structure, as one does in CDT.[18] This has led to the first genuine instance of emergence in nonperturbative quantum gravity. More specifically, there is strong, quantitative evidence from CDT lattice gravity for the dynamical generation of a quantum spacetime with properties that on sufficiently large scales match those of a (semi-)classical de Sitter space, in terms of dimensionality [31, 36], shape and its quantum fluctuations [32, 37] and average curvature [38] (see also the reviews [10–12]).

Before taking a closer look at the nature of these results, and how it reflects both the nonperturbative physics and the corresponding toolbox, we can draw an important conclusion from the discussion above. It is not so much that achieving emergence is subtle and difficult, which is certainly true, but that nonperturbative computational tools have been absolutely essential in informing our current understanding of this phenomenon. It is impossible to guess the dynamical content of a given gravitational path integral or partition function without being able to evaluate it explicitly. For example, the fact that microscopic four-dimensional building blocks do not generically give rise to macroscopic four-dimensional spaces in a continuum limit runs counter to any (Semi-classical) or perturbative intuition and to many practitioners came as a great surprise. However, it merely illustrates the need for reality checks in the form of numerical experiments, to rein in our often speculative ideas about quantum gravity and canalize them in the right direction.

13.9 Lattice quantum gravity: Unlocking the early universe

As argued above, quantum gravity is not simple and we should not expect it to be. It is significantly more complex than and structurally different from non-abelian gauge field theory, and one of the toughest problems theoretical physicists have set themselves. Keeping in mind the time it took to observe gravitational waves – a key prediction of the *classical* theory – and to get a grip on computing their waveforms, we should also not be surprised that the time scale for progress in quantum gravity is long. Fundamental quantum gravity shares the need to go beyond perturbation theory with the nonperturbative sectors of general relativity and QCD, and the same essential need for effective computational tools to tackle this sector directly. In addition, taking into account the lack of experimental guidance in quantum gravity, the case for computation as *the* key to progress in understanding the theory is overwhelming.

The go-to methodology in nonperturbative quantum field theory is a lattice regularization, but this did not succeed initially due to the static and Euclidean nature of the lattices employed, as mentioned earlier. Since these early days, coming up with solutions to these issues and testing their viability has been a continuous, collective effort, culminating in a fully functional lattice formulation "2.0" of quantum gravity. It is based on regularizing the gravitational path integral[19] in terms of CDT, and has opened a measurement window on the *terra incognita* of Planck-scale physics.

Like real experiments, MC experiments[20] are also constrained by the resources available, including computing power, storage capacity and efficiency of the algorithms (see e.g. [39]), and all measurement data are subject to statistical and systematic errors. Just like in lattice QCD, it requires ingenuity to construct observables and set up numerical experiments that yield reliable results for the available lattice sizes, which in typical simulations are of the order of 10^5–10^6 simplicial building blocks.

It should be emphasized that being able to compute *anything* in nonperturbative quantum gravity is an unprecedented situation. It allows us to redirect focus away from formal matters and what the theory *should* be like to the actual physical content of the theory and what it is able to deliver. Considering the status quo of lattice gauge theory and the decades of dedicated work that got us there, it is clear that in lattice quantum gravity much still lies ahead and that it will neither be easy nor happen overnight. Unlike QCD, quantum gravity is still at a much more exploratory stage, with a primary focus on finding new observables that can capture the physics of the largely unknown nonperturbative regime. Further computational optimization will clearly be important, but will be tied to completely different physical questions and observables than in nongravitational lattice theories.

The lattice breakthrough allows us to also address a range of conceptual issues in quantum gravity within a concrete computational framework, rather than based on abstract reasoning alone. Examples are the roles of time, causality, unitarity, topology change and spacetime symmetries at the Planck scale, some of which have already been clarified (see e.g. [11]). Any question one poses has to be formulated as an operationally well-defined experiment that can be conducted in the accessible range of lattice parameters. "Operationally well-defined" means among other things that the observable whose eigenvalues are being measured must be labelling-invariant, which typically implies that it is of the form of a nonlocal spacetime average (see [9], Sec. Q29, for further discussion of observables). Note that this requirement is not met for many semi-classically or perturbatively formulated questions that quantum gravity is expected to provide answers to, including the black hole information loss problem [41].

This situation is qualitatively different from the classical one, where reference systems in the form of local coordinate charts always exist, although they may be non-unique and unphysical. In the nonperturbative realm no such coordinate or other reference systems exist. They also cannot be introduced by hand, in the form of equipotential surfaces of some scalar fields [42] or by adding boundaries, say, because these will be subject to the same quantum fluctuations that prevent the existence of useful coordinate systems in the first place. In other words, the unfamiliar nonlocal character of observables is not due to some shortcoming of the chosen formulation, but is an intrinsic feature of Planckian physics.

The observables investigated in CDT lattice gravity so far provide concrete insights into the type of results one will be able to derive. They will be quantitative, but obviously not of an analytical nature. One could wonder whether we will ever be able to develop an analytical description of this nonperturbative regime. Given the presumed strongly interacting character of gravity at the Planck scale, this seems exceedingly unlikely, but in absence of no-go theorems in nonperturbative quantum gravity, it cannot be excluded in principle. A more likely scenario is that we can theoretically model selected aspects of the theory, based on the input from measurements. A good example are the measurements of the correlator of spatial three-volumes in the de Sitter phase of CDT lattice gravity, which have been used to reverse-engineer an effective cosmological action for the scale factor [43]. Other quantities one can hope to extract in a (near-)Planckian regime are universal parameters associated with the scaling behaviour of specific observables. The already mentioned spectral and Hausdorff dimensions [34, 36] are of this type, and coefficients characterizing the fall-off behaviour of diffeomorphism-invariant two-point functions [44] would be another example.

The recent developments and results on lattices reshape our expectations of what quantum gravity is about and what it means to "solve" it, i.e. what we may learn about it in the foreseeable future with the help of our best computational and theoretical tools. Given that it has already been shown that an extended quantum spacetime with *some* de Sitter-like properties is generated dynamically in lattice gravity, a tantalizing goal is to try to connect this to early-universe physics [45], where the background spacetime for quantum fluctuations is usually *assumed* to resemble a de Sitter universe. It would be spectacular if one could show that this assumption can be justified (or possibly corrected) from first principles. This still requires highly nontrivial investigations, e.g. of the extent to which homogeneity and isotropy are present or "emerge" on larger scales [46], and an analysis of local quantum fluctuations and their correlators, which are the subject of ongoing research. Importantly, there is a clear and concrete path forward, and computation is bound to unlock even more of quantum gravity's nonperturbative secrets.

Notes

1. A perturbative quantization is based on splitting the metric field tensor g into the constant Minkowski metric η and a small perturbation h according to $g_{\mu\nu}(x)=\eta_{\mu\nu}+h_{\mu\nu}(x)$.
2. See [3] for a review of early lattice quantum gravity.
3. Here "solving" is put in inverted commas to indicate that it is not clear a priori what this notion entails in nonperturbative quantum gravity.
4. Potentially confusing for non-experts, such toy models sometimes run under the label "quantum gravity" without highlighting their limited character.
5. A timeline of the main developments of nonperturbative quantum gravity since 1980 can be found in [9], Secs. Q15 and Q22.
6. The spectral dimension of (quantum) spacetime and its so-called volume profile, cf. [12].
7. It depends on at most a small number of parameters that have to be fixed by comparing with real-world experiment or observation.
8. Smooth invertible one-to-one maps of the underlying manifold to itself.
9. A set of conditions a Euclidean (lattice) quantum field theory must satisfy to allow for a rotation back to Lorentzian signature is discussed in [16].
10. Since by construction all spacelike edges and all timelike edges have the same length, there are only two types of geometrically distinct simplicial building blocks, up to time reflection.
11. Four-simplices are the four-dimensional analogues of two-dimensional triangles and three-dimensional tetrahedra.
12. Summaries of the set-up and results of EDT in four dimensions and further references can be found in [3, 12, 28].
13. In Lorentzian signature it is more convenient to work with *squared* lengths.
14. Sub-Planckian means length scales even smaller than the Planck length, sometimes also called trans-Planckian.
15. This excludes analogue gravity models [33], which usually focus on recovering an effective Lorentzian spacetime, rather than a full-fledged theory of gravity.
16. As noted earlier, different Planckian ingredients will typically be associated with different *quantum* theories, but the present argument focuses on the emergence of *classical* structures.
17. This is an "entropic" effect, inasmuch as there are many more ways for the building blocks to form distinct branched polymers than other less degenerate macroscopic structures [35].
18. Note that this causal structure is of course not fixed, but quantum-fluctuates alongside other aspects of the spacetime geometry.
19. The focus here is on pure gravity, but the formalism allows for a straightforward coupling to matter fields if desired, see e.g. [11] for further discussion.
20. Markov chain MC is the method of choice; quantum computing and machine learning tools have been considered, but currently do not offer significant computational improvements [40].

References

[1] Goroff, M. H., and A. Sagnotti. (1985). Quantum Gravity at Two Loops. *Physics Letters B*, 160, 81–86.

[2] Van de Ven, A. E. M. (1992). Two Loop Quantum Gravity. *Nuclear Physics B*, 378, 309–366.

[3] Loll, R. (1998). Discrete Approaches to Quantum Gravity in Four Dimensions. *Living Reviews in Relativity*, 1(13). doi: https://doi.org/10.12942/lrr-1998-1310.12942/lrr-1998-13.

[4] Armas, J. (2021). *Conversations on Quantum Gravity*. Cambridge University Press.

[5] Stephani, H., D. Kramer, M. MacCallum, C. Hoenselaers, and E. Herlt. (2009). *Exact Solutions of Einstein's Field Equations* (2nd ed.). Cambridge University Press.

[6] Baumgarte, T. W., and S. L. Shapiro. (2021). *Numerical Relativity: Starting from Scratch*. Cambridge University Press.

[7] Workman, R. L., et al. (Particle Data Group). (2022). Review of Particle Physics. *Progress of Theoretical and Experimental Physics*, 2022(8), 083C01.

[8] Budd, T. (2023). Lessons from the Mathematics of Two-Dimensional Euclidean Quantum Gravity. In C. Bambi, L. Modesto, and I. L. Shapiro (Eds.), *Handbook of Quantum Gravity* (pp. 3329–3383). Springer.

[9] Loll, R., G. Fabiano, D. Frattulillo, and F. Wagner. (2022). Quantum Gravity in 30 Questions. *Proceedings of Science* (CORFU2021), 316.

[10] Ambjørn, J., A. Görlich, J. Jurkiewicz, and R. Loll. (2012). Nonperturbative Quantum Gravity. *Physics Reports*, 519(4), 127–210.

[11] Loll, R. (2020). Quantum Gravity from Causal Dynamical Triangulations: A Review. *Classical and Quantum Gravity*, 37(1), 013002.

[12] Ambjørn, J., and R. Loll. (2025). Causal Dynamical Triangulations: Gateway to Nonperturbative Quantum Gravity. In M. Bojowald and R. Szabo (Eds.), *Encyclopedia of Mathematical Physics* (2nd ed., Vol. 1, pp. 555–567). Elsevier.

[13] Surya, S. (2019). The Causal Set Approach to Quantum Gravity. *Living Reviews in Relativity*, 22(1).

[14] Ambjørn, J., J. Gizbert-Studnicki, A. Görlich, and D. Németh. (2024). Is Lattice Quantum Gravity Asymptotically Safe? Making Contact Between Causal Dynamical Triangulations and the Functional Renormalization Group. *Physical Review D*, 110, 126006.

[15] Reuter, M., and F. Saueressig. (2019). *Quantum Gravity and the Functional Renormalization Group*. Cambridge University Press.

[16] Montvay, I., and G. Münster. (1994). *Quantum Fields on a Lattice*. Cambridge University Press.

[17] Goldenfeld, N. (1992). *Lectures on Phase Transitions and the Renormalization Group*. Addison-Wesley.

[18] Bambi, C., L. Modesto, and I. Shapiro (Eds.). (2024). *Handbook of Quantum Gravity*. Springer.

[19] Dasgupta, A., and R. Loll. (2001). A Proper Time Cure for the Conformal Sickness in Quantum Gravity. *Nuclear Physics B*, 606, 357–379.

[20] Smolin, L. (1979). Quantum Gravity on a Lattice. *Nuclear Physics B*, 148, 333–372.

[21] Williams, R. (2009). Quantum Regge Calculus. In D. Oriti (Ed.), *Approaches to Quantum Gravity* (pp. 360–377). Cambridge University Press.

[22] Ambjørn, J., B. Durhuus, and T. Jonsson. (1997). *Quantum Geometry: A Statistical Field Theory Approach*. Cambridge University Press.

[23] Ambjørn, J., and R. Loll. (1998). Nonperturbative Lorentzian Quantum Gravity, Causality and Topology Change. *Nuclear Physics B*, 536, 407–434.

[24] Wilson, K. G. (1974). Confinement of Quarks. *Physical Review D*, 10, 2445–2459.

[25] Ambjørn, J., J. Jurkiewicz, and R. Loll. (2001). Dynamically Triangulating Lorentzian Quantum Gravity. *Nuclear Physics B*, 610, 347–382.

[26] David, F. (1985). Planar Diagrams, Two-Dimensional Lattice Gravity and Surface Models. *Nuclear Physics B*, 257, 45–58.

[27] Agishtein, M. E., and A. A. Migdal. (1992). Critical Behavior of Dynamically Triangulated Quantum Gravity in Four Dimensions. *Nuclear Physics B*, 385, 395–412.

[28] Ambjørn, J. (2024). Lattice Quantum Gravity: EDT and CDT. In C. Bambi, L. Modesto, and I. L. Shapiro (Eds.), *Handbook of Quantum Gravity*. Springer.

[29] Ambjørn, J., S. Jordan, J. Jurkiewicz, and R. Loll. (2011). A Second-Order Phase Transition in CDT. *Physical Review Letters*, 107, 211303.

[30] Coumbe, D. N., J. Gizbert-Studnicki, and J. Jurkiewicz. (2016). Exploring the New Phase Transition of CDT. *Journal of High Energy Physics*, 02, 144.

[31] Ambjørn, J., J. Jurkiewicz, and R. Loll. (2004). Emergence of a 4-D World from Causal Quantum Gravity. *Physical Review Letters*, 93, 131301.

[32] Ambjørn, J., A. Görlich, J. Jurkiewicz, and R. Loll. (2008). Planckian Birth of the Quantum de Sitter Universe. *Physical Review Letters*, 100, 091304.

[33] Barceló, C., S. Liberati, and M. Visser. (2011). Analogue Gravity. *Living Reviews in Relativity*, 14, 3. doi: https://doi.org/10.12942/lrr-2011-310.12942/lrr-2011-3.

[34] Ambjørn, J., J. Jurkiewicz, and R. Loll. (2005). The Spectral Dimension of the Universe is Scale-Dependent. *Physical Review Letters*, 95, 171301.
[35] Ambjørn, J., A. Görlich, J. Jurkiewicz, and R. Loll. (2008). CDT – An Entropic Theory of Quantum Gravity. In *Proceedings of CLAQG08 Workshop on Continuum and Lattice Approaches to Quantum Gravity*.
[36] Ambjørn, J., J. Jurkiewicz, and R. Loll. (2005). Reconstructing the Universe. *Physical Review D*, 72, 064014.
[37] Ambjørn, J., A. Görlich, J. Jurkiewicz, and R. Loll. (2008). The Nonperturbative Quantum de Sitter Universe. *Physical Review D*, 78, 063544.
[38] Klitgaard, N., and R. Loll. (2020). How Round is the Quantum de Sitter Universe? *European Physical Journal C*, 80(10), 990.
[39] Brunekreef, J., A. Görlich, and R. Loll. (2024). Simulating CDT Quantum Gravity. *Computer Physics Communications*, 300, 109170.
[40] Clemente, G., and J. Gizbert-Studnicki. Private communication.
[41] Mathur, S. D. (2009). The Information Paradox: A Pedagogical Introduction. *Classical and Quantum Gravity*, 26, 224001.
[42] Ambjørn, J., Z. Drogosz, J. Gizbert-Studnicki, A. Görlich, J. Jurkiewicz, and D. Németh. (2021). Scalar Fields in Causal Dynamical Triangulations. *Classical and Quantum Gravity*, 38, 195030.
[43] Ambjørn, J., J. Gizbert-Studnicki, A. Görlich, and J. Jurkiewicz. (2014). The Effective Action in 4D CDT: The Transfer Matrix Approach. *Journal of High Energy Physics*, 06, 034.
[44] Van der Duin, J., and R. Loll. (2024). Curvature Correlators in Nonperturbative 2D Lorentzian Quantum Gravity. *European Physical Journal C*, 84(7), 759.
[45] Glaser, L., and R. Loll. (2017). CDT and Cosmology. *Comptes Rendus Physique*, 18, 265–274.
[46] Loll, R., and A. Silva. (2023). Measuring the Homogeneity of the Quantum Universe. *Physical Review D*, 107, 086013.

14 Structure and Statistical Properties of the Semiclassical Einstein Equations

Daniel Terno

14.1 Introduction

The Einstein equations of general relativity

$$G_{\mu\nu} = 8\pi T_{\mu\nu} \tag{14.1}$$

are possibly the most perfect expression of classical physics. As summarized in the famous aphorism of John Wheeler, space (represented by the Einstein tensor $G_{\mu\nu} = R_{\mu\nu} - \frac{1}{2}Rg_{\mu\nu}$, with $R_{\mu\nu}$ and R being the Ricci tensor and the scalar, respectively), tells matter how to move, and the matter in the form of the energy-momentum tensor (EMT) $T_{\mu\nu}$ tells space how to curve.

Mathematically, general relativity is the simplest member of a broader family of metric theories of gravity [1, 2]. Both theoretical and observational considerations indicate that general relativity is a low-energy limit of an effective quantum gravity theory [1–4]. Despite this, Eq. (14.1) remains the fundamental tool for exploring gravitational phenomena, ranging from the post-Newtonian corrections of satellite trajectories to astrophysical and cosmological studies.

Predictions of all proposed quantum gravity theories are expressed in classical terms [3, 5]. The observable universe is modeled using classical geometry, which forms the foundation for both the standard cosmological model and discussions of its tensions or alternatives [6–9].

As the rest of physics falls within the scope of quantum theory, astrophysics and cosmology "routinely" [5] combine quantum mechanical descriptions of matter and classical gravity. Quantum mechanics, whether in its nonrelativistic form or as quantum field theory and particle physics, determines

DOI: 10.4324/9781003546139-19

basic parameters—typically expectation values — that characterize matter and fields. When many-body properties of bulk matter become significant, statistical mechanics provides additional methods. These calculations are generally performed in flat spacetime, as the relevant scales are much smaller than the curvature scale. Then, some algorithm that expresses the equivalence principle incorporates flat spacetime results into equations valid in general relativity [5, 9].

From a foundational viewpoint, the absence of experimental evidence for gravitational field quantization (see [10, 11] for discussions on experimental attempts) makes hybrid quantum-classical schemes plausible. These schemes necessitate an interface that defines a mathematically coherent relationship between functions describing geometry (e.g., $G_{\mu\nu}$) and operators characterizing quantum matter [12–14]. A four-level hierarchy of models represents coupling of quantum matter to classical gravity [15]. It begins with the Newton–Schrödinger equation, which describes non-relativistic particles in weak gravitational potentials (level 0), progresses to quantum fields propagating on curved backgrounds (level 1), and includes semiclassical gravity (level 2). Beyond these, it encompasses stochastic semiclassical gravity, effective field theory approaches to matter-gravity systems, and models incorporating a minimal length scale expected from canonical quantum gravity or modified commutation relations.

Arguments by Møller [16] and Rosenfeld [17] suggest using the (renormalized) expectation value of the EMT operator as the source of the Einstein equations, leading to a mean-field quantum hybrid approach (level 2 in the above classification) [4, 15]. This proposal results in the semiclassical Einstein equation (SCE),

$$G_{\mu\nu} = 8\pi\langle\psi|\hat{T}_{\mu\nu}|\psi\rangle_{\text{ren}} \tag{14.2}$$

accompanied by the formal evolution equation,

$$i\hbar\frac{\partial}{\partial t}\psi = \hat{H}[\hat{\phi}, \hat{\pi}, g]\psi \tag{14.3}$$

where the Hamiltonian depends on all matter fields, their conjugate momenta, and the metric $g_{\mu\nu}$ (The Einstein tensor $G_{\mu\nu}$ is derived from this metric). On a given background, renormalization introduces finite terms quadratic in curvature. The coefficients of these terms must be determined, and their inclusion raises the system's order to the fourth. While solving such a system conceptually requires a self-consistent approach, practical calculations are often perturbative or rely on specific geometric backgrounds. In practice it is usually done as a perturbative treatment on a chosen background [18–20]. Alternatively, emphasis may be placed on specific properties of geometry that enable the extraction of structural features of the solutions [21].

Before addressing the implications of particular experiments for semiclassical gravity, two significant foundational issues require attention. First, mixtures of matter states—and consequently mixtures of geometries—introduce additional complexities. Averaging geometric quantities across different spacetimes leads to gauge noncovariant results, which are only meaningful when there is a clear algorithm for resolving gauge freedom in all possible scenarios [20].

Second, hybrid schemes for quantum-classical dynamics can broadly be categorized as reversible (unitary) and irreversible [14]. The former aim to provide a mathematically consistent quantum-classical counterpart to fully quantum unitary and fully classical Hamiltonian theories, without introducing dissipation or diffusion. However, all known reversible schemes are in general inconsistent [14], and there is a substantial reason to suspect this is not merely due to a lack of ingenuity [22]. Moreover, one key insight from the studies of such models is that quantum matter complies with the Heisenberg uncertainty relations only if classical quantities are defined with a certain inherent uncertainty [14, 23] .

A useful perspective on hybrid schemes views them as resulting from applying a classical limit to only one subsystem within the combined quantum-classical system. From this viewpoint, the full quantum description encompasses the entire system, with the hybrid formally derived by introducing two Planck constants, $\hbar_c$ and $\hbar_q$. Setting $\hbar_q = \hbar$ and taking the limit $\hbar_c \to 0$ yields the hybrid dynamics [24].

The SCEs belong to this class. In the discussion below, we first demonstrate that the two formal methods for deriving Eq.(14.2) are equivalent when viewed as derivations of a hybrid equation. This derivation further indicates that, strictly speaking, only a statistical interpretation of the Einstein tensor as $\langle G_{\mu\nu}\rangle_\psi$ is viable. Subsequently, the generalization of Eq.(14.2) becomes straightforward. In this generalized form, the SCEs are more challenging to falsify: they are automatically consistent with the results of the Page–Geilker experiment [25] and make it impossible to distinguish between proper and improper mixtures—a property recently discussed in the context of mean-field theories [26].

In the following we set $c = 1$, occasionally keep $G \neq 1$ and explicitly write the Planck constant.

14.2 Derivation of the quantum-classical hybrid

Despite the reasonable form of Eq. (14.2), its derivation is nontrivial, and its interpretation remains somewhat contentious [3, 4, 20]. This can be compared with the Newton–Schrödinger equation, where the inclusion of the self-gravity term is highly intuitive. However, the Newton–Schrödinger equation is not a one-particle weak-field non-relativistic limit of the SCE but rather an equation governing an effective mean-field wave function in the limit of an infinite number of particles [27].

There are at least two distinct methods by which Eq. (14.2) can be formally derived. To simplify the discussion, we consider the Einstein–Hilbert action for gravity and describe the matter content as scalar fields, possibly conformally coupled [20].

The first method expands a quantum metric and quantum scalar field formally around a classical vacuum solution, deriving the equation of motion for the expected metric by retaining only the tree-level diagrams for gravitons and both the tree-level and one-loop diagrams for the scalar field. Discussion of the two counterterms, whose coefficients must be determined [18, 20, 28], as well as the physical conditions under which this approach is valid [3, 4], is beyond our current scope. However, it is important to note that the loop expansion in quantum field theory is effectively an expansion in powers of $\hbar$: tree diagrams contribute terms of order $\hbar^0$, while one-loop diagrams contribute terms of order $\hbar$.

The second method uses a formal device for considering N non-interacting scalar fields in the same quantum state and performing the loop expansions. The limit $N \longrightarrow \infty$ is then taken under the constraint $GN = \text{const}$. In this expansion, graviton loops are suppressed relative to the matter loops. In the $N \longrightarrow \infty$ limit, fluctuations in the expectation value of the matter EMT become negligible, and Eq. (14.2) emerges as the limiting equation.

We adopt the analysis of [19], which provided the original realization of this approach. Consider the transition amplitude between two states specified by a particular three-geometry, represented as a three-metric ${}^3g$ in a specified gauge, and the values of N scalar field configurations $\vec{\phi} = \{\phi_j\}_{j=1}^N$ on the initial and final hypersurfaces Σ' and Σ'', respectively. This amplitude is expressed as the path integral:

$$\left\langle {}^3g'', \vec{\phi}'' | {}^3g', \vec{\phi}' \right\rangle = \int_{\Sigma'}^{\Sigma''} \breve{D}[g] D[\vec{\phi}] \exp\left(\frac{i}{\hbar}\left(S_g[g] + S_m[g, \vec{\phi}]\right)\right) \tag{14.4}$$

Here, the fields are constrained to take prescribed values on the initial and final surfaces. The measure includes the four terms $\mathcal{F}_\alpha$ that represent the gauge conditions, along with the Faddeev-Popov determinant $\Delta_{\mathcal{F}}$:

$$\breve{D}[g] = D[g] \prod_{\alpha=0}^{4} \delta(\mathcal{F}_\alpha[g]) \Delta_{\mathcal{F}}[g] \tag{14.5}$$

while the infinite gauge volume factor has been omitted.

The gravitational action includes [28] the Einstein–Hilbert term, cosmological, constant, two counterterms and the boundary term that we are not writing out explicitly,

$$S_g = \frac{1}{16\pi G_0} \int d^4x \sqrt{-g} \left(R - 2\Lambda_0 + \alpha_0 R^2 + \beta_0 R_{\mu\nu} R^{\mu\nu}\right) + \text{boundary terms} \tag{14.6}$$

where G_0 and Λ_0 are the bare values of the gravitational constant and the cosmological constant, respectively, and α_0 and β_0 are additional bare coupling constant. The divergent parts of $\langle \hat{T}_{\mu\nu} \rangle_\psi$ are removed by redefinition of these constants. Their finite renormalized values are physical parameters of the theory [20, 28].

The N identical massless conformably coupled real scalar fields,

$$S_m = -\frac{1}{2} \sum_{j=1}^{N} \int d^4x \sqrt{-g} \left(\partial_\mu \phi_j \partial^\mu \phi_j + \tfrac{1}{6} R \phi_j^2 \right) \tag{14.7}$$

but it is immaterial for our argument. The exposition is further simplified if one assumes that the states of scalar fields are described by the same initial and final configurations, $|\phi'\rangle$ and $|\phi''\rangle$, respectively.

The evaluation of the path integral (14.4) proceeds in two step. First, for a given metric and for each field configuration, the functional $Y_{\phi'',\phi'}[g]$ via

$$\exp(iY_{\phi'',\phi'}[g]) := \langle \phi'' | \phi' \rangle_g = \int D[\phi] e^{iS_m[g,\phi]/\hbar}. \tag{14.8}$$

Then the full amplitude becomes

$$\langle {}^3g'', \vec{\phi}'' | {}^3g', \vec{\phi}' \rangle = \int_{\Sigma'}^{\Sigma''} \breve{D}[g] \exp\left(\frac{i}{\hbar} (S_g[g] + NY_{\phi'',\phi'}[g]) \right) \tag{14.9}$$

Taking the $N \to \infty$ limit requires rescaling of the gravitational coupling such as

$$NG =: \kappa = \text{const} \tag{14.10}$$

that brings the amplitude to the form

$$\langle {}^3g'', \vec{\phi}'' | {}^3g', \vec{\phi}' \rangle = \int_{\Sigma'}^{\Sigma''} \breve{D}[g] \exp\left(\frac{iN}{\hbar} (S_g[g] + Y_{\phi'',\phi'}[g]) \right) \tag{14.11}$$

For large N the dominant contribution to the functional integral comes from metrics near the extremum of $\Gamma_{\phi'',\phi'}[g] := S_g[g] + Y_{\phi'',\phi'}[g]$. Here S_g is the classical action, and $Y_{\phi'',\phi'}[g]$ is (analogous to the) effective action, that for non-interacting scalar field is precisely given by the Gaussian expression. For interacting field one-loop expressions provide the leading corrections.

The connection to $1/N$ power counting is particularly transparent in the toy model of [4], whose action is

$$\begin{aligned} S_g + S_m = &-\frac{1}{G} \int d^4x \left(\partial_\mu h \partial^\mu h + h \partial_\mu h \partial^\mu + \ldots \right) \\ &- \tfrac{1}{2} \sum_{j=1}^{N} \left(\partial_\mu \phi_j \partial^\mu \phi_j + \ldots \right) + \sum_{j=1}^{N} \left(h \partial_\mu \phi_j \partial^\mu \phi_j + \ldots \right). \end{aligned} \tag{14.12}$$

The self-coupling graviton term of the order $O(h^3)$, which appears in perturbative gravity beyond the linear approximation, is explicably included.

The computation of the dressed graviton propagator includes several types of diagrams. The limit $N \longrightarrow \infty$ is again needs to be accompanied by the rescaling of Eq. (14.10). The first Feynman diagram is the free graviton propagator, which is now of the order $O(\kappa/N)$. Next, there are N identical diagrams with one loop of matter and two graviton propagators as external legs. Presence of two graviton propagators consigns them to the order $O(\kappa^2/N^2)$; hence the overall contribution can be represented by a single diagram of order $O(\kappa^2/N)$. The combined effect of all diagrams with two loops of matter and three graviton propagators is of the order $O(k^3/N)$, and so on.

The contributions that involve graviton loops are even more suppressed. A diagram with one graviton loop and two graviton legs contains four graviton propagators and two vertices. As the propagators contribute factors $(\kappa/N)^4$ and the vertices $(N/\kappa)^2$, this diagram is of the order $O(\kappa^3/N^2)$. As a result, in the limit $N \longrightarrow \infty$, there are no contributions from the graviton propagators while the matter fields are quantized and contribute accordingly.

The scaling relations of the toy model as well as of both the more rigorous derivations of the semiclassical equation become nearly automatic if we keep $N = 1$ (or just keep any finite number of various matter field), but proceed as appropriate in the derivation of hybrid dynamics. We formally introduce two Planck constants (for gravity and matter fields, respectively) and take the classical limit for the gravitational sector only. It is accomplished by setting $\hbar = N\hbar_g = \text{const}$, hence

$$\begin{aligned}\exp\left(i(S_g + S_m)/\hbar\right) &\longrightarrow \exp\left(i(S_g/\hbar_g + S_m/\hbar)\right)\\ &= \exp\left(i(NS_g + S_m)/\hbar\right)\end{aligned} \tag{14.13}$$

with $N \longrightarrow \infty$ realizing the partial classical limit. Then suppression of the non-classical gravitational contributions becomes obvious.

14.3 Statistical interpretation

The SCE is obtained by calculating expectation values of various correlation functions and then taking the limit $\hbar_g \longrightarrow 0$. Hence, the meaning of the left-hand side of the SCE is:

$$\langle G_{\mu\nu}\rangle_\psi = 8\pi\langle\hat{T}_{\mu\nu}\rangle_\psi^{\text{ren}} \tag{14.14}$$

This interpretation was posited by Ballentine in his commentary on the Page–Geilker experiment [29]. Here, we see that this is a direct consequence of the way the SCE is derived.

It provides a resolution to the question of how the post-measurement state update rules [30] affect the SCE. It is convenient to discuss this as well as the interpretation of the Page–Geilker experiment in terms of the toy model proposed by Unruh [3, 31].

The experiment can be understood as an elaboration of Schrödinger's cat gedankenexperiment. Two macroscopically distinct mass configurations are determined by the state of a decaying unstable particle. Denoting the state where the particle has not yet decayed, with the mass in one configuration, as $|0\rangle \equiv |n\rangle|L\rangle$, and the state of a decayed particle with the other mass configuration as $|1\rangle \equiv |d\rangle|L\rangle$, the quantum state after the approximate exponential decay of the unstable particle evolves as:

$$\psi(t) = \alpha(t)|0\rangle + \beta(t)|1\rangle \tag{14.15}$$

where

$$|\alpha|^2 \approx e^{-\lambda t}, \qquad |\beta|^2 \approx 1 - e^{-\lambda t}. \tag{14.16}$$

A Cavendish torsion balance is employed to measure the gravitational field induced by the mass configurations.

Initially $\langle \hat{T}_{\mu\nu}\rangle(0) = \langle 0|\hat{T}_{\mu\nu}|0\rangle$. Assuming that the exchange term is negligible, which is the case for macroscopically distinct configurations, we have

$$\langle \hat{T}_{\mu\nu}\rangle(t) \approx e^{-\lambda t}\langle 0|\hat{T}_{\mu\nu}|0\rangle + (1 - e^{-\lambda t}\langle 1|\hat{T}_{\mu\nu}|1\rangle) \tag{14.17}$$

Hence, according to the SCE [Eq. (14.2)], the Cavendish balance would follow the dynamics of the expectation value. In the experiment [25], the balance suddenly moved to the position consistent with the distribution L (when it entered the future light cone of the detection) after the counter clicked. Thus, Eq. (14.2) is falsified as an adequate description of gravity [25, 29, 31]. Nevertheless, semiclassical gravity at this level does not provide any information about individual events. The correct SCE (14.14) is still in agreement with the experiment.

In the light of Eq. (14.14) the choice between "no-collapse" (i.e. the many-world interpretation of quantum mechanics) and contradiction with the Bianchi identity [25, 32] does not arise. It is forced by the observation that in general if

$$\langle \psi|\hat{T}^{\mu}_{\ \nu}|\psi\rangle_{;\mu} = 0 \tag{14.18}$$

the measurement-induced discontinuity in the state description will produce the corresponding discontinuity in $G^{\mu}_{\ \nu\mu}$. However, this problem does not arise with Eq. (14.14), and neither a possibility of a superluminal communication.

This clarification does not matter much in actual applications. The SCE is usually used in astrophysical and cosmological problems that do not involve quantum states is a superposition of substantially different EMT distributions

[5]. Nevertheless, this statistical interpretation trivially matches the results of [25]. It also gives a clear indication why stochastic gravity is necessary if one wants to capture the effect of fluctuations up to the second order [4].

If this is the case, then the nonlinear relations between curvature and metric make its precise identification from the knowledge of $\langle G_{\mu\nu}\rangle_\psi$ impossible. Similar ambiguities in the averaging procedure are source of trouble for cosmology [7, 8]. The equations below can both describe the application of some averaging procedure to inhomogeneous geometry and evaluate exsections for a random process that is generating geometries (under assumption of the appropriate gauge fixing that ensures compatibility of different situations). Averaging both sides of Eq. (14.1) results in

$$\langle G^\mu{}_\nu\rangle = \langle R^\mu{}_\nu\rangle - \tfrac{1}{2}\delta^\mu_\nu\langle g^{\lambda\rho}R_{\lambda\rho}\rangle = 8\pi\langle T^\mu{}_\nu\rangle \tag{14.19}$$

One can focus on the averaged metric introduces $\bar{g}_{\mu\nu} := \langle g_{\mu\nu}\rangle$, and introduce

$$\delta g_{\mu\nu} := g_{\mu\nu} - \bar{g}_{\mu\nu} \tag{14.20}$$

Then it is possible to define the connection $\bar{\Gamma}^\lambda{}_{\mu\nu}$ and other objects that are based on the averaged metric. The averaged Einstein equations can be written as

$$\bar{G}_{\mu\nu} + \delta G_{\mu\nu} = 8\pi\langle T_{\mu\nu}\rangle \tag{14.21}$$

where non-zero value of the correction term indicates that the Einstein tensor of the averaged metric is not in general the average of the Einstein tensors, $G[\langle g\rangle] \neq \langle G[g]\rangle$.

Most general states in quantum theory are mixed states—convex combinations (i.e., the weighted averages) of some pure states that are the extreme points of the set of all quantum states. In the SCE framework the geometric quantities are certain real-valued functionals of pure quantum states. Hence the standard rules of quantum mechanics lead to the formal expression

$$\langle G_{\mu\nu}\rangle_\rho = 8\pi\mathrm{tr}\left(\rho\hat{T}_{\mu\nu}\right)_{\mathrm{ren}} = 8\pi\sum_i \langle\psi|\hat{T}_{\mu\nu}|\psi\rangle_{\mathrm{ren}} \tag{14.22}$$

where the mixed state is

$$\rho = \sum_i w_i|\psi_i\rangle\langle\psi_i|, \qquad w_i \geqslant 0. \tag{14.23}$$

As discussed above there is no direct relationship between $\langle G_{\mu\nu}\rangle_\rho$ and $\bar{g}_{\mu\nu} = \langle g_{\mu\nu}\rangle_\rho$. This is so even if the EMT of each of the mixture components has low variance, so $\langle\delta g\rangle_{\psi_i} \approx 0$. However, in this case

$$\langle g_{\mu\nu}\rangle = \sum_i w_i\langle g_{\mu\nu}\rangle_{\psi_i} \tag{14.24}$$

does not have to correspond to any of the geometries in the ensemble.

A formal expression for the averaged geometric quantities, such as Eq. (14.22), can be given a more operationally meaningful form if all quantities are described relative to a reference frame that is constructed according to a predefined algorithm using relational quantities. Statistical moments of invariant quantities, such as independent curvature scalars [33], can be calculated directly.

This interpretation of the SCE satisfies the requirement of operational indistinguishability between proper and improper mixtures [30] that was analyzed by Fedida and Kent [26]. The mass configuration can be decided either as a result of a random process (a proper mixture) or as tracing out of the auxiliary degrees of freedom (an improper mixture). In terms of the Unruh model the matter density matrix is in both cases

$$\rho = |\alpha|^2 |L\rangle\langle L| + |\beta|^2 |R\rangle\langle R| \tag{14.25}$$

In the example of Page-Geilker experiment the proper mixture corresponds to the ensemble of the post-measurement configurations (when the results are not revealed), and the improper mixture results from tracing out the unstable particle. Assume that Eq. (14.2) holds. Once the Cavendish balance is in the future light cone of the random choice of the matter configuration, it will respond in one of the two possible distinct ways. However, in case of improper mixture, the balance will behave as (time-dependent) weighted average of the two responses.

Such behavior would allow to distinguish between two types of mixtures. However, as the only valid prediction of the SCE is the expectation value $\langle G_{\mu\nu}\rangle_\rho$, such differentiation is impossible. In addition, in the Newtonian limit if we assume that each of the two matter configurations r results in a low dispersion value of $\langle \hat{T}_{00}\rangle_{L,R}$ (i.e. the states are approximate eigenvalues of the Hamiltonian [3, 31]), we will have that the Newtonian gravitational potential satisfies

$$\langle \varphi\rangle_\rho \approx |\alpha|^2 \varphi_L + (1 - |\alpha|^2)\varphi_R \tag{14.26}$$

where $\varphi_{L,R}$ are the gravitational potentials that correspond to the respective mass distributions.

14.4 Discussion

We have shown that the two formal derivations of the SCE are equivalent when viewed as derivations of the quantum-classical hybrid with $\hbar_g \to 0$. A key consequence of this derivation is that the Einstein tensor is fundamentally a stochastic quantity, and the SCE predicts only its expectation value. Sharp (low-dispersion) predictions are possible only under special conditions. As a result, unlike Eq. (14.2), it does not produce any of the undesirable effects discussed in [25, 26, 32]. This analysis also highlights the necessity of stochastic gravity [4], even under the assumption that all gravitational field fluctuations originate from fluctuations in the quantum matter.

The SCE lacks matter-gravity entanglement and can, in principle, be falsified in experiments where the gravitational field is used to establish entanglement between two matter subsystems with negligible other interactions [10]. However, further investigation is required to determine how its low-energy implications affect the interpretation of tabletop experiments.

Acknowledgments

Useful discussion with David Ahn, Samuel Fedida, Masahiro Hotta, Adrian Kent, and Sebastian Murk are gratefully acknowledged.

References

[1] Belenchia, A., M. Letizia, S. Liberati, and E. Di Casola. (2018). Higher-Order Theories of Gravity: Diagnosis, Extraction and Reformulation via Non-Metric Extra Degrees of Freedom—A Review. *Reports on Progress in Physics*, 81, 036001.

[2] De Felice, A., and S. Tsujikawa. (2010). $f(R)$ Theories. *Living Reviews in Relativity*, 13, 3. http://www.livingreviews.org/lrr-2010-3lrr-2010-3.

[3] Kiefer, C. (2012). *Quantum Gravity* (3rd ed.). Oxford University Press.

[4] Hu, B.-L., and E. Verdaguer. (2020). *Semiclassical and Stochastic Gravity: Quantum Field Effects on Curved Spacetime*. Cambridge University Press.

[5] Wallace, D. (2022). Quantum Gravity at Low Energies. *Studies in History and Philosophy of Science*, 94, 31.

[6] Abdalla, E., et al. (2022). Cosmology Intertwined: A Review of the Particle Physics, Astrophysics, and Cosmology Associated with the Cosmological Tensions and Anomalies. *Journal of High Energy Astrophysics*, 34, 49.

[7] Alure, P. K., et al. (2023). Is the Observable Universe Consistent with the Cosmological Principle? *Classical and Quantum Gravity*, 40, 094001.

[8] Wiltshire, D. L. (2011). What Is Dust?—Physical Foundations of the Averaging Problem in Cosmology. *Classical and Quantum Gravity*, 28, 164006.

[9] Padmanabhan, T. (2020). *Theoretical Astrophysics Volume I: Astrophysical Processes*. Cambridge University Press.

[10] Marletto, C., and V. Vedral. (2024). Quantum-Information Methods for Quantum Gravity Laboratory-Based Tests. *arXiv:2410.07262*.

[11] Tobar, G., S. K. Manikandan, T. Beitel, and I. Pikovski. (2024). Detecting Single Gravitons with Quantum Sensing. *Nature Communications*, 15, 7229.

[12] Boucher, W., and J. Traschen. (1988). Semiclassical Physics and Quantum Fluctuations. *Physical Review D*, 37, 3522.

[13] Peres, A., and D. R. Terno. (2001). Hybrid Classical-Quantum Dynamics. *Physical Review A*, 63, 022101.

[14] Terno, D. R. (2023). Classical-Quantum Hybrid Models. *arXiv:2309.05014*. Encyclopedia of Mathematical Physics 2nd edition, Volume 2 doi:10.1016/B978-0-323-95703-8.00105-1

[15] Hu, B.-L. (2014). Gravitational Decoherence, Alternative Quantum Theories and Semiclassical Gravity. *Journal of Physics: Conference Series*, 504, 012021.

[16] Møller, C. (1962). In A. Lichnerowicz and M. A. Tonnelat (Eds.), *Les Théories Relativistes de la Gravitation* (p. 15). Editions du CNRS.

[17] Rosenfeld, L. (1963). On Quantization of Fields. *Nuclear Physics*, 40, 353.

[18] Parker, L., and D. Toms. (2009). *Quantum Field Theory in Curved Spacetime*. Cambridge University Press.

[19] Hartle, J. B., and G. T. Horowitz. (1981). Ground-State Expectation Value of the Metric in the $1/N$ or Semiclassical Approximation to Quantum Gravity. *Physical Review D*, 24, 257.

[20] Flanagan, É. É., and R. M. Wald. (1996). Does Back Reaction Enforce the Averaged Null Energy Condition in Semiclassical Gravity? *Physical Review D*, 54, 6233.

[21] Murk, S., R. B. Mann, and D. R. Terno. (2022). Black Holes and Their Horizons in Semiclassical and Modified Theories of Gravity. *International Journal of Modern Physics D*, 31, 2230015.

[22] Galley, T. D., F. Giacomini, and J. H. Selby. (2023). Any Consistent Coupling between Classical Gravity and Quantum Matter is Fundamentally Irreversible. *Quantum*, 7, 1142.

[23] Barceló, C., R. Carballo-Rubio, L. J. Garay, and R. Gómez-Escalante. (2012). Hybrid Classical–Quantum Formulations Ask for Hybrid Notions. *Physical Review A*, 86, 042120.

[24] Caro, J., and L. L. Salcedo. (1999). Impediments to Mixing Classical and Quantum Dynamics. *Physical Review A*, 60, 842.

[25] Page, D. N., and C. D. Geilker. (1981). Indirect Evidence for Quantum Gravity. *Physical Review Letters*, 47, 979.

[26] Fedida, S., and A. Kent. (2024). Mixture Equivalence Principles and Post-Quantum Theories of Gravity. *arXiv:2412.12288*. DOI: https://doi.org/10.1103/pttr-6kj7 Phys. Rev. D 111, 126016 (2025)

[27] Anastopoulos, C., and B.-L. Hu. (2014). Problems with the Newton–Schrödinger Equations. *New Journal of Physics*, 16, 085007.

[28] Ford, L. H. (2005). Spacetime in Semiclassical Gravity. *arXiv:gr-qc/0504096*. https://doi.org/10.48550/arXiv.gr-qc/0504096.

[29] Ballentine, L. E. (1982). Comment on "Indirect Evidence for Quantum Gravity". *Physical Review Letters*, 48, 522.

[30] Busch, P., P. Lahti, J.-P. Pellonpää, and K. Ylinen. (2016). *Quantum Measurement*. Springer.

[31] Unruh, W. G. (1984). Steps towards a Quantum Theory of Gravity. In S. M. Christensen (Ed.), *Quantum Theory of Gravity* (p. 234). Adam Hilger.

[32] Eppley, K., and E. Hannah. (1977). The Necessity of Quantizing the Gravitational Field. *Foundations of Physics*, 7, 51.

[33] Stephani, H., D. Kramer, M. MacCallum, C. Hoenselaers, and E. Herlt. (2003). *Exact Solutions of Einstein's Field Equations* (2nd ed.). Cambridge University Press.

Index

Made in the USA
Las Vegas, NV
04 May 2026

fe65812a-095d-4960-bc8f-05bafb1bbdccR01